WIRELESS NETWORKS

WIRELESS NETWORKS

P. Nicopolitidis
Aristotle University, Greece

M. S. Obaidat
Monmouth University, USA

G. I. Papadimitriou
Aristotle University, Greece

A. S. Pomportsis
Aristotle University, Greece

JOHN WILEY & SONS, LTD

Other Wiley Editorial Offices

John Wiley & Sons Inc.,
111 River Street, Hoboken, NJ 07030, USA

Jossey-Bass, 989 Market Street, San Francisco,
CA 94103–1741, USA

Wiley-VCH Verlag GmbH,
Boschstr. 12, D–69469 Weinheim, Germany

John Wiley & Sons Australia Ltd, 33 Park Road,
Milton, Queensland 4064, Australia

John Wiley & Sons (Asia) Pte Ltd, 2 Clementi Loop 02–01,
Jin Xing Distripark, Singapore 129809

John Wiley & Sons Canada Ltd, 22 Worcester Road,
Etobicoke, Ontario, Canada M9W 1L1

British Library Cataloguing in Publication Data

A catalogue record for this book is available from the British Library

ISBN 0-470-84529-5

Typeset in 10/12pt Times by Deerpark Publishing Services Ltd, Shannon, Ireland.
Printed and bound in Great Britain by T.J. International Limited, Padstow, Cornwall.
This book is printed on acid-free paper responsibly manufactured from sustainable forestry in which at least two trees are planted for each one used for paper production.

To My Parents
  Petros Nicopolitidis

To My Mother and the Memory of My Late Father
  Mohammad Salameh Obaidat

To My Parents Zoi and Ilias,
To My Wife Maria and our Children
  Georgios I. Papadimitriou

To My Sons Sergios and George
  Andreas S. Pomportsis

Contents

3   First Generation (1G) Cellular Systems   95

Preface

The field of wireless networks has witnessed tremendous growth in recent years and it has become one of the fastest growing segments of the telecommunications industry. Wireless communication systems, such as cellular, cordless and satellite phones as well as wireless local area networks (WLANs) have found widespread use and have become an essential tool to many people in every-day life. The popularity of wireless networks is so great that we will soon reach the point where the number of worldwide wireless subscribers will be higher than the number of wireline subscribers. This popularity of wireless communication systems is due to its advantages compared to wireline systems. The most important of these advantages is the freedom from cables, which enables the 3A paradigm: communication anywhere, anytime, with anyone. For example, by dialing a friend or colleague's mobile phone number, one is able to contact him in a variety of geographical locations, thus overcoming the disability of fixed telephony.

This book aims to provide in-depth coverage of the wireless technological alternatives offered today. In Chapter 1, a short introduction to wireless networks is made.

In Chapter 2, background knowledge regarding wireless communications is provided. Issues such as electromagnetic wave propagation, modulation, multiple access for wireless systems, etc. are discussed Readers who are already familiar with these issues may skip this chapter.

In Chapter 3, the first generation of cellular systems is discussed. Such systems are still used nowadays, nevertheless they are far from being at the edge of technology. Chapter 3 discusses two representative first generation systems, the Advanced Mobile Phone System (AMPS) and the Nordic Mobile Telephony (NMT) system.

In Chapter 4, the second generation of cellular systems is discussed. The era of mobile telephony as we understand it today, is dominated by second generation cellular standards. Chapter 4 discusses several such systems, such as D-AMPS, cdmaOne and the Global system for Mobile Communications (GSM). Moreover, data transmission over 2G systems is discussed by covering the so-called 2.5G systems, such as the General Packet Radio Service (GPRS), cdmaTwo, etc. Finally, Chapter 4 discusses Cordless Telephony (CT) including the the Digital European Cordless Telecommunications Standard (DECT) and the Personal Handyphone System (PHS) standards.

Chapter 5 discusses the third generation of cellular systems. These are the successors of second generation systems. They are currently starting to be deployed and promise data rates up to 2 Mbps. The three different third generation air-interface standards (Enhanced Data

Rates for GSM Evolution (EDGE), cdma2000 and wideband CDMA (WCDMA)) are discussed.

Chapter 6 provides a vision of 4G and beyond mobile and wireless systems. Such systems target the market of 2010 and beyond, aiming to offer data rates of at least 50 Mbps. Due to the large time window to their deployment, both the telecommunications scene and the services offered by 4G systems and beyond are not yet known and as a result aims for these systems may be changing over time.

Chapter 7 discusses satellite-based wireless systems. After discussing the characteristics of the various satellite orbits, Chapter 7 covers the VSAT, Iridium and Globalstar systems and discusses a number of issues relating to satellite-based Internet access.

Chapter 8 discusses fixed wireless systems. The main points of this chapter are the well-known Multichannel Multipoint Distribution Service (MMDS) and Local Multipoint Distribution Service (LMDS).

Chapter 9 covers wireless local area networks. It discusses the design goals for wireless local area networks, the different options for using a physical layer and the MAC protocols of two wireless local area network standards, IEEE 802.11 and ETSI HIPERLAN 1. Furthermore, it discusses the latest developments in the field of wireless local area networks.

Chapter 10 is devoted to Wireless Asynchronous Transfer Mode (WATM). After providing a brief introduction to ATM, it discusses WATM and HIPELRAN 2, an ATM-compatible wireless local area network. The chapter also provides a section on wireless ad-hoc routing protocols.

Chapter 11 describes Personal Area Networks (PANs). The concept of a PAN differs from that of other types of data networks in terms of size, performance and cost. PANs target applications that demand short-range communications. After a brief introduction, Chapter 11 covers the Bluetooth and HomeRF PAN standards.

Chapter 12 discusses security issues in wireless networks. Security is a crucial point in all kinds of networks but is even more crucial in wireless networks due to the fact that wireless transmission cannot generally be confined to a certain geographical area.

Chapter 13 deals with the basics of simulation modeling and its application to wireless networking. It discusses the basic issues involved in the development of a simulator and presents several simulation studies of wireless network systems.

Finally, Chapter 14 discusses several economical issues relating to wireless networks. It is reported that although voice telephony will continue to be a significant application, the wireless-Internet combination will shift the nature of wireless systems from today's voice-oriented wireless systems towards data-centric ones. The impacts of this change on the key players in the wireless networking world are discussed. Furthermore, the chapter covers charging issues in the wireless networks.

We would like to thank the reviewers of the original book proposal for their constructive suggestions. Also, we would like to thank our students for some feedback that we received while trying the manuscript in class. Many thanks to Wiley's editors and editorial assistants for their outstanding work.

1

Introduction to Wireless Networks

Although it has history of more than a century, wireless transmission has found widespread use in communication systems only in the last 15–20 years. Currently the field of wireless communications is one of the fastest growing segments of the telecommunications industry. Wireless communication systems, such as cellular, cordless and satellite phones as well as wireless local area networks (WLANs) have found widespread use and have become an essential tool in many people's every-day life, both professional and personal. To gain insight into the wireless market momentum, it is sufficient to mention that it is expected that the number of worldwide wireless subscribers in the years to come will be well over the number of wireline subscribers. This popularity of wireless communication systems is due to its advantages compared to wireline systems. The most important of these advantages are mobility and cost savings.

Mobile networks are by definition wireless, however as we will see later, the opposite is not always true. Mobility lifts the requirement for a fixed point of connection to the network and enables users to physically move while using their appliance with obvious advantages for the user. Consider, for example, the case of a cellular telephone user: he or she is able to move almost everywhere while maintaining the potential to communicate with all his/her colleagues, friends and family. From the point of view of these people, mobility is also highly beneficial: the mobile user can be contacted by dialing the very same number irrespective of the user's physical location; he or she could be either walking down the same street as the caller or be thousands of miles away. The same advantage also holds for other wireless systems. Cordless phone users are able to move inside their homes without having to carry the wire together with the phone. In other cases, several professionals, such as doctors, police officers and salesman use wireless networking so that they can be free to move within their workplace while using their appliances to wirelessly connect (e.g., through a WLAN) to their institution's network.

Wireless networks are also useful in reducing networking costs in several cases. This stems from the fact that an overall installation of a wireless network requires significantly less cabling than a wired one, or no cabling at all. This fact can be extremely useful:

- *Network deployment in difficult to wire areas.* Such is the case for cable placement in rivers, oceans, etc. Another example of this situation is the asbestos found in old buildings. Inhalation of asbestos particles is very dangerous and thus either special precaution must

be taken when deploying cables or the asbestos must be removed. Unfortunately, both
solutions increase the total cost of cable deployment.

- *Prohibition of cable deployment.* This is the situation in network deployment in several
  cases, such as historical buildings.
- *Deployment of a temporary network.* In this case, cable deployment does not make sense,
  since the network will be used for a short time period.

Deployment of a wireless solution, such as a WLAN, is an extremely cost-efficient solution
for the scenarios described above. Furthermore, deployment of a wireless network takes
significantly less time compared to the deployment of a wired one. The reason is the same:
no cable is installed.

In this introductory chapter we briefly overview the evolution of wireless networks, from
the early days of pioneers like Samuel Morse and Guglielmo Marconi to the big family of
today's wireless communications systems. We then proceed to briefly highlight the major
technical challenges in implementing wireless networks and conclude with an overview of
the subjects described in the book.

1.1 Evolution of Wireless Networks

Wireless transmission dates back into the history of mankind. Even in ancient times, people
used primitive communication systems, which can be categorized as wireless. Examples are
smoke signals, flashing mirrors, flags, fires, etc. It is reported that the ancient Greeks utilized a
communication system comprising a collection of observation stations on hilltops, with each
station visible from its neighboring one. Upon receiving a message from a neighboring
station, the station personnel repeated the message in order to relay it to the next neighboring
station. Using this system messages were exchanged between pairs of stations far apart from
one another. Such systems were also employed by other civilizations.

However, it is more logical to assume that the origin of wireless networks, as we understand them today, starts with the first radio transmission. This took place in 1895, a few years
after another major breakthrough: the invention of the telephone. In this year, Guglielmo
Marconi demonstrated the first radio-based wireless transmission between the Isle of Wight
and a tugboat 18 miles away. Six years later, Marconi successfully transmitted a radio signal
across the Atlantic Ocean from Cornwall to Newfoundland and in 1902 the first bidirectional
communication across the Atlantic Ocean was established. Over the years that followed
Marconi's pioneering activities, radio-based transmission continued to evolve. The origins
of radio-based telephony date back to 1915, when the first radio-based conversation was
established between ships.

1.1.1 Early Mobile Telephony

In 1946, the first public mobile telephone system, known as Mobile Telephone System
(MTS), was introduced in 25 cities in the United States. Due to technological limitations,
the mobile transceivers of MTS were very big and could be carried only by vehicles. Thus, it
was used for car-based mobile telephony. MTS was an analog system, meaning that it
processed voice information as a continuous waveform. This waveform was then used to
modulate/demodulate the RF carrier. The system was half-duplex, meaning that at a specific

time the user could either speak or listen. To switch between the two modes, users had to push a specific button on the terminal.

MTS utilized a Base Station (BS) with a single high-power transmitter that covered the entire operating area of the system. If extension to a neighboring area was needed, another BS had to be installed for that area. However, since these BSs utilized the same frequencies, they needed to be sufficiently apart from one another so as not to cause interference to each other. Due to power limitations, mobile units transmitted not directly to the BS but to receiving sites scattered along the system's operating area. These receiving sites were connected to the BS and relayed voice calls to it. In order to place a call from a fixed phone to an MTS terminal, the caller first called a special number to connect to an MTS operator. The caller informed the operator of the mobile subscriber's number. Then the operator searched for an idle channel in order to relay the call to the mobile terminal. When a mobile user wanted to place a call, an idle channel (if available) was seized through which an MTS operator was notified to place the call to a specific fixed telephone. Thus, in MTS calls were switched manually.

Major limitations of MTS were the manual switching of calls and the fact that a very limited number of channels was available: In most cases, the system provided support for three channels, meaning that only three voice calls could be served at the same time in a specific area.

An enhancement of MTS, called Improved Mobile Telephone System (IMTS), was put into operation in the 1960s. IMTS utilized automatic call switching and full-duplex support, thus eliminating the intermediation of the operator in a call and the need for the push-to-talk button. Furthermore, IMTS utilized 23 channels.

1.1.2 Analog Cellular Telephony

IMTS used the spectrum inefficiently, thus providing a small capacity. Moreover, the fact that the large power of BS transmitters caused interference to adjacent systems plus the problem of limited capacity quickly made the system impractical. A solution to this problem was found during the 1950s and 1960s by researchers at AT&T Bell Laboratories, through the use of the cellular concept, which would bring about a revolution in the area of mobile telephony a few decades later. It is interesting to note that this revolution took a lot of people by surprise, even at AT&T. They estimated that only one million cellular customers would exist by the end of the century; however today, there are over 100 million wireless customers in the United States alone.

Originally proposed in 1947 by D.H. Ring, the cellular concept [1] replaces high-coverage BSs with a number of low-coverage stations. The area of coverage of each such BS is called a 'cell'. Thus, the operating area of the system was divided into a set of adjacent, non-overlapping cells. The available spectrum is partitioned into channels and each cell uses its own set of channels. Neighboring cells use different sets of channels in order to avoid interference and the same channel sets are reused at cells away from one another. This concept is known as frequency reuse and allows a certain channel to be used in more than one cell, thus increasing the efficiency of spectrum use. Each BS is connected via wires to a device known as the Mobile Switching Center (MSC). MSCs are interconnected via wires, either directly between each other or through a second-level MSC. Second-level MSCs might be interconnected via a third-level MSC and so on. MSCs are also responsible for assigning channel sets to the various cells.

The low coverage of the transmitters of each cell leads to the need to support user movements between cells without significant degradation of ongoing voice calls. However, this issue, known today as handover, could not be solved at the time the cellular concept was proposed and had to wait until the development of the microprocessor, efficient remote-controlled Radio Frequency (RF) synthesizers and switching centers.

The first generation of cellular systems (1G systems) [2] was designed in the late 1960s and, due to regulatory delays, their deployment started in the early 1980s. These systems can be thought of as descendants of MTS/IMTS since they were of also analog systems. The first service trial of a fully operational analog cellular system was deployed in Chicago in 1978. The first commercial analog system in the United States, known as Advanced Mobile Phone System (AMPS), went operational in 1982 offering only voice transmission. Similar systems were used in other parts of the world, such as the Total Access Communication System (TACS) in the United Kingdom, Italy, Spain, Austria, Ireland, MCS-L1 in Japan and Nordic Mobile Telephony (NMT) in several other countries. AMPS is still popular in the United States but analog systems are rarely used elsewhere nowadays. All these standards utilize frequency modulation (FM) for speech and perform handover decisions for a mobile at the BSs based on the power received at the BSs near the mobile. The available spectrum within each cell is partitioned into a number of channels and each call is assigned a dedicated pair of channels. Communication within the wired part of the system, which also connects with the Packet Switched Telephone Network (PSTN), uses a packet-switched network.

1.1.3 Digital Cellular Telephony

Analog cellular systems were the first step for the mobile telephony industry. Despite their significant success, they had a number of disadvantages that limited their performance. These disadvantages were alleviated by the second generation of cellular systems (2G systems) [2], which represent data digitally. This is done by passing voice signals through an Analog to Digital (A/D) converter and using the resulting bitstream to modulate an RF carrier. At the receiver, the reverse procedure is performed.

Compared to analog systems, digital systems have a number of advantages:

- Digitized traffic can easily be encrypted in order to provide privacy and security. Encrypted signals cannot be intercepted and overheard by unauthorized parties (at least not without very powerful equipment). Powerful encryption is not possible in analog systems, which most of the time transmit data without any protection. Thus, both conversations and network signaling can be easily intercepted. In fact, this has been a significant problem in 1G systems since in many cases eavesdroppers picked up user's identification numbers and used them illegally to make calls.
- Analog data representation made 1G systems susceptible to interference, leading to a highly variable quality of voice calls. In digital systems, it is possible to apply error detection and error correction techniques to the voice bitstream. These techniques make the transmitted signal more robust, since the receiver can detect and correct bit errors. Thus, these techniques lead to clear signals with little or no corruption, which of course translates into better call qualities. Furthermore, digital data can be compressed, which increases the efficiency of spectrum use.
- In analog systems, each RF carrier is dedicated to a single user, regardless of whether the

user is active (speaking) or not (idle within the call). In digital systems, each RF carrier is shared by more than one user, either by using different time slots or different codes per user. Slots or codes are assigned to users only when they have traffic (either voice or data) to send.

A number of 2G systems have been deployed in various parts of the world. Most of them include support for messaging services, such as the well-known Short Message Service (SMS) and a number of other services, such as caller identification. 2G systems can also send data, although at very low speeds (around 10 kbps). However, recently operators are offering upgrades to their 2G systems. These upgrades, also known as 2.5G solutions, support higher data speeds.

1.1.3.1 GSM

Throughout Europe, a new part of the spectrum in the area around 900 MHz has been made available for 2G systems. This allocation was followed later by allocation of frequencies at the 1800 MHz band. 2G activities in Europe were initiated in 1982 with the formation of a study group that aimed to specify a common pan-European standard. Its name was 'Groupe Speciale Mobile' (later renamed Global System for Mobile Communications). GSM [3], which comes from the initials of the group's name, was the resulting standard. Nowadays, it is the most popular 2G technology; by 1999 it had 1 million new subscribers every week. This popularity is not only due to its performance, but also due to the fact that it is the only 2G standard in Europe. This can be thought of as an advantage, since it simplifies roaming of subscribers between different operators and countries.

The first commercial deployment of GSM was made in 1992 and used the 900 MHz band. The system that uses the 1800 MHz band is known as DCS 1800 but it is essentially GSM. GSM can also operate in the 1900 MHz band used in America for several digital networks and in the 450 MHz band in order to provide a migration path from the 1G NMT standard that uses this band to 2G systems.

As far as operation is concerned, GSM defines a number of frequency channels, which are organized into frames and are in turn divided into time slots. The exact structure of GSM channels is described later in the book; here we just mention that slots are used to construct both channels for user traffic and control operations, such as handover control, registration, call setup, etc. User traffic can be either voice or low rate data, around 14.4 kbps.

1.1.3.2 HSCSD and GPRS

Another advantage of GSM is its support for several extension technologies that achieve higher rates for data applications. Two such technologies are High Speed Circuit Switched Data (HSCSD) and General Packet Radio Service (GPRS). HSCSD is a very simple upgrade to GSM. Contrary to GSM, it gives more than one time slot per frame to a user; hence the increased data rates. HSCD allows a phone to use two, three or four slots per frame to achieve rates of 57.6, 43.2 and 28.8 kbps, respectively. Support for asymmetric links is also provided, meaning that the downlink rate can be different than that of the uplink. A problem of HSCSD is the fact that it decreases battery life, due to the fact that increased slot use makes terminals spend more time in transmission and reception modes. However, due to the fact that reception requires significantly less consumption than transmission, HSCSD can be efficient for web browsing, which entails much more downloading than uploading.

GPRS operation is based on the same principle as that of HSCSD: allocation of more slots within a frame. However, the difference is that GPRS is packet-switched, whereas GSM and HSCSD are circuit-switched. This means that a GSM or HSCSD terminal that browses the Internet at 14.4 kbps occupies a 14.4 kbps GSM/HSCSD circuit for the entire duration of the connection, despite the fact that most of the time is spent reading (thus downloading) Web pages rather than sending (thus uploading) information. Therefore, significant system capacity is lost. GPRS uses bandwidth on demand (in the case of the above example, only when the user downloads a new page). In GPRS, a single 14.4 kbps link can be shared by more than one user, provided of course that users do not simultaneously try to use the link at this speed; rather, each user is assigned a very low rate connection which can for short periods use additional capacity to deliver web pages. GPRS terminals support a variety of rates, ranging from 14.4 to 115.2 kbps, both in symmetric and asymmetric configurations.

1.1.3.3 D-AMPS

In contrast to Europe, where GSM was the only 2G standard to be deployed, in the United States more than one 2G system is in use. In 1993, a time-slot-based system known as IS-54, which provided a three-fold increase in the system capacity over AMPS, was deployed. An enhancement of IS-54, IS-136 was introduced in 1996 and supported additional features. These standards are also known as the Digital AMPS (D-AMPS) family. D-AMPS also supports low-rate data, with typical ranges around 3 kbps. Similar to HSCSD and GRPS in GSM, an enhancement of D-AMPS for data, D-AMPS+ offers increased rates, ranging from 9.6 to 19.2 kbps. These are obviously smaller than those supported by GSM extensions. Finally, another extension that offers the ability to send data is Cellular Digital Packet Data (CDPD). This is a packet switching overlay to both AMPS and D-AMPS, offering the same speeds with D-AMPS+. Its advantages are that it is cheaper than D-AMPS+ and that it is the only way to offer data support in an analog AMPS network.

1.1.3.4 IS-95

In 1993, IS-95, another 2G system also known as cdmaOne, was standardized and the first commercial systems were deployed in South Korea and Hong Kong in 1995, followed by deployment in the United States in 1996. IS-95 utilizes Code Division Multiple Access (CDMA). In IS-95, multiple mobiles in a cell whose signals are distinguished by spreading them with different codes, simultaneously use a frequency channel. Thus, neighboring cells can use the same frequencies, unlike all other standards discussed so far. IS-95 is incompatible with IS-136 and its deployment in the United States started in 1995. Both IS-136 and IS-95 operate in the same bands with AMPS. IS-95 is designed to support dual-mode terminals that can operate either under an IS-95 or an AMPS network. IS-95 supports data traffic at rates of 4.8 and 14.4 kbps. An extension of IS-95, known as IS-95b or cdmaTwo, offers support for 115.2 kbps by letting each phone use eight different codes to perform eight simultaneous transmissions.

1.1.4 Cordless Phones

Cordless telephones first appeared in the 1970s and since then have experienced a significant growth. They were originally designed to provide mobility within small coverage areas, such as homes and offices. Cordless telephones comprise a portable handset, which communicates with a BS connected to the Public Switched Telephone Network (PSTN). Thus, cordless telephones primarily aim to replace the cord of conventional telephones with a wireless link.

Early cordless telephones were analog. This fact resulted in poor call quality, since handsets were subject to interference. This situation changed with the introduction of the first generation of digital cordless telephones, which offer voice quality equal to that of wired phones.

Although the first generation of digital cordless telephones was very successful, it lacked a number of useful features, such as the ability for a handset to be used outside of a home or office. This feature was provided by the second generation of digital cordless telephones. These are also known as telepoint systems and allow users to use their cordless handsets in places such as train stations, busy streets, etc. The advantages of telepoint over cellular phones were significant in areas where cellular BSs could not be reached (such as subway stations). If a number of appropriate telepoint BSs were installed in these places, a cordless phone within range of such a BS could register with the telepoint service provider and be used to make a call. However, the telepoint system was not without problems. One such problem was the fact that telepoint users could only place and not receive calls. A second problem was that roaming between telepoint BSs was not supported and consequently users needed to remain in range of a single telepoint BS until their call was complete. Telepoint systems were deployed in the United Kingdom where they failed commercially. Nevertheless, in the mid-1990s, they faired better in Asian countries due to the fact that they could also be used for other services (such as dial-up in Japan). However, due to the rising competition by the more advanced cellular systems, telepoint is nowadays a declining business.

The evolution of digital cordless phones led to the DECT system. This is a European cordless phone standard that provides support for mobility. Specifically, a building can be equipped with multiple DECT BSs that connected to a Private Brach Exchange (PBX). In such an environment, a user carrying a DECT cordless handset can roam from the coverage area of one BS to that of another BS without call disruption. This is possible as DECT provides support for handing off calls between BSs. In this sense, DECT can be thought of as a cellular system. DECT, which has so far found widespread use only in Europe, also supports telepoint services.

A standard similar to DECT is being used in Japan. This is known as the Personal Handyphone System (PHS). It also supports handoff between BSs. Both DECT and PHS support two-way 32 kbps connections, utilize TDMA for medium access and operate in the 1900 MHz band.

1.1.5 Wireless Data Systems

The cellular telephony family is primarily oriented towards voice transmission. However, since wireless data systems are used for transmission of data, they have been digital from the beginning. These systems are characterized by bursty transmissions: unless there is a packet to transmit, terminals remain idle. The first wireless data system was developed in 1971 at the

University of Hawaii under the research project ALOHANET. The idea of the project was to offer bi-directional communications between computers spread over four islands and a central computer on the island of Oahu without the use of phone lines. ALOHA utilized a star topology with the central computer acting as a hub. Any two computers could communicate with each other by relaying their transmissions through the hub. As will be seen in later chapters, network efficiency was low; however, the system's advantage was its simplicity. Although mobility was not part of ALOHA, it was the basis for today's mobile wireless data systems.

1.1.5.1 Wide Area Data Systems

These systems offer low speeds for support of services such as messaging, e-mail and paging. Below, we briefly summarize several wide area data systems. A more thorough discussion is given in Ref. [4].

- *Paging systems.*These are one-way cell-based systems that offer very low-rate data transmission towards the mobile user. The first paging systems transmitted a single bit of information in order to notify users that someone wanted to contact them. Then, paging messages were augmented and could transfer small messages to users, such as the telephone number of the person to contact or small text messages. Paging systems work by broadcasting the page message from many BSs, both terrestrial and satellite. Terrestrial systems typically cover small areas whereas satellites provide nationwide coverage. It is obvious that since the paging message is broadcasted, there is no need to locate mobile users or route traffic. Since transmission is made at high power levels, receivers can be built without sophisticated hardware, which of course translates into lower manufacturing costs and device size. In the United States, two-way pagers have also appeared. However, in this case mobile units increase in size and weight, and battery time decreases. The latter fact is obviously due to the requirement for a powerful transmitter in the mobile unit capable of producing signals strong enough to reach distant BSs. Paging systems were very popular for many years, however, their popularity has started to decline due to the availability of the more advanced cellular phones. Thus, paging companies have started to offer services at lower prices in order to compete with the cellular industry.
- *Mobitex.*This is a packet-switched system developed by Ericsson for telemetry applications. It offers very good coverage in many regions of the world and rates of 8 kbps. In Mobitex, coverage is provided by a system comprising BSs mounted on towers, rooftops, etc. These BSs are the lower layer of a hierarchical network architecture. Medium access in Mobitex is performed through an ALOHA-like protocol. In 1998, some systems were built for the United States market that offered low-speed Internet access via Mobitex.
- *Ardis.* This circuit-switched system was developed by Motorola and IBM. Two versions of Ardis, which is also known as DataTAC, exist: Mobile Data Communications 4800 (MDC4800) with a speed of 4.8 kbps and Radio Data Link Access Protocol (RD-LAP), which offers speeds of 19.2 kbps while maintaining compatibility with MDC4800. As in Mobitex, coverage is provided by a few BSs mounted on towers, rooftops, etc., and these BSs are connected to a backbone network. Medium access is also carried out through an ALOHA-like protocol.
- *Multicellular Data Network (MCDN).* This is a system developed by Metricom and is also

known as Ricochet. MCDN was designed for Internet access and thus offers significantly higher speeds than the above systems, up to 76 kbps. Coverage is provided through a dense system of cells of radius up to 500 m. Cell BSs are mounted close to street level, for example, on lampposts. User data is relayed through BSs to an access point that links the system to a wired network. MCDN is characterized by round-trip delay variability, ranging from 0.2 to 10 s, a fact that makes it inefficient for voice traffic. Since cells are very scattered, coverage of an entire country is difficult, since it would demand some millions of BS installations. Finally, the fact that MCDN demands spectrum in the area around the 900 MHz band makes its adoption difficult in countries where these bands are already in use. Such is the case in Europe, where the 900 MHz band is used by GSM. Moving MCDN to the 2.4 GHz band which is license-free in Europe would make cells even smaller. This would result in a cost increase due to the need to install more BSs.

1.1.5.2 Wireless Local Area Networks (WLANS)

WLANs [2,5,6] are used to provide high-speed data within a relatively small region, such as a small building or campus. WLAN growth commenced in the mid-1980s and was triggered by the US Federal Communications Commission (FCC) decision to authorize license-free use of the Industrial, Scientific and Medical (ISM) bands. However, these bands are likely to be subject to significant interference, thus the FCC sets a limit on the power per unit bandwidth for systems utilizing ISM bands. Since this decision of the FCC, there has been a substantial growth in the area of WLANs. In the early years, however, lack of standards enabled the appearance of many proprietary products thus dividing the market into several, possibly incompatible parts.

The first attempt to define a standard was made in the late 1980s by IEEE Working Group 802.4, which was responsible for the development of the token-passing bus access method. The group decided that token passing was an inefficient method to control a wireless network and suggested the development of an alternative standard. As a result, the Executive Committee of IEEE Project 802 decided to establish Working Group IEEE 802.11, which has been responsible since then for the definition of physical and MAC sub-layer standards for WLANs. The first 802.11 standard offered data rates up to 2 Mbps using either spread spectrum transmission in the ISM bands or infrared transmission. In September 1999, two supplements to the original standard were approved by the IEEE Standards Board. The first standard, 802.11b, extends the performance of the existing 2.4 GHz physical layer, with potential data rates up to 11 Mbps. The second standard, 802.11a aims to provide a new, higher data rate (from 20 to 54 Mbps) physical layer in the 5 GHz ISM band. All these variants use the same Medium Access Control (MAC) protocol, known as Distributed Foundation Wireless MAC (DFWMAC). This is a protocol belonging in the family of Carrier Sense Multiple Access protocols tailored to the wireless environment. IEEE 802.11 is often referred to as wireless Ethernet and can operate either in an ad hoc or in a centralized mode. An ad hoc WLAN is a peer-to-peer network that is set up in order to serve a temporary need. No networking infrastructure needs to be present and network control is distributed along the network nodes. An infrastructure WLAN makes use of a higher speed wired or wireless backbone. In such a topology, mobile nodes access the wireless channel under the coordination of a Base Station (BS), which can also interface the WLAN to a fixed network backbone.

In addition to IEEE 802.11, another WLAN standard, High Performance European Radio LAN (HIPERLAN), was developed by group RES10 of the European Telecommunications Standards Institute (ETSI) as a Pan-European standard for high speed WLANs. The HIPER-LAN 1 standard covers the physical and MAC layers, offering data rates between 2 and 25 Mbps by using narrowband radio modulation in the 5.2 GHz band. HIPERLAN 1 also utilizes a CSMA-like protocol. Despite the fact that it offers higher data rates than most 802.11 variants, it is less popular than 802.11 due to the latter's much larger installed base. Like IEEE 802.11, HIPERLAN 1 can operate either in an ad hoc mode or with the supervision of a BS that provides access to a wired network backbone.

1.1.5.3 Wireless ATM (WATM)

In 1996 the ATM Forum approved a study group devoted to WATM. WATM [7,8] aims to combine the advantages of freedom of movement of wireless networks with the statistical multiplexing (flexible bandwidth allocation) and QoS guarantees supported by traditional ATM networks. The latter properties, which are needed in order to support multimedia applications over the wireless medium, are not supported in conventional LANs due to the fact that these were created for asynchronous data traffic. Over the years, research led to a number of WATM prototypes.

An effort towards development of a WLAN system offering the capabilities of WATM is HIPERLAN 2 [9,10]. This is a connection-oriented system compatible with ATM, which uses fixed size packets and offers high speed wireless access (up to 54 Mbps at the physical layer) to a variety of networks. Its connection-oriented nature supports applications that demand QoS.

1.1.5.4 Personal Area Networks (PANs)

PANs are the next step down from LANs and target applications that demand very short-range communications (typically a few meters). Early research for PANs was carried out in 1996. However, the first attempt to define a standard for PANs dates back to an Ericsson project in 1994, which aimed to find a solution for wireless communication between mobile phones and related accessories (e.g. hands-free kits). This project was named Bluetooth [11,12] (after the name of the king that united the Viking tribes). It is now an open industry standard that is adopted by more than 100 companies and many Bluetooth products have started to appear in the market. Its most recent version was released in 2001. Bluetooth operates in the 2.4 MHz ISM band; it supports 64 kbps voice channels and asynchronous data channels with rates ranging up to 721 kbps. Supported ranges of operation are 10 m (at 1 mW transmission power) and 100 meters (at 1 mW transmission power).

Another PAN project is HomeRF [13]; the latest version was released in 2001. This version offers 32 kbps voice connections and data rates up to 10 Mbps. HomeRF also operates in the 2.4 MHz band and supported ranges around 50 m. However, Bluetooth seems to have more industry backing than HomeRF.

In 1999, IEEE also joined the area of PAN standardization with the formation of the 802.15 Working Group [14,15]. Due to the fact that Bluetooth and HomeRF preceded the initiative of IEEE, a target of the 802.15 Working Group will be to achieve interoperability with these projects.

1.1.6 Fixed Wireless Links

Contrary to the wireless systems presented so far (and later on), fixed wireless systems lack the capability of mobility. Such systems are typically used to provide high speeds in the local loop, also known as the last mile. This is the link that connects a user to a backbone network, such as the Internet. Thus, fixed wireless links are competing with technologies such as fiber optics and Digital Subscriber Line (DSL).

Fixed wireless systems are either point-to-point or point-to-multipoint systems. In the first case, the company that offers the service uses a separate antenna transceiver for each user whereas in the second case one antenna transceiver is used to provide links to many users. Point-to-multipoint is the most popular form of providing fixed wireless connectivity, since many users can connect to the same antenna transceiver. Companies offering point-to-multipoint services place various antennas in an area, thus forming some kind of cellular structure. However, these are different from the cells of conventional cellular systems, since cells do not overlap, the same frequency is reused at each cell and no handoff is provided since users are fixed. The most common fixed wireless systems are presented below and are typically used for high-speed Internet access:

- *ISM-band systems.* These are systems that utilize the 2.4 GHz ISM band. Transmission is performed by using spread spectrum technology. Specifically, many such systems actually operate using the IEEE 11 Mbps 802.11b standard, which utilizes spread spectrum technology. ISM-band systems are typically organized into cells of 8 km radius. The maximum capacity offered within a cell is 11 Mbps although most of the time capacity is between 2 and 6 Mbps. In point-to-multipoint systems this capacity is shared among the cell's users.
- *MMDS.* Multipoint Multichannel Distribution System (MMDS) utilizes the spectrum originally used for analog television broadcasting. This spectrum is in the bands between 2.1 and 2.7 GHz. Such systems are typically organized into cells of 45 km. These higher ranges are possible due to the fact that in licensed bands, transmission at a higher power is permitted. The maximum capacity of an MMDS cell is 36 Mbps and is shared between the users of the cell. MMDS supports asymmetric links with a downlink up to 5 Mbps and an uplink up to 256 kbps.
- *LMDS.* Local Multipoint Distribution System (LMDS) utilizes higher frequencies (around 30 GHz) and thus smaller cells (typically 1-2 km) than MMDS. It offers a maximum cell capacity of 155 Mbps.

1.1.7 Satellite Communication Systems

The era of satellite systems began in 1957 with the launch of Sputnik by the Soviet Union. However, the communication capabilities of Sputnik were very limited. The first real communication satellite was the AT&T Telstar 1, which was launched by NASA in 1962. Telstar 1 was enhanced in 1963 by its successor, Telstar 2. From the Telstar era to today, satellite communications [16] have enjoyed an enormous growth offering services such as data, paging, voice, TV broadcasting, Internet access and a number of mobile services.

Satellite orbits belong to three different categories. In ascending order of height, these are the circular Low Earth Orbit (LEO), Medium Earth Orbit (MEO) and Geosynchronous Earth Orbit (GEO) categories at distances in the ranges of 100–1000 km, 5000–15 000 km and approximately 36 000 km, respectively. There also exist satellites that utilize elliptical orbits. These try to combine the low propagation delay property of LEO systems and the stability of GEO systems.

The trend nowadays is towards use of LEO orbits, which enable small propagation delays and construction of simple and light ground mobile units. A number of LEO systems have appeared, such as Globalstar and Iridium. They offer voice and data services at rates up to 10 kbps through a dense constellation of LEO satellites.

1.1.8 Third Generation Cellular Systems and Beyond

Despite their great success and market acceptance, 2G systems are limited in terms of maximum data rate. While this fact is not a limiting factor for the voice quality offered, it makes 2G systems practically useless for the increased requirements of future mobile data applications. In future years, people will want to be able to use their mobile platforms for a variety of services, ranging from simple voice calls, web browsing and reading e-mail to more bandwidth hungry services such as video conferencing, real-time and bursty-traffic applications. To illustrate the inefficiency of 2G systems for capacity-demanding applications, consider a simple transfer of a 2 MB presentation. Such a transfer would take approximately 28 minutes employing the 9.6 kbps GSM data transmission. It is clear that future services cannot be realized over the present 2G systems.

In order to provide for efficient support of such services, work on the Third Generation (3G) of cellular systems [17–19] was initiated by the International Telecommunication Union (ITU) in 1992. The outcome of the standardization effort, called International Mobile Telecommunications 2000 (IMT-2000), comprises a number of different 3G standards. These standards are as follows:

- EDGE, a TDMA-based system that evolves from GSM and IS-136, offering data rates up to 473 kbps and backwards compatibility with GSM/IS-136;
- cdma2000, a fully backwards-compatible descendant of IS-95 that supports data rates up to 2 Mbps;
- WCDMA, a CDMA-based system that introduces a new 5-MHz wide channel structure, capable of offering speeds up to 2 Mbps.

As far as the future of wireless networks is concerned, it is envisioned that evolution will be towards an integrated system, which will produce a common packet-switched (possibly IP-based) platform for wireless systems. This is the aim of the Fourth Generation (4G) of cellular networks [20–22], which targets the market of 2010 and beyond. The unified platform envisioned for 4G wireless networks will provide transparent integration with the wired networks and enable users to seamlessly access multimedia content such as voice, data and video irrespective of the access methods of the various wireless networks involved. However, due to the length of time until their deployment, several issues relating to future 4G networks are not so clear and are heavily dependent on the evolution of the telecommunications market and society in general.

1.2 Challenges

The use of wireless transmission and the mobility of most wireless systems give rise to a number challenges that must be addressed in order to develop efficient wireless systems. The most important of these challenges are summarized below.

1.2.1 Wireless Medium Unreliability

Contrary to wireline, the wireless medium is highly unreliable. This is due to the fact that wireless signals are subject to significant attenuation and distortion due to a number of issues, such as reflections from objects in the signal's path, relative movement of transmitter and receiver, etc. The way wireless signals are distorted is difficult to predict, since distortions are generally of random nature. Thus, wireless systems must be designed with this fact in mind. Procedures for hiding the impairments of the wireless links from high-layer protocols and applications as well as development of models for predicting wireless channel behavior would be highly beneficial.

1.2.2 Spectrum Use

As will be seen in later chapters, the spectrum for wireless systems is a scarce resource and must be regulated in an efficient way. Considering the fact that the spectrum is also an expensive resource, it can be seen that efficient regulation by the corresponding organizations, such as the FCC in the United States and ETSI in Europe, would be highly beneficial. If companies have the ability to get a return on investments in spectrum licenses, reluctance to deploy wireless systems will disappear, with an obvious advantage for the proliferation of wireless systems.

Finally, the scarcity of spectrum gives rise to the need for technologies that can either squeeze more system capacity over a given band, or utilize the high bands that are typically less crowded. The challenge for the latter option is the development of equipment at the same cost and performance as equipment used for lower band systems.

1.2.3 Power Management

Hosts in wireline networks are not subject to power limitations, whereas those in wireless networks are typically powered by batteries. Since batteries have a finite storage capacity, users have to regularly recharge their devices. Therefore, systems with increased time of operation between successive recharges are needed so as not to burden the users with frequent recharging. Furthermore, mobile networks need to utilize batteries that do not weigh a lot in order to enable terminal mobility. Since battery technology is not as advanced as many would like, the burden falls on developing power-efficient electronic devices. Power consumption of dynamic components is proportional to CV^2F, where C is the capacitance of the circuit, V is the voltage swing and F is the clock frequency [23]. Thus, to provide power efficiency, one or more of the following are needed: (a) greater levels of VLSI integration to reduce capacitance; (b) lower voltage circuits must be developed; (c) clock frequency must be reduced.

Alternatively, wireless systems can be built in such a way that most of the processing is carried out in fixed parts of the network (such as BSs in cellular systems), which are not power-limited. Finally, the same concerns for power management also affect the software for wireless networks: efficient software can power down device parts when those are idle.

1.2.4 Security

Although security is a concern for wireline networks as well, it is even more important in the wireless case. This is due to the fact that wireless transmissions can be picked up far more easily than transmissions in a wire. While in a wireline network physical access to the wire is needed to intercept a transmission, in a wireless network transmissions can be intercepted even hundreds or thousands of meters away from the transmitter. Sufficient levels of security should thus be provided if applications such as electronic banking and commerce are to be deployed over wireless networks.

1.2.5 Location/Routing

Most wireless networks are also mobile. Thus, contrary to wireline networks, the assumption of a static network topology cannot be made. Rather, the topology of a wireless network is likely to change in time. In cellular networks, mobility gives rise to the capability of roaming between cells. In this case, efficient techniques are needed for (a) locating mobile terminals and (b) support ongoing calls during inter-cell terminal movements (handoffs). In the case of ad hoc data networks, routing protocols that take into account dynamic network topologies are needed.

1.2.6 Interfacing with Wired Networks

The development of protocols and interfaces that allow mobile terminals to connect to a wired network backbone is significant. This would give mobile users the potential to use their portable devices for purposes such as file-transfer, reading email and Internet access.

1.2.7 Health Concerns

The increasing popularity of wireless networks has raised a lot of concerns regarding their impact on human health. This is an important issue but it is not within the scope of this book. Nevertheless, it is worth mentioning that most concerns are targeted at cellular phones. This is because the transmission power of cellular phones is typically higher than that of wireless systems, such as WLANs and PANs. Moreover, contrary to WLANs and PANs, (a) cellular phones are operated at close proximity to the human brain, (b) when used for voice calls they emit radiation for the entire duration of the call (this is not the case with WLANs, where transmissions are typically of bursty nature; thus terminals transmit for short time durations).

Although microwave and radio transmissions are not as dangerous as higher-band radiation (such as gamma and X-rays), prolonged use of microwaves can possibly affect the human brain. One such effect of prolonged use of microwaves is a rise in temperature (as intentionally achieved in microwave ovens). A number of studies have appeared in the medical literature; however a final answer has yet to be given to the question of health concerns: Some people say that it is not yet proven that wireless networks cause health problems. Nevertheless, others say that the studies so far have not proved the opposite. Overall, the situation somewhat resembles the early days of electricity, when many people believed that the presence of wires carrying electricity in their homes would damage their health. Obviously, after so many years it can be seen that such a fear was greatly overestimated.

Similarly, a lot of time will have to pass before we possess such knowledge on the health concerns regarding wireless networks.

1.3 Overview

1.3.1 Chapter 2: Wireless Communications Principles and Fundamentals

In order to provide a background on the area of wireless networks, Chapter 2 describes fundamental issues relating to wireless communications. The main issues covered are as follows:

- A description of the various bands of the electromagnetic spectrum and their properties is given. In increasing order of frequency, these are the radio, microwave, infrared, visible light, ultraviolet, X-ray and gamma-ray bands. The higher a band's frequency, the more data it can carry; however, the bands above visible light are rarely used due to the fact that they are difficult to modulate and dangerous to living creatures. The most commonly used bands for commercial communication systems fall within the microwave range.
- The fact that spectrum is a scarce resource is identified. Thus, spectrum use must abide by some form of regulation in order to ensure interference-free operation. The three main approaches for regulating spectrum, comparative bidding, lottery and auctions are discussed.
- The physical phenomena that govern wireless signal propagation are discussed. These include free space loss, Doppler shift and the propagation phenomena, which cause multipath propagation and shadowing. These are reflection, diffraction and scattering. The implications of these phenomena on signal reception are discussed. The characteristics of signal propagation in rural and urban situations (e.g. city streets) and the differences between indoor and outdoor signal propagation are examined.
- The advantages of digital over analog data representation are given. These are increased transmission reliability, efficient use of spectrum and security.
- Voice coding, which is used to convert voice from its analog form to a digital form that will be transmitted over a digital wireless network, such as a 2G network, is discussed. The discussion starts with the process of PCM conversion of an analog signal to a digital signal and proceeds to vocoders and hybrid codecs, which try to reduce the bit rate required for digitized voice transmission. Vocoders work by digitally encoding not the actual voice signals but rather the mechanics of voice production. Vocoders can achieve rates as low as 1.2–2.4 kbps at the expense however of producing 'mechanized' voice signals. Hybrid codecs transmit both vocoding and PCM voice information in an effort to overcome this problem.
- The most common modulation techniques are presented along with examples describing each of them. Analog modulation techniques include Amplitude Modulation (AM) and Frequency Modulation (FM). The digital modulation techniques are Amplitude Shift Keying (ASK), Frequency Shift Keying (FSK), Phase Shift Keying (PSK) and Quadrate Amplitude Modulation (QAM).
- Multiple access techniques for wireless networks are presented. These are (a) Frequency Division Multiple Access (FDMA), which separates users in the frequency domain, (b) Time Division Multiple Access (TDMA), which separates users in the time domain, (c)

Code Division Multiple Access (CDMA), in which users are separated by using different codes each, (d) ALOHA and Carrier Sense Multiple Access (CSMA), which are random access protocols and (e) Randomly Addressed Polling (RAP) and Group RAP (GRAP) as examples of polling protocols.

- An overview of techniques that increase the performance of wireless systems by combating the impairments of wireless links is given. These include antenna diversity, multi-antenna transmission, coding, equalization, power control and multicarrier modulation.
- The concept of cellular networks, which is extensively used by commercial systems in order to increase the efficiency of spectrum usage, is discussed, along with the associated issues of frequency reuse and location/handoff.
- The basic issues of ad hoc networks, which are wireless networks having no central administration, are discussed. Network topology determination, connectivity maintenance and packet routing in ad hoc wireless networks are discussed. Next, the semi ad hoc concept is presented.
- Two different approaches for delivering data to mobile clients have been presented. These are the push and pull approaches.
- Finally, an introductory overview of the basic techniques and interactions at the different network layers is made with the help of the OSI reference model.

1.3.2 Chapter 3: First Generation (1G) Cellular Systems

Chapter 3 discusses the first generation cellular systems. The era of cellular telephony as we understand it today began with the introduction of the these 1G systems. 1G systems served mobile telephone calls via analog transmission of voice traffic. Despite the fact that 1G systems are considered technologically primitive today the fact remains that a significant number of people still use analog cellular phones and an analog cellular infrastructure is found throughout North America and other parts of the world. Furthermore, they have found use as a basis for the development of several second generation systems. An example of this is D-AMPS, which is a 2G system evolving from AMPS. This chapter describes:

- *The Advanced Mobile Phone System (AMPS)*. AMPS divides the frequency spectrum into several channels, each 30 kHz wide. These channels are either speech or control channels. Speech channels utilize Frequency Modulation (FM), while control channels can use Binary Frequency Shift Keying (BFSK) at a rate of 10-kb/s. Both data messages and frequency tones are used for AMPS control signaling and two operators can be collocated in the same geographical area.
- *The Nordic Mobile Telephony (NMTS) system*. Two versions of NMT exist. The first operates in the area around 450 MHz and the second operates in the area around 900 MHz. These variants are known as NMT 450 and NMT 900, respectively.

1.3.3 Chapter 4: Second Generation (2G) Cellular Systems

The era of mobile telephony as we understand it today is dominated by second generation cellular standards. Chapter 4 describes several 2G standards:

- D-AMPS, the 2G TDMA system that is used in North America, descended from the 1G

AMPS, is described. D-AMPS operates at 800 MHz as an overlay over the analog AMPS network. It maintains the 30-kHz channel spacing of AMPS and uses AMPS carriers to deploy digital channels. Each such digital channel can support three times the users that are be supported by AMPS with the same carrier. Digital channels are organized into frames, with each frame comprising six slots. The actual channel a user sees is comprised of one or two slots within each frame. D-AMPS can be seen as an overlay on AMPS that 'steals' some carriers and changes them to carry digital traffic. Obviously, this does not affect the underlying AMPS network, since an AMPS MS can continue to operate. IS-136 is a descendant of D-AMPS that also operates in the 800 MHz bands. However, upgrades to the 1900 band are planned. While D-AMPS maintains the analog channels of AMPS, IS-136 is a fully digital standard. Both D-AMPS and its successor IS-136 support voice as well as data services. Supported speeds for data services are up to 9.6 kbps.

- IS-95, which is the only 2G system based on CDMA is discussed. It is a fully digital standard that operates in the 800 MHz band, like AMPS. In IS-95, multiple mobiles in a cell whose signals are distinguished by spreading them with different codes simultaneously use a frequency channel. Thus, neighboring cells can use the same frequencies, unlike all other standards discussed so far. IS-95 supports data traffic at rates of 4.8 and 14.4 kbps.

- The widely used Global System for Mobile Communications (GSM) is described. Four variants of GSM exist, operating at 900 MHz, 1800 MHz, 1900 MHZ and 450 MHz. It is a fully digital standard and manages channel access via a TDD mechanism that splits the available bandwidth in the time domain. The resulting access method is actually a hierarchy of slots, frames, multiframes and hyperframes. Apart from voice services, GSM also offers data transfer services. The speeds a user sees are typically round 9.6 kbps.

- IS-41, which is actually not a 2G standard but rather a protocol that operates on the network side of North American cellular networks is discussed.

- Approaches for data transmission over 2G systems, including GRPS, HSCSD, cdmaTwo and D-AMPS + are discussed. HSCSD is a simple upgrade to GSM that lets each MS to be allocated up to four slots within each frame thus resulting in maximum speeds up to 57.6 kbps. The problem of HSCSD stems from its circuit-switched nature. GPRS also allocates up to eight slots to each MS thus resulting in maximum speeds up to 115.2 kbps. However, it has the advantage of being packet-switched. CdmaTwo is an upgrade of cdmaOne that lets a MS use up to eight spreading codes. This is equivalent to performing more than one CDMA transmission, thus resulting in speeds up to 115.2 kbps. Furthermore, the problems faced by TCP in a wireless environment, mobileIP, an extension of the Internet Protocol (IP) that supports terminal mobility and the Wireless Access Protocol (WAP) are discussed.

- Cordless Telephony (CT) including analog and digital CT, the Digital European Cordless Telecommunications Standard (DECT) and Personal Handyphone System (PHS) standards are discussed.

1.3.4 Chapter 5: Third Generation (3G) Cellular Systems

Chapter 5 discusses 3G mobile and wireless networks. The goal of 3G wireless networks is to provide efficient support for both voice and high bit-rate data services (ranging from 144 kbps

to 2 Mbps) in order to remove the deficiency of 2G systems for supporting bandwidth-hungry
data services. The main issues covered in Chapter 5 are:

- The fact that assignment of new bands for 3G systems has proven to be a difficult task is
  identified. Apart from bands already regulated for 3G, a number of additional bands that
  can be used for 3G are mentioned. The development and commercial use of efficient
  technologies that can alleviate problems due to non-uniform worldwide spectrum regulation and spectrum shortage will be highly beneficial. Technologies that achieve this, such
  as software radio and multi-user detection, are described.
- A description of the service classes that will be offered by 3G systems is given. These are
  (a) voice and audio for the support of voice/audio traffic, (b) wireless messaging, offering
  multimedia-capable messaging, (c) switched data, supporting dial-up access, (d) medium
  multimedia, enabling web browsing, (e) high multimedia, for high-speed Internet access
  and (d) interactive high multimedia, that will offer the maximum speeds possible. Some of
  the 3G applications that will probably be popular among the user community are
  presented.
- Standardization procedures and the outcome of the standardization effort, IMT-2000,
  which comprises three different 3G standards for the air-interface are discussed. These
  standards are (a) EDGE, a TDMA-based system that evolves from GSM and IS-136,
  offering data rates up to 473 kbps and backwards compatibility with GSM/IS-136, (b)
  cdma2000, a fully backwards-compatible descendant of IS-95 that supports data rates up
  to 2 Mbps and (c) WCDMA, a CDMA-based system that introduces a new 5-MHz wide
  channel structure, capable of offering speeds up to 2 Mbps.
- Finally, possible architectures for the fixed parts of future 3G cellular networks are
  discussed.

1.3.5 Chapter 6: Future Trends: Fourth Generation (4G) Systems and Beyond

Chapter 6 provides a vision of 4G mobile and wireless systems. Such systems target the
market of 2010 and beyond, aiming to offer support to mobile applications demanding data
rates of at least 50 Mbps. Due to the large time window until their deployment, both the
telecommunications scene and the services offered by 4G and future systems are not yet
known and, as a result, the aims for these systems may change over time. However, as 3G
systems move from research to the implementation stage, 4G and future systems will be an
extremely interesting field of research on future generation wireless systems. The main issues
covered by Chapter 6 are:

- 4G systems aim to provide a common IP-based platform for the multiple mobile and
  wireless systems offering higher data rates. The desired properties of 4G systems are
  identified. Furthermore, OFDM, a promising technology for providing high data rates,
  is presented.
- A description of possible applications and service classes that will dominate the 4G
  market, as these emerge from ongoing research, are presented. These include (a) tele-
  presence, (b) information access, (c) inter-machine communication, (d) intelligent shopping and (e) location-based services.
- A discussion on the challenge of predicting the future of wireless networks is made. Many

issues of these systems are not so clear and are dependent on the evolution of the tele-communications market and society in general. Three different scenarios for the future generations of wireless networks are presented, along with possible research issues for each scenario.

1.3.6 Chapter 7: Satellite Networks

Chapter 7 discusses satellite-based wireless systems. Although facing competition from terrestrial technologies and having faced market problems, satellite-based systems seem to be promising for offering voice and especially Internet services to users scattered around the world. The main issues covered by Chapter 7 are:

- The characteristics of satellite communications and the three bands mainly used for satellite communication networks are discussed. Possible applications of satellite communications are presented. These include voice telephony, use in cellular systems, connectivity for aircraft passengers, Global Positioning Systems (GPS) and Internet access.
- The various possible orbits of satellite systems and their characteristics are described. These include the circular LEO (100–1000 km), MEO (5000–15 000 km), GEO (approximately 36 000 km) and elliptical orbits. LEO and MEO are characterized by relatively short propagation delays. They form constellations that orbit the Earth at speeds greater than its rotation speed. GEO systems experience higher propagation delays than LEO/MEO systems. However, they have the advantage of rotating at a speed equal to that of the Earth's rotation, and thus a GEO satellite appears fixed at a certain point in the sky. Elliptical orbits try to combine the low propagation delay property of LEO systems and the stability of GEO systems.
- The VSAT approach along with its topology and operation are presented. VSAT systems are especially useful for interconnecting large numbers of users residing in remote areas. They can operate either by using an Earth Station (ES) as a 'hub', or by using an intelligent satellite that incorporates the hub's functionality.
- The Iridium and Globalstar, voice-oriented satellite systems are described. Iridium, which was abandoned in 2000 for economical reasons, targets worldwide coverage through a LEO constellation of 66 satellites orbiting at 11 different planes with six satellites per plane. Satellites are able to communicate with each other through Inter-Satellite links (ISLs). Globalstar is a relatively simple system, which demands the presence of a Globalstar ground unit (gateway) in range of the satellite that serves the user. ISLs are not supported in Globalstar.
- Finally, a number of issues relating to satellite-based Internet access, including possible architectures, routing and transport issues, are discussed.

1.3.7 Chapter 8: Fixed Wireless Access Systems

Chapter 8 discusses fixed wireless systems. The US Federal Communications Commission (FCC) has licensed wireless broadband services at four locations in the radio spectrum: the Multichannel Multipoint Distribution Service (MMDS), Digital Electronic Messaging Service (DEMS), Local Multipoint Distribution Service (LMDS), and Microwave Service.

LMDS and MMDS are discussed in this chapter. MMDS networks utilize a single omni-directional central antenna that can provide MMDS service to an area faster and with a much smaller investment than other broadband services. One MMDS supercell can cover an area of about 3850 square miles. However, it is not easy to obtain line-of-sight, which may affect as many as 60% of households. Local Multipoint Distribution Service (LMDS) requires easy deployment. It was developed to provide a radio-based delivery service for a wide variety of broadband services. Due to the huge spectrum available, LMDS can provide high speed services with data rates reaching 155 Mbps. However, LMDS requires small cell sizes due to the high frequency at which they operate. Therefore, the average LMDS cell can cover between 12.6 and 28.3 square miles. The service provider can choose to launch its system at a pace to match its individual business plan without sacrificing quality of service (QoS). More-over, LMDS subscribers will be able to utilize a rooftop or window-based antenna to receive signals from a radio base station.

1.3.8 Chapter 9: Wireless Local Area Networks

Chapter 9 discusses Wireless Local Area Networks (WLANs). The main issues covered by Chapter 9 are:

- The two types of WLAN topologies used today (ad hoc and infrastructure) are presented. A number of requirements for WLAN systems are presented.
- The five current physical layer alternatives for WLANs, which are based either on infrared (IR) or microwave transmission, are presented. The IR-based physical layer provides the advantages of greater security and potentially higher data rates, however, not many IR-based products exist. Microwave alternatives include Frequency Hopping Spread Spectrum Modulation, Direct Sequence Spread Spectrum Modulation, Narrowband Modulation and Orthogonal Frequency Division Multiplexing (OFDM). The Spread Spectrum and the OFDM approaches offer superior performance in the presence of fading which is the dominant propagation characteristic of wireless transmission. The Spread Spectrum techniques trade off bandwidth for this superiority, offering moderate data rates. Narrowband modulation, on the other hand, can potentially offer higher data rates than Spread Spectrum, but are subject to increased performance degradation due to interference. The OFDM approach is a form of multicarrier modulation that achieves high data rates.
- The two WLAN MAC standards available today, IEEE 802.11 and HIPERLAN 1, which both employ contention-based CSMA-like MAC algorithms are presented. The 802.11 MAC includes a mechanism that combats the hidden terminal problem whereas such a technique is not included in the HIPERLAN 1 standard. The latter includes a mechanism for multi-hop network support, effectively increasing the network's operating area. However, it pays the price of reduced overall performance compared to the single hop case. The way both of the standards try to support time-bounded services is described. Power saving and security in both standards are also discussed.
- The latest developments in the WLAN area are discussed. These include the 802.11a and 802.11b standards, which are physical layer enhancements of 802.11 that provide high data rates. Furthermore, the aims of the ongoing work within Task Groups d, e, f, g, h, i of Working Group 802.11 are reported.

1.3.9 Chapter 10: WATM and Wireless Ad Hoc Routing

Chapter 10 describes Wireless Asynchronous Transfer Mode (WATM). WATM combines the advantages of wired ATM networks and wireless networks. These are the flexible bandwidth allocation offered through the statistical multiplexing capability of ATM and the freedom of terminal movement offered by wireless networks. This combination will enable implementation of QoS demanding applications over the wireless medium. The main issues covered by Chapter 10 are:

- A brief introduction to ATM is made in order to enable the discussion on WATM. The implementation challenges for WATM are discussed.
- The protocol stack for WATM (physical, MAC and Data Link Control (DLC) layer functionality) is described. Typical bit rates for WATM at the physical layer are in the region of 25 Mbps. Nevertheless, higher PHY speeds are possible and WATM projects under development have succeeded in achieving data rates of 155 Mbps. A number of requirements for an efficient MAC protocol for WATM are presented.
- The issues of location management and handoff in wireless ATM networks are discussed.
- Next, Chapter 10 describes HIPERLAN 2, an ATM compatible WLAN standard developed by ETSI. Contrary to WLAN protocols, HIPERLAN 2 is connection oriented and ATM compatible. HIPERLAN 2 will support speeds up to 25 Mbps at the DLC layer.
- Finally, a number of routing protocols for multihop ad hoc wireless networks are presented. These routing protocols fall into two families: table-driven and on-demand. In table-driven protocols, each network node maintains one or more routing tables, which are used to store the routes from this node to all other network nodes. In on-demand routing protocols, a route is established only when required for a network connection.

1.3.10 Chapter 11: Personal Area Networks (PANS)

Chapter 11 describes Personal Area Networks (PANs). The concept of a PAN differs from that of other types of data networks in terms of size, performance and cost. PANs target applications that demand short-range communications. The main issues covered by Chapter 11 are:

- After a brief introduction to PANs and ongoing PAN-related activities by IEEE Working Group 802.15, PAN concerns and possible PAN applications are highlighted.
- Bluetooth, an industry initiative to develop a de facto, open, standard for PANs is presented. Bluetooth aims to meet the communication needs of all mobile computing and communication devices located in a reduced geographical space, ranging up to 100 m. The Bluetooth specification 1.1, which comprises two parts, core and profiles, is discussed. The core specification defines the layers of the Bluetooth protocol stack. Profiles aim to ensure interoperability between Bluetooth devices The Bluetooth radio channel, which operates in the 2.4 GHz ISM band using Frequency Hopping Spread Spectrum modulation is discussed, followed by a discussion on the way Bluetooth devices connect to form small networks, known as piconets. Piconet interconnections are known as scatternets. Bluetooth supports both voice and asynchronous data channels. Voice channels are 64 kbps each, whereas asynchronous data channels are either asymmetric, with a maximum data rate of 721 kbps in one direction and 57.6 in the other, or symmetric with a

432 kbps maximum rate in both directions. Furthermore, power management and security services of Bluetooth are discussed.

- HomeRF is then presented. HomeRF aims to enable interoperable wireless voice and data networking within the home at ranges higher than those of Bluetooth. Version 1.2 of HomeRF supported speeds at upper layers of 1.6 and 0.8 Mbps, a little higher than the Bluetooth rates. However, version 2.0 provides for rates up to 10 Mbps by using wider (5 MHz) channels in the ISM band through Frequency Hopping Spread Spectrum. This makes it more suitable than Bluetooth for transmitting music, audio, video and other high data applications. However, Bluetooth seems to have more industry backing. Furthermore, due to its complexity (hybrid MAC, using CSMA/CA, higher capability physical layer), HomeRF devices are more expensive than Bluetooth devices. The operation of the HomeRF MAC layer resembles that of IEEE 802.11. Finally, issues regarding system synchronization, power management and security in HomeRF are also discussed.

1.3.11 Chapter 12: Security Issues in Wireless Systems

Almost all wireless networks are at risk of compromise. Unfortunately fixing the problem is not a straightforward procedure. This chapter discusses security issues in wireless networks. It has been found that all IEEE 802.11 wireless networks deployed have security problems [20]. Among the effective interim short-term solution is the use of a WEP with a robust key management system, VPNs schemes and high level security schemes such as IPSec. Although these schemes do not completely resolve the problem, they can be used until the IEEE 802.11 standard committee establishes new effective encapsulation algorithms. Basically, there is no wireless technology that is better than another for all applications. Each has its own advantages and drawbacks. Despite the fact that wireless LANs are not completely secure, their ease of use has always been considered a key factor in their amazing widespread success. Biometric-based security schemes have great potential to secure and authenticate access to all types of networks including wireless networks.

1.3.12 Chapter 13: Simulation of Wireless Network Systems

This chapter deals with the basics of simulation modeling and its application to wireless networking. It starts by introducing the fundamentals of discrete-event simulation, the basic building block of any simulation program (simulator), simulation methodology. Then the commonly used distributions, their major characteristics, and applications are surveyed. The techniques used to generate and test random numbers are presented. Then the techniques used to generate random variates (observations) are presented and the variates that can be generated by each of these techniques are investigated. Finally, the chapter concludes by presenting four examples of the simulation of wireless network systems. These examples cover the performance evaluation of a simple IEEE 802.11 WLAN, simulation of QoS in IEEE 802.11 WLAN system, simulation comparison of the TRAP and RAP wireless LANs protocols and simulation of the Topology Broadcast Based on Reverse-Path Forwarding (TBRPF) protocol using an 802.11 WLAN-based MObile ad hoc NETwork (MONET) model.

1.3.13 Chapter 14: Economics of Wireless Networks

Chapter 14 discusses several economical issues relating to wireless networks. It is reported that although voice telephony will continue to be a significant application, the wireless-Internet combination will shift the nature of wireless systems from today's voice-oriented wireless systems towards data-centric systems. The impact of this change on the key players in wireless networking world is discussed. Furthermore, charging issues in wireless networks are discussed.

WWW Resources

1. *www.palowireless.com*: this web site contains information on a number of systems presented in this book.
2. *www.telecomwriting.com*: this web site includes information on the history and evolution of wireless networks, ranging from the early work on wireless transmission in the 19th century to present cellular mobile systems.

References

[1] McDonald V. H. The Cellular Concept, *Bell Systems Technology Journal*, January, 1979, 15-49.
[2] Padgett J. E., Gunther C. G. and Hattori T. Overview of Personal Communications, *IEEE Communications Magazine*, January, 1995, 28–41.
[3] Rahnema M. Overview of the GSM System and Protocol Architecture, *IEEE Communications Magazine*, April, 1993, 92–100.
[4] Salkintzis A. K. A Survey of Mobile Data Networks, *IEEE Communications Surveys*, Third Quarter, 2(3), 1999, 2–18.
[5] Bantz D. F. and Bauchot F. J. Wireless LAN Design Alternatives, *IEEE Network*, March/April, 1994, 43–53.
[6] Nicopolitidis P., Papadimitriou G. I. and Pomportsis A. S. Design Alternatives for Wireless Local Area Networks, *International Journal of Communication Systems*, Wiley, February, 2001, 1–42.
[7] Awater G. A. and Kruys J. Wireless ATM–an Overview, *Mobile Networks and Applications*, 1, 1996, 235–243.
[8] Raychaudhuri D. Wireless ATM Networks: Technology Status and Future Directions, in *Proceedings of the IEEE*, October, 1999, 1790–1806.
[9] Jush J. K., Malmgren G., Schramm P. and Torsner J. HIPERLAN Type 2 for Broadband Wireless Communication, *Ericsson Review*, 2, 2000.
[10] Johnsson M. HiperLAN/2 - The Broadband Radio Transmission Technology Operating in the 5 GHz Frequency Band, HiperLAN/2 Global Forum, 1999, Version 1.0.
[11] Bhagwat P. Bluetooth: Technology for Short–Range Wireless Apps, *IEEE Internet Computing*, May/June, 2001, 96–103.
[12] Haartsen J. The Bluetooth Radio System, *IEEE Personal Communications*, February, 2000, 28–36.
[13] Lansford J. and Bahl P. The Design and Implementation of HomeRF: a Radio Frequency Wireless Networking Standard for the Connected Home, *Proceedings of the IEEE*, October, 2000, 1662–1676.
[14] IEEE Project 802.15. http://www.ieee802.org/15.
[15] Heile B., Gifford I. and Siep T. IEEE 802 Perspectives, The IEEE P802.15 Working Group for Wireless Personal Area Networks, *IEEE Network*, July, 1999.
[16] Satellite Communications-A Continuing Revolution, *IEEE Aerospace & Electronic Systems Magazine, Jubilee Issue*, October, 2000, 95–107.
[17] Ojanpera T. and Prasad R. An Overview of Third Generation Wireless Personal Communications: A European Perspective, *IEEE Personal Communications*, December, 1998, 59–65.
[18] Sarikaya B. Packet Mode in Wireless Networks: Overview of Transition to Third Generation, *IEEE Communications Magazine*, September, 2000, 164–172.

[19] Nilsson M. Third-Generation Radio Access Standards, *Ericsson Review*, 3, 1999.

[20] Mohr W. Development of Mobile Communications Systems Beyond Third Generation, *Wireless Personal Communications, Kluwer*, June, 2001, 191–207.

[21] Varshney U. and Jain R. Issues in Emerging 4G Wireless Networks, *IEEE Computer*, June, 2001, 94–96.

[22] Flament M., Gessler F., Lagergren F., Queseth O., Stridh R., Unbehaun M., Wu J. and Zander J. Key Research Issues in 4th Generation Wireless Infrastructures, in *Proceedings of the PCC Workshop*, Stockholm, Sweden, 1998.

[23] Forman G. H. and Zahorjan J. The Challenges of Mobile Computing, *IEEE Computer*, April, 1994, 38–46.

2

Wireless Communications Principles and Fundamentals

2.1 Introduction

Wireless networks, as the name suggests, utilize wireless transmission for exchange of information. The exact form of wireless transmission can vary. For example, most people are accustomed to using remote control devices that employ infrared transmission. However, the dominant form of wireless transmission is radio-based transmission. Radio technology is not new, it has a history of over a century and its basic principles remain the same with those in its early stage of development.

In order to explain wireless transmission, an explanation of electromagnetic wave propagation must be given. A great deal of theory accompanies the way in which electromagnetic waves propagate. In the early years of radio transmission (at the end of the nineteenth century) scientists believed that electromagnetic waves needed some short of medium in order to propagate, since it seemed very strange to them that waves could propagate through a vacuum. Therefore the notion of the ether was introduced which was thought as an invisible medium that filled the universe. However, this idea was later abandoned as experiments indicated that ether does not exist. Some years later, in 1905 Albert Einstein developed a theory which explained that electromagnetic waves comprised very small particles which often behaved like waves. These particles were called photons and the theory explained the physics of wave propagation using photons. Einstein's theory stated that the number of photons determines the wave's amplitude whereas the photons' energy determines the wave's frequency. Thus, the question that arises is what exactly is radiation made of, waves or photons. A century after Einstein, an answer has yet to be given and both approaches are used. Usually, lower frequency radiation is explained using waves whereas photons are used for higher frequency light transmission systems.

Wireless transmission plays an important role in the design of wireless communication systems and networks. As a result, the majority of these systems' characteristics stem from the nature of wireless transmission. As was briefly mentioned in the previous chapter, the primary disadvantage of wireless transmission, compared to wired transmission, is its increased bit error rate. The bit error rates (BER)[1] experienced over a wireless link can be as high as 10^{-3} whereas typical BERs of wired links are around 10^{-10}. The primary reason for

---

[1] A BER equal to 10^{-x} means that 1 out of 10^x received bits is received with an error, that is, with its value inverted.

the increased BER is atmospheric noise, physical obstructions found in the signal's path, multipath propagation and interference from other systems.

Another important aspect in which wireless communication systems differ from wired systems, is the fact that in wired systems, signal transmissions are confined within the wire. Contrary to this, for a wireless system one cannot assume an exact geographical location in which the propagation of signals will be confined. This means that neighboring wireless systems that use the same waveband will interfere with one another. To solve this problem, wavebands are assigned after licensing procedures. Licensing involves governments, operators, corporations and other parties, making it a controversial procedure as most of the times someone is bound to complain about the way wavebands have been assigned.

Licensing makes the wireless spectrum a finite resource, which must be used as efficiently as possible. Thus, wireless systems have to achieve the highest performance possible over a waveband of specific width. Therefore, such systems should be designed in a way that they offer a physical layer able to combat the deficiencies of wireless links. Significant work has been done in this direction with techniques such as diversity, coding and equalization able to offer a relatively clean channel to upper layers of wireless systems. Furthermore, the cellular concept offers the ability to reuse parts of the spectrum, leading to increased overall performance and efficient use of the spectrum.

2.1.1 Scope of the Chapter

The remainder of this chapter describes the fundamental issues related to wireless transmission systems. Section 2.2 describes the various bands of the electromagnetic spectrum and discusses the way spectrum is licensed. Section 2.3 describes the physical phenomena that govern wireless propagation and a basic wireless propagation model. Section 2.4 describes and compares analog and digital radio transmission. Section 2.5 describes the basic modulation techniques that are used in wireless communication systems while Section 2.6 describes the basic categories of multiple access techniques. Section 2.7 provides an overview of diversity, smart antennae, multiantenna transmission, coding, equalization, power control and multicarrier modulation, which are all techniques that increase the performance over a wireless link. Section 2.8 introduces the cellular concept, while Section 2.9 describes the ad hoc and semi ad hoc concepts. Section 2.10 describes and compares packet-mode and circuit-mode wireless services. Section 2.11 presents and compares two approaches for delivering data to mobile clients, the pull and push approaches. Section 2.12 provides an overview of the basic techniques and interactions between the different layers of a wireless network. The chapter ends with a brief summary in Section 2.13.

2.2 The Electromagnetic Spectrum

Electromagnetic waves were predicted by the British physicist James Maxwell in 1865 and observed by the German physicist Heinrich Hertz in 1887. These waves are created by the movement of electrons and have the ability to propagate through space. Using appropriate antennas, transmission and reception of electromagnetic waves through space becomes feasible. This is the base for all wireless communications.

Electromagnetic waves are generated through generation of an electromagnetic field. Such a field is created whenever the speed of an electrical charge is changed. Transmitters are

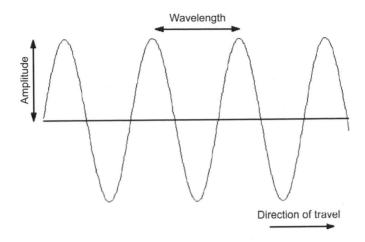

Figure 2.1 Wavelength and amplitude of an electromagnetic wave

based on this principle: in order to generate an electromagnetic wave, a transmitter vibrates electrons, which are the particles that orbit all atoms and contain electricity. The speed of electron vibration determines the wave's frequency, which is the fundamental characteristic of an electromagnetic wave. It states how many times the wave is repeated in one second and is measured in hertz (to honor Heinrich Hertz). Higher vibration speeds for electrons produce higher frequency waves. Reception of a wave works in the same way, by examining values of electrical signals that are induced to the receiver's antenna by the incoming wave.

Another fundamental characteristic of an electromagnetic wave is its wavelength. This refers to the distance between two consecutive maximum or minimum peaks of the electromagnetic wave and is measured in meters. The wavelength of a periodic sine wave is shown in Figure 2.1, which also shows the wave's amplitude. The amplitude of an electromagnetic wave is the height from the axis to a wave peak and represents the strength of the wave's transmission. It is measured in volts or watts.

The wavelength λ and frequency f of an electromagnetic wave are related according to the following equation:

$$c = \lambda f \tag{2.1}$$

where c is a constant representing the speed of light. The constant nature of c means that given the wavelength, the frequency of a wave can be determined and vice versa. Thus, waves can be described in terms of their wavelength or frequency with the latter option being the trend nowadays. The equation holds for propagation in a vacuum, since passing through any material lowers this speed. However, passing through the atmosphere does not cause significant speed reduction and thus the above equation is a very good approximation for electromagnetic wave propagation inside the earth's atmosphere.

2.2.1 Transmission Bands and their Characteristics

The complete range of electromagnetic radiation is known as the electromagnetic spectrum. It comprises a number of parts called bands. Bands, however, do not exist naturally. They are

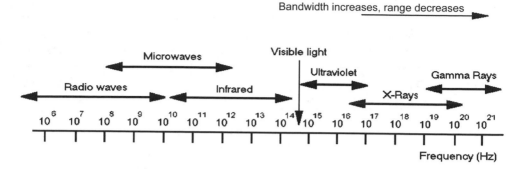

Figure 2.2   The electromagnetic spectrum

used in order to explain the different properties of various spectrum parts. As a result, there is not a clear distinction between some bands of the electromagnetic spectrum. This can be seen in Figure 2.2, which shows the electromagnetic spectrum and its classification into several bands.

As can be seen from the figure, frequency is measured on a logarithmic scale. This means that by moving from one point to another on the axis, frequency is increased by a factor of 10. Thus, higher bands have more bandwidth and can carry more data. However, the bands above visible light are rarely used in wireless communication systems due to the fact that they are difficult to modulate and are dangerous to living creatures. Another difference between the spectrum bands relates to the attenuation they suffer. Higher frequency signals typically have a shorter range than lower frequency signals as higher frequency signals are more easily blocked by obstacles. An example of this is the fact that light cannot penetrate walls, while radio signals can.

The various bands of the spectrum are briefly summarized below in increasing order of frequency. Of these, the most important for commercial communication systems are the radio and microwave bands.

- *Radio.* Radio waves occupy the lowest part of the spectrum, down to several kilohertz. They were the first to be applied for wireless communications (Gugliemo Marconi sent the first radio message across the Atlantic Ocean in the early 1900s). Lower frequency radio bands have lower bandwidth than higher frequency bands. Thus, modern wireless communications systems favor the use of high frequency radio bands for fast data services while lower frequency radio bands are limited to TV and radio broadcasting. However, higher frequency radio signals have a shorter range as mentioned above. This is the reason that radio stations in the Long Wavelength (LW) band are easily heard over many countries whereas Very High Frequency (VHF) stations can only cover regions about the size of a city. Nevertheless, reduced range is a potential advantage for wireless networking systems, since it enables frequency reuse. This will be seen later in this chapter when the cellular concept is covered. The LW, VHF and other portions of the radio band of the spectrum are shown in Figure 2.3. The HF band has the unique characteristic that enables worldwide transmission although having a relatively high frequency. This is due to the fact that HF signals are reflected off the ionosphere and can thus travel over very large distances.

Frequency	Band name	Applications
< 3 KHz	Extremely Low Frequency (ELF)	Submarine communications
3 KHz -30 KHz	Very Low Frequency (VLF)	Marine communications
20 KHz -300 KHz	Low Frequency (LF)or Long Wave (LW)	AM radio
300 KHz -3 MHz	Medium Frequency (MF) or Medium Wave (MW)	AM radio
3 MHz - 30 MHz	High Frequency (HF) or Short Wave (HW)	AM radio
30 MHz -300 MHz	Very High Frequency (VHF)	FM Radio-TV
300 MHz - 3 GHz	Ultra High Frequency (UHF)	TV-cellular telephony
3 GHz - 30 GHz	Super High Frequency (SHF)	Satellites
30 GHz - 300 GHz	Extra High Frequency (EHF)	Satellites-radars

Figure 2.3   The various radio bands and their common use

Although not very reliable, this was the only way to communicate overseas before the satellite era.

- *Microwaves.* The high frequency radio bands (UHF, SHF and EHF) are referred to as microwaves. Microwaves get their name from the fact that they have small wavelengths compared to the other radio waves. Microwaves have a large number of applications in wireless communications which stem from their high bandwidth. However, they have the disadvantage of being easily attenuated by objects found in their path. The commonly used parts of the microwave spectrum are shown in Figure 2.4.
- *Infrared (IR).* IR radiation is located below the spectrum of red visible light. Such rays are emitted by very hot objects and the frequency depends on the temperature of the emitting body. When absorbed, the temperature increases. IR radiation is also emitted by the human body and night vision is based on this fact. It also finds use in some wireless communica-

Frequency	Band name	Applications
0.4 GHz - 1.5 GHz	L	Broadcasting-cellular
1.5 GHz - 5.2 GHz	S	Cellular
3.9 GHz - 6.2 GHz	C	Satellites
5.2 GHz - 10.9 GHz	X	Fixed wireless-satellite
10.9 GHz - 36 GHz	K	Fixed wireless-satellite
36 GHz - 46 GHz	Q	Fixed wireless
46 GHz -56 GHz	V	Future satellite
56 GHz -100 GHz	W	Future cellular

Figure 2.4   The various microwave bands and their common use

tion systems. An example is the infrared-based IEEE 802.11 WLAN covered in Chapter 9. Furthermore, other communication systems exchange information either by diffused IR transmission or point-to-point infrared links.

- *Visible light.* The tiny part of the spectrum between UV and Infrared (IR) in Figure 2.4 represents the visible part of the electromagnetic spectrum.
- *Ultraviolet (UV).* In terms of frequency, UV is the next band in the spectrum. Such rays can be produced by the sun and ultraviolet lamps. UV radiation is also dangerous to humans.
- *X-Rays.* X-Rays, also known as Rontgen rays, are characterized by shorter frequency than gamma rays. X-Rays are also dangerous to human health as they can easily penetrate body cells. Today, they find use in medical applications, the most well known being the examination of possible broken bones.
- *Gamma rays.* Gamma rays occupy the highest part of the electromagnetic spectrum having the highest frequency. These kinds of radiation carries very large amounts of energy and are usually emitted by radioactive material such as cobalt-60 and cesium-137. Gamma rays can easily penetrate the human body and its cells and are thus very dangerous to human life. Consequently, they are not suitable for wireless communication systems and their use is confined to certain medical applications. Due to their increased potential for penetration, gamma rays are also used by engineers to look for cracks in pipes and aircraft parts.

Signal transmission in bands lower than visible light are generally not considered as harmful (e.g. UV, X and gamma rays). However, they are not entirely safe, since any kind of radiation causes increase in temperature. Recall the way microwave ovens work: Their goal is for food molecules to absorb microwaves which cause heat and help the food to cook quickly.

2.2.2 Spectrum Regulation

The fact that wireless networks do not use specific mediums for signal propagation (such as cables) means that the wireless medium can essentially be shared by arbitrarily many systems. Thus, wireless systems must operate without excessive interference from one another. Consequently, the spectrum needs to be regulated in a manner that ensures limited interference.

Regulation is commonly handled inside each country by government-controlled national organizations although lately there has been a trend for international cooperation on this subject. An international organization responsible for worldwide spectrum regulation is the International Telecommunications Union (ITU). ITU has regulated the spectrum since the start of the century by issuing guidelines that state the spectrum parts that can be used by certain applications. These guidelines should be followed by national regulation organizations in order to allow use of the same equipment in any part of the world. However, following the ITU guidelines is not mandatory. For spectrum regulation purposes, the ITU splits the world into three parts: (i) the American continent; (ii) Europe, Africa and the former Soviet union; and (iii) the rest of Asia and Oceania. Every couple of years the ITU holds a World Radiocommunication Conference (WRC) to discuss spectrum regulation issues by taking into account industry and consumer needs as well as social issues. Almost any interested member (e.g. scientists and radio amateurs) can attend the conference, although most of the time attendees are mainly government agencies and industry people. The latest WRC was held in 2000 in which spectrum regulation for the Third Generation (3G) of wireless networks was discussed. 3G wireless networks are covered in Chapter 5.

Several operators that offer wireless services often exist inside each country. National regulation organizations should decide how to license the available spectrum to operators. This is a troublesome activity that entails political and sociological issues apart from technological issues. Furthermore, the actual policies of national regulation organizations differ. For example, the Federal Communications Commission (FCC), the national regulator inside the United States licenses spectrum to operators without limiting them on the type of service to deploy over this spectrum. On the other hand, the spectrum regulator of the European Union does impose such a limitation. This helps growth of a specific type of service, an example being the success of the Global System Mobile (GSM) communications inside Europe (GSM is described in Chapter 4). In the last year, the trend of licensing spectrum for specific services is being followed by other countries too, an example being the licensing by many countries of a specific part in the 2 GHz band for 3G services.

Until now, three main approaches for spectrum licensing have been used: comparative bidding, lottery and auction. Apart from these, the ITU has also reserved some parts of the spectrum that can be used internationally without licensing. These are around the 2.4 GHz band and are commonly used by WLAN and Personal Area Networks (PANs). These are covered in Chapters 9 and 11, respectively. Parts of the 900 MHz and 5 GHz bands are also available for use without licensing in the United States and Canada.

2.2.2.1 Comparative Bidding

This is the oldest method of spectrum licensing. Each company that is interested in becoming an operator forms a proposal that describes the types of services it will offer. The various interested companies submit their proposals to the regulating agency which then grades them according to the extent that they fulfill certain criteria, such as pricing, technology, etc., in an effort to select those applications that serve the public interest in the best way. However, the problem with this method is the fact that government-controlled national regulators may not be completely impartial and may favor some companies over others due to political or economic reasons. When a very large number of companies declare interest for a specific license, the comparative bidding method is likely to be accompanied by long delays until service deployment. Regulating organizations will need more time to study and evaluate the submitted proposals. This increases costs of both governments and candidate operators. In the late 1980s, the FCC sometimes needed more than three years to evaluate proposals. Comparative bidding is not thought to be a popular method for spectrum licensing nowadays. Nevertheless, inside the European Union, Norway, Sweden, Finland, Denmark, France and Spain used it for licensing spectrum for 3G services.

2.2.2.2 Lottery

This method aims to alleviate the disadvantages of comparative bidding. Potential operators submit their proposals to the regulators, which then give licenses to applicants that win the lottery. This method obviously is not accompanied by delays. However, it has the disadvantage that public interest is not taken into account. Furthermore, it attracts the interest of speculator companies that do not posses the ability to become operators. Such companies may enter the lottery and if they manage to get the license, they resell it to companies that lost the lottery but nevertheless have the potential to offer services using the license. In such cases, service deployment delays may also occur as speculators may take their time in order to achieve the best possible price for their license.

2.2.2.3. Auction

This method is based on the fact that spectrum is a scarce, and therefore expensive, resource. Auctioning essentially allows governments to sell licenses to potential operators. In order to sell a specific license, government issues a call for interested companies to join the auction and the company that makes the highest bid gets the license. Although expensive to companies, auction provides important revenue to governments and forces operators to use the spectrum as efficiently as possible. Spectrum auctions were initiated by the government of New Zealand in 1989 with the difference that spectrum was not sold. Rather, for a period of for two decades, it was leased to the highest bidder who was free to use it for offering services or lease it to another company.

Despite being more efficient than comparative bidding and lotteries, auction also has some disadvantages. The high prices paid for spectrum force companies passed on high charges to the consumers. It is possible that the companies' income from deployed services is over-estimated. As a result companies may not be able to get enough money to pay for the license and go bankrupt. This is the reason why most regulating agencies nowadays tend to ask for all the money in advance when giving a license to the highest bidder.

Since 1989 auction has been used by other countries as well. In 1993, FCC abandoned lotteries and adopted auction as the method for giving spectrum licenses. In 2000 auction was used for licensing 3G spectrum in the United Kingdom resulting in 40 billion dollars of revenue to the British government, ten times more than expected. Auctioning of 3G spectrum was also used inside the European Union by Holland, Germany, Belgium and Austria. Italy and Ireland used a combination of auction and comparative bidding with the winners of comparative bidding entering an auction in order to compete for 3G licenses.

2.3 Wireless Propagation Characteristics and Modeling

2.3.1 The Physics of Propagation

An important issue in wireless communications is of course the amount of information that can be carried over a wireless channel, in terms of bit rate. According to information theory, an upper bound on the bit rate W of any channel of bandwidth H Hz whose signal to thermal noise ratio is S/N, is given by Shannon's formula:

$$W = H\log_2\left(1 + \frac{S}{N}\right) \tag{2.2}$$

Equation (2.2) applies to any transmission media, including wireless transmission. However, as already mentioned, Equation (2.2) gives only the maximum bit rate that can be achieved on a channel. In real wireless channels the bit rates achieved can be significantly lower, since apart from the thermal noise, there exist a number of impairments on the wireless channels that cause reception errors and thus lower the achievable bit rates. Most of these impairments stem from the physics of wave propagation. Understanding of the wave propagation mechanism is thus of increased importance, since it provides a means for predicting the coverage area of a transmitter and the interference experienced at the receiver. Although the mechanism that governs propagation of electromagnetic waves through space is of increased complexity, it can generally be attributed to the following phenomena: free space path loss, Doppler Shift which is caused by station mobility and the propagation mechanisms of reflection, scattering and diffraction which cause signal fading.

2.3.1.1 Free Space Path Loss

This accounts for signal attenuation due to distance between the transmitter and the receiver. In free space, the received power is proportional to r^{-2}, where r is the distance between the transmitter and the receiver. However, this rule is rarely used as the propagation phenomena described later significantly impact the quality of signal reception.

2.3.1.2 Doppler Shift

Station mobility gives rise to the phenomenon of Doppler shift. A typical example of this phenomenon is the change in the sound of an ambulance passing by. Doppler shift is caused when a signal transmitter and receiver are moving relative to one another. In such a situation the frequency of the received signal will not be the same as that of the source. When they are moving towards each other the frequency of the received signal is higher than that of the source, and when they are moving away from each other the frequency decreases. This phenomenon becomes important when developing mobile radio systems.

2.3.1.3 Propagation Mechanisms and Slow/Fast Fading

As mentioned above, electromagnetic waves generally experience three propagation mechanisms: reflection, scattering and diffraction. Reflection occurs when an electromagnetic wave falls on an object with dimensions very large compared to the wave's wavelength. Scattering occurs when the signal is obstructed by objects with dimensions in the order of the wavelength of the electromagnetic wave. This phenomenon causes the energy of the signal to be transmitted over different directions and is the most difficult to predict. Finally, diffraction, also known as shadowing, occurs when an electromagnetic wave falls on an impenetrable object. In this case, secondary waves are formed behind the obstructing body despite the lack of line-of-sight (LOS) between the transmitter and the receiver. However, these waves have less power than the original one. The amount of diffraction is dependent on the radio frequency used, with low frequency signals diffracting more than high frequency signals. Thus, high frequency signals, especially, Ultra High Frequencies (UHF), and microwave signals require LOS for adequate signal strength. Shadowed areas are often large, resulting in the rate of change of the signal power being slow. Thus, shadowing is also referred to as slow fading. Reflection scattering and diffraction are shown in Figure 2.5.

In a wireless channel, the signal from the transmitter may be reflected from objects (such as hills, buildings, etc.) resulting in echoes of the signal propagating over different paths with

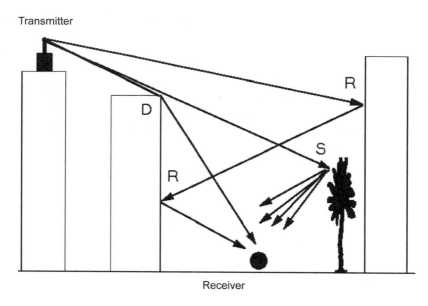

Figure 2.5   Reflection (R), diffraction (D) and scattering (S)

different path lengths. This phenomenon is known as multipath propagation and can possibly lead to fluctuations in received signal power. This is due to the fact that echoes travel a larger distance due to reflections and they arrive at the receiver after the original signal. Therefore, the receiver sees the original signal followed by echoes that possibly distort the reception of the original signal by causing small-scale fluctuations in the received signal. The time duration between the reception of the first signal and the reception of the last echo is known as the channel's delay spread.

Because these small-scale fluctuations are experienced over very short distances (typically at half wavelength distances), multipath fading is also referred to either as fast fading or small-scale fading. When a LOS exists between the receiver and the transmitter, this kind of fading is known as Ricean fading. When a LOS does not exist, it is known as Rayleigh fading. Multipath fading causes the received signal power to vary rapidly even by three or four orders of magnitude when the receiver moves by only a fraction of the signal's wavelength. These fluctuations are due to the fact that the echoes of the signal arrive with different phases at the receiver and thus their sum behaves like a noise signal. When the path lengths followed by echoes differ by a multiple of half of the signal's wavelength, arriving signals may partially or totally cancel each other. Partial signal cancellation at the receiver due to multipath propagation is shown in Figure 2.6. Despite the rapid small-scale fluctuations due to multipath propagation, the average received signal power, which is computed over receiver movements of 10–40 wavelengths and used by the mobile receiver in roaming and power control decisions, is characterized by very small variations in the large scale, as shown in Figure 2.7, and decreases only when the transmitter moves away from the receiver over significantly large distances.

Multipath propagation can lead to the presence of energy from a previous symbol during the detection time of the current symbol which has catastrophic effects at signal reception.

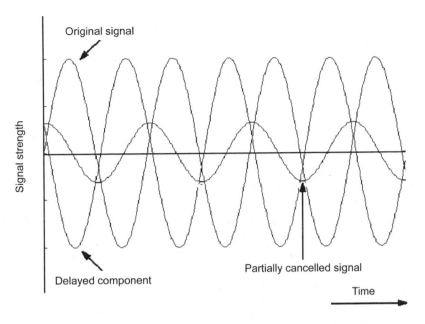

Figure 2.6   Partial signal cancellation due to multipath propagation

This is known as intersymbol interference (ISI) and occurs when the delay spread of a channel is comparable to symbol detection time [1]. This criterion is equivalent to

$$B > B_c \tag{2.3}$$

where B is the transmitted signal bandwidth (equivalently, the transmitted symbol rate), and B_c is the channel's coherence bandwidth, which is the frequency band over which the fading of different frequency components of the channel is essentially the same. When Equation

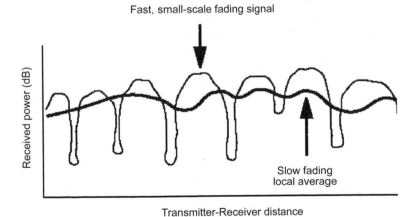

Figure 2.7   Variation of signal level according to transmitter–receiver distance

(2.3) applies, the channel is said to be frequency selective or wideband, otherwise it is said to be flat or narrowband. The fading type is known as frequency selective or flat, respectively.

The zones affected by multipath fading tend to be small, multiple areas of space where periodic attenuation of a received signal is experienced. In other words, the received signal strength will fluctuate, causing a momentary, but repetitive, degradation in quality.

2.3.2 Wireless Propagation Modeling

As can be seen from the above discussion, in a wireless system, the actual signal arriving at a receiver is the sum of components that derive from several difficult to predict propagation phenomena. Thus, the need for a model that predicts the signal arriving at the receiver arises. Such models are known as propagation models [2] and are essentially a set of mathematical expressions, algorithms and diagrams that predict the propagation of a signal in a given environment. Propagation models are either empirical (also known as statistical), theoretical (also known as deterministic) or a combination of the above.

Empirical models describe the radio characteristics of an environment based on measurements made in several other environments. An obvious advantage of empirical models is the fact that they implicitly take into account all the factors that affect signal propagation albeit these might not be separately identified. Furthermore, such models are computationally efficient. However, the accuracy of empirical models is affected by the accuracy of the measurements that are used. Moreover, the accuracy of such models depends on the similarity of the environment where the measurements were made and the environment to be analyzed.

Theoretical models base their predictions not on measurements but on principles of wave theory. Consequently, theoretical models are independent of measurements in specific environments and thus their predictions are more accurate for a wide range of different environments. However, their disadvantage is the fact that they are expressed by algorithms that are very complex and thus computationally inefficient. For that reason, theoretical models are often used only in indoor and small outdoor areas where they obviously provide greater accuracy than empirical models.

In terms of the radio environment they describe, propagation models can be categorized into indoor and outdoor models. Moreover outdoor models are subdivided into macrocell models describing propagation over large outdoor areas and microcell models describing propagation over small outdoor areas (typically city blocks). A large number of propagation models have been proposed but detailed presentation is outside the scope of this chapter. The interested reader is referred to corresponding technical papers [2]. In the remainder of this section we describe the behavior of outdoor macrocell/microcell and indoor environments and we describe how propagation occurs in these situations and the factors that affect it.

2.3.2.1 Macrocells

The concept of the cell is described later, however for the purposes of this discussion, a macrocell is considered to be a relatively large area that is under the coverage of a BS. Macrocells were the basis for organization of the first generation of cellular systems. As a result, the need to predict the received signal power arose first for macrocells.

When free space loss was discussed, it was mentioned that although in free space, the received power is proportional to r^{-2}, where r is the distance between the transmitter and the

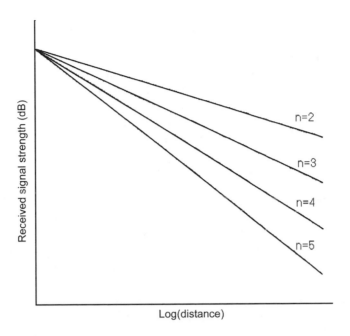

Figure 2.8 Log-log form of Equation (2.4)

receiver; this rule, however, is rarely used as the other propagation phenomena affect received signal power. In real situations a good estimator for the received signal strength $P(r)$ for a distance r between the transmitter and the receiver is given by

$$P(r) = kr^{-n} \tag{2.4}$$

where k is a constant and the exponent n is a parameter that describes the environment. A value of $n = 2$ describes propagation into free space, while values of n between 2 and 4 are used for modeling macrocells. The form of Equation (2.4) in a log-log scale is shown in Figure 2.8.

The same power law model also applies to path loss. Thus, the average path loss at a distance r is (in dB)[2]

$$PL(r) = PL(r_0) + 10n\log\left(\frac{r}{r_0}\right) \tag{2.5}$$

where r_0 is a reference distance that must be appropriately selected and is typically 1 km for macrocells. However, the path loss model of Equation (2.5) does not take into account the fact that for a certain transmitter–receiver distance, different path loss values are possible due to the fact that shadowing may occur in some locations and not in others. To take this fact into account, Equation (2.5) now becomes [3]

---

[2] When we say that the relative strength of signal X, $P(X)$ to that of signal Y, $P(Y)$ is D dB then $D = 10\log(P(X)/P(Y))$. Thus dB is a convention used to measure the relative strength of two signals and has no physical meaning, since the relative strength of two signals is just a number.

$$PL(r) = PL(r_0) + 10n\log\left(\frac{r}{r_0}\right) + X_\sigma \tag{2.6}$$

where $X\sigma$ is a zero-mean Gaussian-distributed random variable with standard deviation σ.

Macrocells were the basis for the first generation of cellular systems. The first propagation model for such systems was made by Okomura and was based on comprehensive measurements of Japanese environments. The model of Okomura was later enhanced by Hata by transforming it into parametric formulas. These works produced results that confirm the above path loss model and although strictly empirical, they have proven to be robust not only for Japanese environments but in other environments as well.

2.3.2.2 Microcells

Microcells cover much smaller regions than macrocells. Propagation in microcells differs significantly from that observed in macrocells. The smaller area of a microcell results in smaller delay spreads. Microcells are most commonly used in densely populated areas such as parts of a city. The model of Equation (2.6) also describes path loss in microcells, with a typical r_0 value of 100 m.

Andersen et al. [3] mention the concept of a 'street microcell', which is shown in Figure 2.9. This kind of microcell is created by placing transmitter antennas lower than surrounding buildings. Thus, most of the signal power propagates along streets. Even in this case nearby buildings play an important role regarding received signal quality. Assuming the situation of

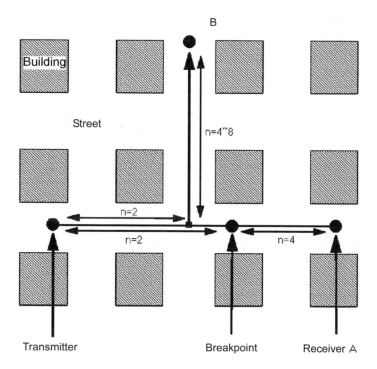

Figure 2.9   Path loss situations in a street microcell

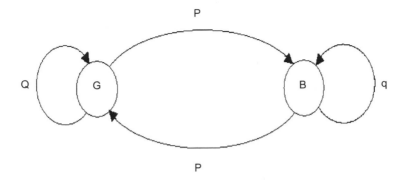

Figure 2.11   Transition diagram of a Markov chain

2.3.3 Bit Error Rate (BER) Modeling of Wireless Channels

Although there are a number electromagnetic wave propagation impairments, such as free-space loss and thermal noise, fading is the primary cause of reception errors in wireless communications. In the previous paragraphs, the discussion was made in terms of received signal strength. However, in most cases one is interested in viewing the effects of wireless propagation impairments from a higher point of view: the way in which bit errors occur.

Wireless channels are more prone to bit errors than wired channels. Apart from the higher BER of wireless channels compared to wired channels, measurements also indicate a difference in the pattern of bit error occurrence. In contrast to the random nature of bit error occurrence in wired channels, bit errors over wireless channels occur in bursts and Markov chain model approximations have been shown to be adequate for wireless channel bit error modeling [4]. Such models comprise two states, a good (G) and a bad (B) state, and parameters that define the transition procedure between the two states. State G is error free, thus bit errors only occur in state B. Future states are independent of past states and depend only on the present state. In other words, the model is memoryless. Figure 2.11 depicts the transition diagram of a Markov chain. P is the probability of the channel state transiting from state G to state B, p defines the probability of transition from state B to state G, Q and q the probabilities of the channel remaining in states G and B, respectively. Obviously $Q = 1 - P$ and $q = 1 - p$. In state B, bit errors are assumed to occur with probability h. Values for the model parameters are obtained through statistical measurements of particular channels. These values are different for different channels and physical environments. Markov chain models can efficiently approximate the behavior of a wireless channel and are widely used in simulations of wireless systems.

2.4 Analog and Digital Data Transmission

An important parameter of message relaying between a source and a destination is whether the message is analog or digital. These terms relate to the nature of the message and can characterize either the transmitted data or the form of the actual signal used to carry the message. Thus, we have analog and digital data, as well as analog and digital signals. Analog and digital signal representations are shown in Figures 2.12 and 2.13, respectively. The

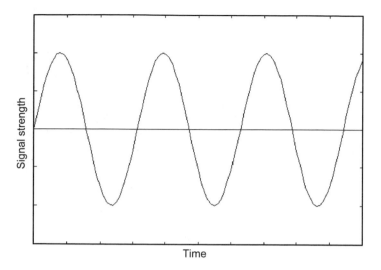

Figure 2.12   Analog signal

difference is obvious: analog signals take continuous values in time whereas digital ones change between certain levels at specific time positions. In the following we discuss and compare analog and digital data representation, while the basic modulation methods for wireless networks, which are used to transmit the signal over the wireless medium, are discussed in Section 2.5.

The vast majority of the early radio communication systems concerned sound transmission. Television transmission comprises two analog components, corresponding to sound and image. Moreover, the only service offered by early cellular systems (e.g. Advanced Mobile

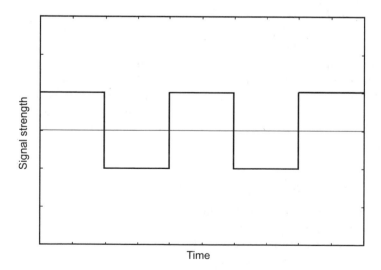

Figure 2.13   Digital signal

Phone System, AMPS) was voice conversation. Thus, all these systems represented the information to be transmitted in an analog form since the physical nature of both sound and image is analog. However, modern wireless systems are increasingly being used for computer data communications, such as file transfer. The natural form of such data is digital, thus digital representation is used. There is a trend towards digital representation of analog data, which stems from the inherent advantages of digital over analog technology. These advantages are briefly summarized below:

- *Transmission reliability.* Transmission of a message through a medium is generally degraded by noise, which is more or less present in all communication mediums. As mentioned earlier, noise causes bit errors and BERs of wireless channels are significantly higher than those of wired channels. The digital representation of a message increases the tolerance of a wireless system to noise. This is due to the fact that, as seen from Figure 2.13, a digital signal is not continuous but rather comprises a number of levels. As a result, in order for noise to alter the message content, it has to be strong enough to change the signal level to another one. Furthermore, digital messages can be accompanied by additional bits, called checksum bits. The actual content of these bits is based on error detecting/correcting algorithms and the procedure is known as Forward Error Correction (FEC). An error detection algorithm works by appending extra bits to a binary message in a way that the receiver can use the received bits and determine whether or not a bit error has occurred and thus, request a retransmission if needed. Error correction algorithms work in the same way, however, in this case the receiver has the ability not only to detect but also to correct bit errors. The Hamming code is a widely known technique used both for error correction and detection.
- *Efficient use of spectrum.* The above mentioned increased noise tolerance of digital representation helps increase the amount of information that can be transmitted using a wireless channel. This is because less errors are likely to occur due to the applied coding. Thus, for a given amount of spectrum and a certain time period, more information can be transmitted by using digital representation – a fact that results to a more efficient use of the spectrum. Furthermore, digital data can be compressed easily which increases spectrum efficiency even more.
- *Security.* Wireless channels are probably the most easy to eavesdrop on, therefore security is a crucial issue in such systems. Analog systems can be provided with a certain level of security, however, these have proved easy to crack. Digital data, on the other hand, can be easily and efficiently encrypted even up to a point that makes unauthorized decryption of the message almost impossible. Furthermore, encryption does not come at any expense to the spectral efficiency of the system, meaning than an encrypted message can be transmitted over the same bandwidth required for unencrypted transmission of the same message.

2.4.1 Voice Coding

While the trend in modern wireless networks is towards data communications, the demand for voice-related services such as traditional mobile phone calls is expected to continue to exist. Thus voice needs to be converted from its analog form to a digital form that will be transmitted over the digital wireless network. The devices that perform this operation are known as

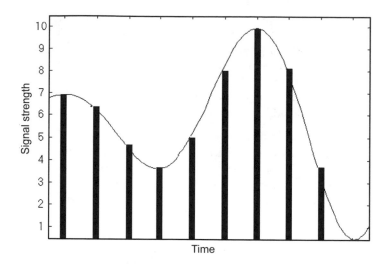

Figure 2.14 PAM pulses created by sampling of the analog signal

codecs (coder/decoder) and have been used mainly in mobile phones. Codecs aim to convert voice into a digital bit stream that has the lowest possible bit rate while maintaining an acceptable quality.

A codec can convert an analog speech signal to its digital representation by sampling the analog signal at regular time intervals. This method is known as Pulse Code Modulation (PCM) and is used in codecs of PSTN and CD systems. There is a direct relationship between the number of samples per second, W, and the width, H, of the analog signal we want to digitize. This is given in the following equation, which tells us that when we want to digitize an analog signal of width, H, there is no point in sampling faster than W:

$$W = 2H \text{ bps} \tag{2.7}$$

The process of PCM conversion of an analog signal to a digital one comprises three stages:

- *Sampling of the analog signal.* This produces a series of samples, known as Pulse Amplitude Modulation (PAM) pulses, with amplitude proportional to the original signal. The PAM pulses produced after sampling of an analog signal are shown in Figure 2.14.
- *Quantizing.* This is essentially the splitting of the effective amplitude range of the analog signal to V levels which are used for approximating the PAM pulses. These V levels (known as quantizing levels) are selected as the median values between various equally spaced signal levels. The quantization of the PAM pulses of Figure 2.14 is shown in Figure 2.15. Quantization obviously distorts the original signal since some information is lost due to approximation. The more the quantizing levels, the less the distortion since the approximation with many levels is more precise. Good voice digitization by PCM is achieved for 128 quantization levels. The distortion due to quantization is known as quantizing noise and is given by the following formula [5]:

$$\frac{S}{N} = 6V + 1.8 \text{ dB} \tag{2.8}$$

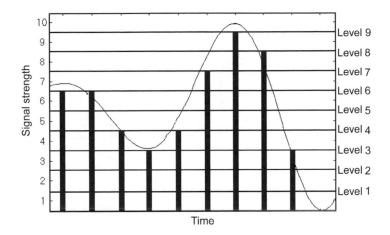

Figure 2.15   PCM pulses produced by quantization

- *Binary encoding.* This is encoding of the quantized values of PAM to binary format, which forms the output of the PCM system and will be used to modulate the signal to be transmitted. For the quantized PAM pulses of Figure 2.15 four bits are used per PCM sample coding (since nine levels can be encoded by four bits) the binary output is 0110011001000011010001111001100 00011.

PCM demands relatively high bit rates and is thus not very useful for wireless communications systems, such as mobile phones. A number of techniques exist that are refinements of PCM and try both to increase voice quality and decrease the output bit rate. PCM with nonlinear encoding takes into account the fact that PCM will produce a largely distorted signal when the effective amplitude of the sampled analog signal is relatively small compared to the amplitude covered by the PCM quantizing levels. Therefore, nonlinear encoding use more levels for such signals – a fact that reduces quantizing noise. For voice signals 24–30 dB S/N improvements have been achieved. Differential PCM (DPCM) outputs the binary representation of the difference between consecutive PCM samples rather than the samples themselves. When x bits are used for encoding the differences, the method is known as x-bit DPCM. The method for $x = 1$ is known as Delta modulation. DPCM schemes obviously reduce the bit rate produced if the differences between samples can be encoded using less bits than those required for encoding the actual samples. However, DPCM techniques have poor performance when steep changes occur in the analog signal. Adaptive DPCM (ADPCM) tries to predict the value of a sample based on previous sample values. ADPCM helps reduce the bit rate down to 16 kbps while still maintaining acceptable voice quality. The following chapters show that 16 kbps is still a large value for mobile phones, however, prediction is used in conjunction with other techniques in mobile phone codecs to lower the bit rate.

2.4.1.1 Vocoders and hybrid codecs

In an effort to reduce the bit rate required for voice transmission, engineers have exploited the actual structure and operation of human speech production organs and the devices that work

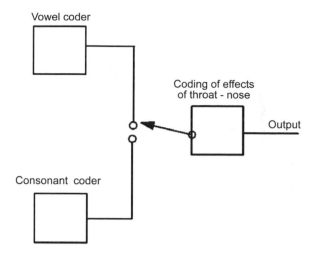

Figure 2.16   Vocoder structure

based on this are known as vocoders. Vocoders, which were initially only an attempt to synthesize speech, work by encoding not the actual voice signals but rather by modeling the mechanics of how sounds are produced (such as mouth movement, voice pitch, etc.). By encoding and transmitting this information the signal can be reconstructed at the receiver.

A simple vocoder diagram is shown in Figure 2.16. It comprises three parts:

- the part responsible for coding vowel sounds, which are attributed to the vocal cords;
- the part responsible for coding consonant sounds, which are produced by lips, teeth, etc.;
- the part that is responsible for coding the effects of the throat and nose on the speech signal.

Vocoders are very useful since they achieve voice transfer with a low bit rate. 'Full-rate' vocoders produce a compressed voice signal of 13 kbps while half-rate vocoders sacrifice some quality and achieve a rate of 8 kbps. Furthermore, there are vocoders that can serve bandwidth-limited scenarios, such as military and space communications. Over the low bandwidth channels of such applications, these vocoders can achieve voice transmission with very low bit rates, as low as 1.2–2.4 kbps. However, the voice produced is not very 'natural' and has a somewhat 'artificial' quality. In some cases it is even difficult to tell who is actually speaking. Hybrid codecs try to overcome this problem by transmitting both vocoding and PCM voice information while also making sure that sounds that are inaudible to the human ear are not transmitted. An example of such a sound is that of a quiet musical instrument in the background of a loud one. Furthermore, codecs that vary the bit rate according to the characteristics of speech sounds have been produced.

2.5 Modulation Techniques for Wireless Systems

In the previous section we covered analog and digital data representation. Whether in analog or digital format, data has to be converted into electromagnetic waves in order to be sent over a wireless channel. The techniques used to perform this are known as modulation techniques

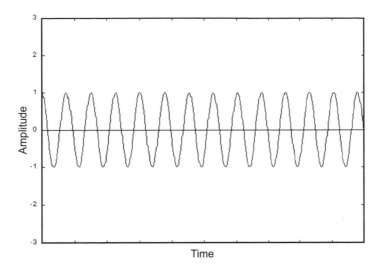

Figure 2.17   Carrier wave

and operate by altering certain properties of a radio wave, known as the carrier wave, which has the frequency of the wireless channel used for communication. Although the properties that are varied are the same both for analog and digital modulation, the nature of the data to be transmitted (analog or digital), directly impacts the output of modulation. Thus, we categorize modulation techniques into analog and digital and present the most common ones in the following subsections.

2.5.1 Analog Modulation

In order for analog data to be transmitted, analog modulation techniques are used. Analog modulation works by impressing the analog signal containing the data on a carrier wave with this impression aiming to change a property of the carrier wave. The most well known analog modulation techniques are Amplitude Modulation (AM) and Frequency Modulation (FM). These work by altering the amplitude and frequency of the carrier wave, respectively. AM and FM have found extensive use in radio broadcasting and are still widely used in these areas.

2.5.1.1 Amplitude Modulation (AM)

As mentioned above, AM works by superimposing the analog information signal $x(t)$ on the carrier wave $c(t)$. The modulated signal $s(t)$ is thus produced by adding $s(t)$ to the product of $s(t)$ and $x(t)$. Mathematically, AM is expressed by the following equation:

$$s(t) = (1 + x(t))\cos(2\pi f t) \tag{2.9}$$

where f is the frequency of the carrier wave and $c(t) = cos(2\pi f t)$ is the carrier wave.

AM results in a wave of an amplitude varying according to the amplitude of the analog information signal $x(t)$. Figures 2.17–2.19 show a carrier wave of amplitude twice that of the

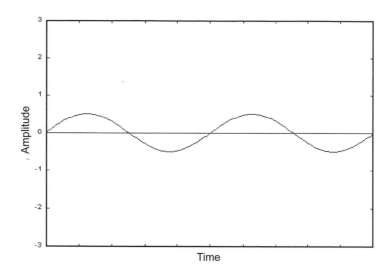

Figure 2.18   Analog information signal

analog information signal, the analog information signal and the result of AM modulation of the signal, respectively.

From Figure 2.19 one can see that the analog information signal can be easily decoded at the receiver by 'following' either the positive or negative peaks of the AM signal. However, this is not possible in cases where the ratio n of the maximum amplitude of the information signal $x(t)$ to that of the carrier $c(t)$ is higher than 1. In this case, decoding is more difficult, as 'following' either the positive or negative peaks of the amplitude-modulated signal does not give $x(t)$ but rather its absolute value, $|x(t)|$. Thus, the information signal is received distorted.

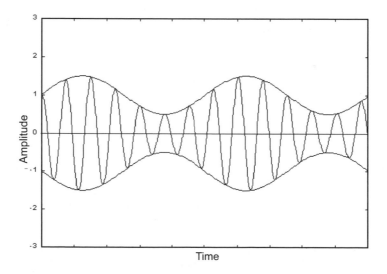

Figure 2.19   Result of AM

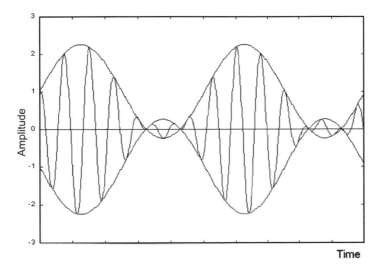

Figure 2.20 Result of AM when $n > 1$

This is shown in Figure 2.20, which depicts the AM signal produced by modulating the carrier wave of Figure 2.17 with an analog signal having twice the amplitude of the carrier. The same problem would also occur if we tried to modulate $c(t)$ by performing only a multiplication with $x(t)$.

2.5.1.2 Frequency Modulation (FM)

In FM, the information signal is used to alter the frequency of the carrier wave rather than its amplitude. This makes FM more resistant to noise than AM, since most of the times noise affects the amplitude of a signal rather than its frequency. FM can be expressed mathematically as

$$s(t) = A\left(\cos 2\pi f + \int^t x(t)dt\right) \qquad (2.10)$$

where A is the amplitude of the carrier wave $c(t)$, f is its frequency and $x(t)$ is the analog information signal. Figure 2.21 shows the output signal of FM for the carrier wave and information signal shown in Figures 2.17 and 2.18, respectively. Apart from conventional analog radio broadcasting, known to most people as FM radio, FM is used in first generation cellular systems, like the AMPS standard which is covered in Chapter 3.

2.5.2 Digital Modulation

Digital modulation techniques work by converting a bit string (digital data) to a suitable continuous time waveform. As in the case of analog modulation, digital modulation also alters a property of a carrier wave. However, in digital modulation these changes occur at discrete time intervals rather than in a continuous manner. The number of such changes over one second is known as the signal's baud rate which is generally different to the bit rate, as will be seen later.

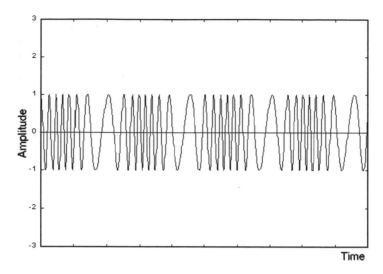

Figure 2.21   Result of FM

The most popular digital modulation techniques are Amplitude Shift Keying (ASK), two-level (binary) and four-level Frequency Shift Keying (FSK), Phase Shift Keying (PSK) and its variants. These are described below.

2.5.2.1 Amplitude Shift Keying (ASK)

The output of ASK for transmission of a binary string x, works as follows. Transmission of a binary 1 is represented by the presence of a carrier for a specific time interval, whereas transmission of a binary 0 is represented by a carrier absence for the same interval. Thus, for a cosine carrier of amplitude A and frequency f, we have

$$s(t) = \begin{cases} A\cos(2\pi ft), & \text{for binary 1} \\ 0 & \text{for binary 0} \end{cases} \tag{2.11}$$

The result of ASK modulation of the binary string of Figure 2.22 using the carrier of Figure 2.17, is shown in Figure 2.23.

2.5.2.2 Frequency Shift Keying (FSK)

The output of FSK for transmission of a binary string x, works as follows. Assuming a carrier of frequency f and a small frequency offset k, transmission of a binary 1 is represented by the presence of a carrier of frequency $f + k$ for a specific time interval, whereas transmission of a binary 0 is represented by a carrier of frequency $f - k$ for the same interval. Thus, for a cosine carrier of amplitude A and frequency f, we have

$$s(t) = \begin{cases} A\cos(2\pi(f + k)t), & \text{for binary 1} \\ A\cos(2\pi(f - k)t), & \text{for binary 0} \end{cases} \tag{2.12}$$

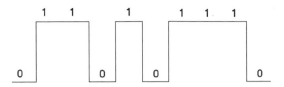

Figure 2.22   Binary string

Since two frequency levels are used, this technique is also known as two-level or binary FSK (BFSK). The result of BFSK modulation of the binary string of Figure 2.23 using the carrier of Figure 2.17 is shown in Figure 2.24.

In BFSK, every frequency shift encodes one bit. By defining more offsets for the frequency deviation, FSK can transmit more information with a single frequency shift. For example, four-level FSK:

$$s(t) = \begin{cases} A\cos(2\pi(f + 2k)t), & \text{for binary } 10 \\ A\cos(2\pi(f + k)t), & \text{for binary } 11 \\ A\cos(2\pi(f - k)t), & \text{for binary } 01 \\ A\cos(2\pi(f - 2k)t), & \text{for binary } 00 \end{cases} \qquad (2.13)$$

can transmit two bits per frequency shift. In this case, the bit rate achieved by the FSK signal is twice its baud rate since each state of the carrier encodes two bits. Higher level FSK modulation is also possible.

FSK is used in a number of wireless communication systems. For example, BFSK and four-level FSK are used in the physical layer of the 802.11 WLAN standard.

2.5.2.3 Phase Shift Keying (PSK)

The output of PSK for transmission of a binary string x, works as follows. Assuming a carrier of frequency f, transmission of a binary 0 is represented by the presence of the carrier for a specific time interval, whereas transmission of a binary 1 is represented by the presence of the carrier signal with a phase difference of π radians, for the same interval. Thus, for a cosine

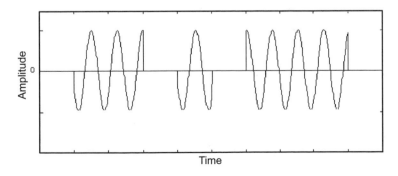

Figure 2.23   Result of ASK modulation of the binary string of Figure 2.19

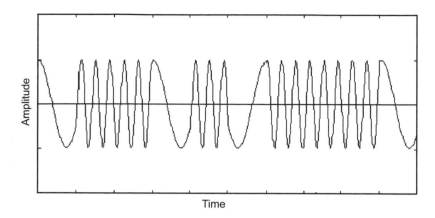

Time

Figure 2.24   Result of BFSK modulation of the binary string of Figure 2.22

carrier of amplitude A and frequency f, we have

$$s(t) = \begin{cases} A\cos(2\pi ft + \pi), & \text{for binary 1} \\ A\cos(2\pi ft), & \text{for binary 0} \end{cases} \qquad (2.14)$$

Since a single phase difference, this technique is also known as two-level or binary PSK (BPSK). The result of BPSK modulation of the binary string of Figure 2.22 using the carrier of Figure 2.17 is shown in Figure 2.25.

In BPSK, every phase representation encodes one bit. By defining more offsets for the frequency deviation, PSK can transmit more information with a single frequency shift. For example, quadrate (four level) PSK (QPSK) uses four different phases, separated by $\pi/2$ radians:

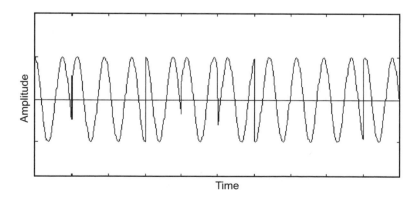

Time

Figure 2.25   Result of BPSK modulation of the binary string of Figure 2.22

$$s(t) = \begin{cases} A\cos\left(2\pi ft + \dfrac{\pi}{4}\right), & \text{for binary } 10 \\[2mm] A\cos\left(2\pi ft + \dfrac{3\pi}{4}\right), & \text{for binary } 11 \\[2mm] A\cos\left(2\pi ft + \dfrac{5\pi}{4}\right), & \text{for binary } 01 \\[2mm] A\cos\left(2\pi ft + \dfrac{7\pi}{4}\right), & \text{for binary } 00 \end{cases} \qquad (2.15)$$

QPSK can thus transmit two bits per frequency shift. In this case, the bit rate achieved by the QPSK signal is twice its baud rate since each state of the carrier encodes two bits. The obvious advantage of QPSK and other four-level modulation schemes makes them suitable choices for many cellular environments.

A number of techniques exist that are essentially PSK variations:

- *Differential PSK (DPSK)*. This is a variant of PSK. In DPSK a binary 1 is represented by changing the phase of the carrier wave relative to the phase of the previous symbol. On the other hand, a binary 0 is represented by a carrier wave having the same phase as the carrier used for transmission of the previous binary symbol. One can see that DPSK provides for self-clocking since phase changes are guaranteed for long runs of 1s.
- *π/4-shifted PSK*. This is another four-level PSK technique that provides self-clocking. π/4-shifted PSK codes pairs of bits by varying the phase of the carrier relative to the phase of the carrier used for the preceding pair of bits, according to Figure 2.26. It can easily be seen that there is always a phase change between consecutive bit transmissions. This can be seen for the transmission of 101001 in Figure 2.27. π/4-shifted PSK has found use in a number of systems, such as the cellular IS-54 standard which is covered in Chapter 4.

Pair of bits	Phase change
00	π/4
01	5π/4
10	-π/4
11	-5π/4

Figure 2.26   Phase changes for π/4-shifted PSK

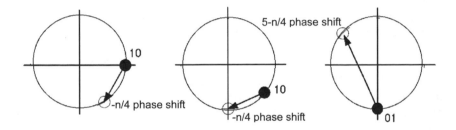

Figure 2.27   π/4-PSK operation

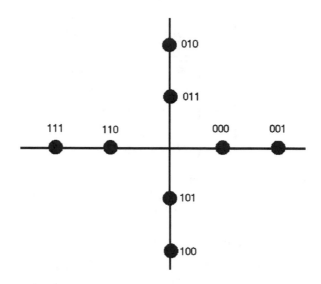

Figure 2.28   8-level QAM constellation encoding 3 bits/baud

- *Quadrate Amplitude Modulation (QAM)*. In QAM both the amplitude of the carrier and its phase are altered. Taking for example QPSK and assuming that we are able to code the four different phases with two different amplitude values, we have eight different combinations which can effectively code three bits per sample (i.e. bit rate = 3 × baud rate). For various QAM schemes these sets of combinations are known as constellation patterns. The constellation pattern for the system mentioned above is shown in Figure 2.28. By using a larger number of phase changes/amplitude combinations, the bit rate/baud rate ratio increases and we can thus get more spectrum-efficient modulation techniques. Thus, higher level QAM schemes have been developed, such as 16-QAM and 64-QAM which use 16 and 64 different numbers of phase changes/amplitude combinations, respectively. However, such techniques are more susceptible to noise, since a larger number of combinations means that these combinations are close to one another and thus noise can change the signal more easily.

2.6 Multiple Access for Wireless Systems

As in all kinds of networks, nodes in a wireless network have to share a common medium for signal transmission. Multiple Access Control (MAC) protocols are algorithms that define the manner in which the wireless medium is shared by the participating nodes. This is done in a way that maximizes overall system performance. MAC protocols for wireless networks can be roughly divided into three categories: Fixed assignment (e.g. TDMA, FDMA), random access (e.g. ALOHA, CSMA/CA) and demand assignment protocols (e.g. polling). The large number of MAC protocols for wireless networks that have appeared in the corresponding scientific literature (a good overview appears in Ref. [6]) demands a large amount of space for a comprehensive review of such protocols. In this section, we present some basic wireless MAC protocols.

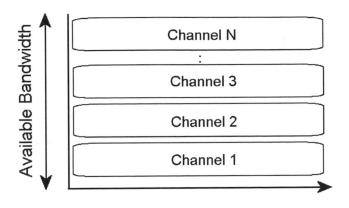

Figure 2.29   Illustration of FDMA

2.6.1 Frequency Division Multiple Access (FDMA)

In order to accommodate various nodes inside the same wireless network, FDMA divides the available spectrum into subbands each of which are used by one or more users. FDMA is shown in Figure 2.29. Using FDMA, each user is allocated a dedicated channel (subband), different in frequency from the subbands allocated to other users. Over the dedicated subband the user exchanges information. When the number of users is small relative to the number of channels, this allocation can be static, however, for many users dynamic channel allocation schemes are necessary.

In cellular systems, channel allocations typically occur in pairs. Thus, for each active mobile user, two channels are allocated, one for the traffic from the user to the Base Station (BS) and one for the traffic from the BS to the user. The frequency of the first channel is known as the uplink (or reverse link) and that of the second channel is known as the downlink (or forward link). For an uplink/downlink pair, uplink channels typically operate on a lower frequency than the downlink channel in an effort to preserve energy at the mobile nodes. This is because higher frequencies suffer greater attenuation than lower frequencies and consequently demand increased transmission power to compensate for the loss. By using low frequency channels for the uplink, mobile nodes can operate at lower power levels and thus preserve energy.

Due to the fact that pairs of uplink/downlink channels are allocated by regulation agencies, most of the time they are of the same bandwidth. This makes FDMA relatively inefficient since in most systems the traffic on the downlink is much heavier than that in the uplink. Thus, the bandwidth of the uplink channel is not fully used. Consider, for example, the case of web browsing through a mobile device. The traffic from the BS to the mobile node is much heavier, since it contains the downloaded web pages, whereas the uplink is used only for conveying short user commands, such as mouse clicks.

The biggest problem with FDMA is the fact that channels cannot be very close to one another. This is because transmitters that operate at a channel's main band also output some energy on sidebands of the channel. Thus, the frequency channels must be separated by guard bands in order to eliminate inter-channel interference. The existence of guard bands, however, lowers the utilization of the available spectrum, as can be seen in Figure 2.30 for

System	Uplink operating frequency range	Downlink operating frequency range	Channel bandwidth	Usable channel bandwidth
AMPS	824 MHz - 849 MHz	869 MHz - 894 MHz	30 KHz	24 KHz
NMT	453 MHz - 457.5 MHz 890 MHz - 915 MHz	463 MHz -467.7 MHz 935 MHz - 960 MHz	25 KHz	9.4 KHz

Figure 2.30   Total and usable channel bandwidths for AMPS and NMT systems

the first generation AMPS and Nordic Mobile Telephony (NMT) systems covered in Chapter 3.

2.6.2 Time Division Multiple Access (TDMA)

In TDMA [7], the available bandwidth is shared in the time domain, rather than in the frequency domain. TDMA is the technology of choice for a wide range of second generation cellular systems such as GSM, IS-54 and DECT which are covered in Chapter 4. TDMA divides a band into several time slots and the resulting structure is known as the TDMA frame. In this, each active node is assigned one (or more) slots for transmission of its traffic. Nodes are notified of the slot number that has been assigned to them, so they know how much to wait within the TDMA frame before transmission. For example, if the bandwidth is spread into N slots, a specific node that has been assigned one slot has to wait for $N-1$ slots between its successive transmissions. Uplink and downlink channels in TDMA can either occur in different frequency bands (FDD-TDMA) or time-multiplexed in the same band (TDD-TDMA). The latter technique obviously has the advantage of easy trading uplink to downlink bandwidth for supporting asymmetrical traffic patterns. Figures 2.31 and 2.32 show the structure of FDD-TDMA and TDD-TDMA, respectively.

TDMA is essentially a half-duplex technique, since for a pair of communicating nodes, at a specific time, only one of the nodes can transmit. Nevertheless, slot duration is so small that the illusion of two-way communication is created. The short slot duration, however, imposes strict synchronization problems in TDMA systems. This is due to the fact that if nodes are far

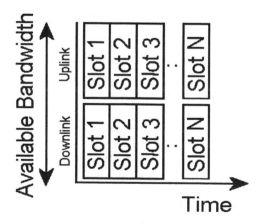

Figure 2.31   Illustration of FDD-TDMA

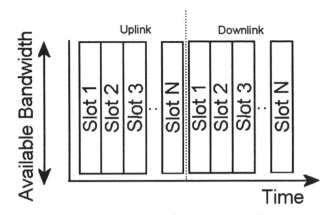

Figure 2.32 Illustration of TDD-TDMA

from one another, the propagation delay can cause a node to miss its turn. This is the case in GSM. Each GSM slot lasts 577 μs, which poses a limit of 35 km on the range of GSM antennas. If this range were to exceed 35 km, the propagation delay becomes large relative to the slot duration, thus resulting in the GSM phone losing its slot. In order to protect inter-slot interference due to different propagation paths to mobiles being assigned adjacent slots, TDMA systems use guard intervals in the time domain to ensure proper operation. Furthermore, the short slot duration means that the guard interval and control information (synchronization, etc.) may be a significant overhead for the system. One could argue that this overhead could be made lower by increasing the slot size. Although this is true, it would lead to increased delay which may not be acceptable for delay-sensitive applications such as voice calls.

Dynamic TDMA schemes allocate slots to nodes according to traffic demands. They have the advantage of adaptation to changing traffic patterns. Three such schemes are outlined below [8]:

- The first scheme was devised by Binder. In this scheme, it is assumed that the number of stations is lower than the number of slots, thus each station can be assigned a specific slot. The remaining slots are not assigned to anyone. According to their traffic demands, stations can contend for the remaining slots using slotted ALOHA, which is presented in the next paragraph. If a station wants to use a remaining slot to transmit information, it does so at the start of the slot. Furthermore, a station can use the home slot of another station if it monitors this home slot to be idle during the previous TDMA frame, a fact that means that the slot owner has no traffic. When the owner wants to use its slot, it transmits at the start of the slot. Obviously a collision occurs which notifies other stations that the slot's owner has traffic to transmit. Consequently, during the next TDMA frame, these stations defer from using that slot which can thus be used by its owner.
- The second scheme was devised by Crowther. In this scheme, it is assumed that the number of stations is unknown and can be variable. Thus, slots are not assigned to stations which contend for every available slot using ALOHA. When a station manages to capture a slot, it transmits a frame. Stations that hear this transmission understand that the station has successfully captured the slot and defer from using it in the next TDMA frame. Thus, a station that captures a slot is free to use it in the next TDMA frame as well.

- The third scheme, due to Roberts, tries to minimize the bandwidth loss due to collisions. Thus, a special slot (reservation slot) in the TDMA frame is split into smaller subslots which are used to resolve contention for slots. Specifically, each station that wants to use a slot transmits a registration request in a random subslot of the reservation slot. Slots are assigned in ascending order. Thus, the first successful reservation assigns the first data slot of the TDMA frame, the second successful reservation assigns the second data slot, etc. Stations are assumed to possess knowledge of the number of slots already assigned, so if the reservation of a station is completed without a collision, the station is assigned the next available slot.

2.6.3 Code Division Multiple Access (CDMA)

As seen above, FDMA accommodates nodes in different frequency subbands whereas TDMA accommodates them in different time parts. The third medium access technique, CDMA [9], follows a different approach. Instead of sharing the available bandwidth either in frequency or time, it places all nodes in the same bandwidth at the same time. The transmission of various users are separated through a unique code that has been assigned to each user.

CDMA has its origins in spread spectrum, a technique originally developed during World War II. The purpose of spread spectrum was to avoid jamming or interception of narrowband communications by the enemy. Thus, the idea of spread spectrum was essentially to use a large number of narrowband channels over which a transmitter hops at specific time intervals. Using this method any enemy that listened to a specific narrowband channel manages to receive only a small part of the message. Of course, the spreading of the transmission over the channels is performed in a random pattern defined by a seed, which is known both to the receiver and transmitter so that they can establish communication. Using this scheme the enemy could still detect the transmission, but jamming or eavesdropping is impossible without knowledge of the seed.

This form of spread spectrum is known as Frequency Hopping Spread Spectrum (FHSS) and although not used as a MAC technique, it has found application in several systems, such as an option for transmission the physical layer of IEEE 802.11 WLAN. This can be justified by the fact that spread spectrum provides a form of resistance to fading: If the transmission is spread over a large bandwidth, different spectral components inside this bandwidth fade independently, thus fading affects only a part of the transmission. On the other hand, if narrowband transmission was used and the narrowband channel was affected by fading, a large portion of the message would be lost. FHSS is revisited in Chapter 9.

CDMA is often used to refer to the second form of spread spectrum, Direct Sequence Spread Spectrum (DSSS), which is used in all CDMA-based cellular telephony systems. CDMA can be understood by considering the example of various conversations using different languages taking place in the same room. In such a case, people that understand a certain language listen to that conversation and reject everything else as noise.

The same principle applies in CDMA. All nodes are assigned a specific n-bit code. The value of parameter n is known as the system's chip rate. The various codes assigned to nodes are orthogonal to one another, meaning that the normalized inner product[3] of the vector representations of any pair of codes equals zero. Furthermore, the normalized inner product

---

[3] The normalized inner product P of two vectors A and B is essentially the cosine of the angle formed between A and B. Thus, it is a metric of similarity of the two vectors, since for A and B being orthogonal, $P = 0$.

Station codes	Transmissions	Decoding traffic of station:
C_A: 1 0 1 0	A: Bit 1 => 1 -1 1 -1	A: $S*C_A$=(-1 -1 3 -1)*(1 -1 1 -1)/4=1 => binary 1.
C_B: 1 0 0 1	B: Bit 0 => -1 1 1 -1	B: $S*C_B$=(-1 -1 3 -1)*(1 -1 -1 1)/4=-1 => binary 0.
C_C: 0 0 1 1	C: Bit 1 => -1 -1 1 1	C: $S*C_C$=(-1 -1 3 -1)*(-1 -1 1 1)/4=1 => binary 1.

	S= -1 -1 3 -1	

Figure 2.33 CDMA operation

of the vector representation of any code with itself and the one's complement of itself equals 1 and -1, respectively. Nodes can transmit simultaneously using their code and this code is used to extract the user's traffic at the receiver. The way in which codes are used for transmission is as follows. If a user wants to transmit a binary one, it transmits its code, whereas for transmission of a binary zero it transmits the one's complement of its code. Assuming that users' transmissions add linearly, the receiver can extract the transmission of a specific transmitter by correlating the aggregate received signal with the transmitter's code. Due to the use of the n-bit code, the transmission of a signal using CDMA occupies n times the bandwidth that would be occupied by narrowband transmission of the same signal at the same symbol rate. Thus, CDMA spreads the transmission over a large amount of bandwidth and this provides resistance to multipath interference, as in the FHSS case. This is the reason that, apart from an channel access mechanism, CDMA has found application in several systems as a method of combating multipath interference. Such a situation is the use of CDMA as an option for transmission the physical layer of IEEE 802.11 WLAN.

An example of CDMA is shown in Figure 2.33 where we map the transmission of the ones and zeros in stations' codes to $+1$ and -1, respectively. For three users, A, B and C and $n = 4$, the figure shows the users' codes C_A, C_B and C_C, these stations bit transmissions and the way that recovery of a specific station's signal is made.

CDMA makes the assumption that the signals of various users reach the receiver with the same power. However, in wireless systems this is not always true. Due to the different attenuation suffered by signals following different propagation paths, the power level of two different mobiles may be different at the BS of a cellular system. This is known as the near-far problem and is solved by properly controlling mobile transmission power so that the signal levels of the various mobile nodes are the same at the BS. This method is known as power control and is described in the next section. Furthermore, as FDMA and TDMA, CDMA demands synchronization between transmitters and receivers. This is achieved by assigning a specific code for transmission of a large sequence by the transmitter. This signal is known as the pilot signal and is used by the receiver for synchronizing with the transmitter.

2.6.4 ALOHA-Carrier Sense Multiple Access (CSMA)

The ALOHA protocol is related to one of the first attempts to design a wireless network. It was the MAC protocol used in the research project ALOHANET which took place in 1971 at the University of Hawaii. The idea of the project was to offer bi-directional communications without the use of phone lines between computers spread over four islands and a central computer based on the island of Oahu. Although ALOHA can be applied to wired networks as well, its origin relates to wireless networks.

The principle of ALOHA is fairly simple: Whenever a station has a packet to transmit, it does so instantaneously. If the station is among a few active stations within the network, the chances are that its transmission will be successful. If, however, the number of stations is relatively large, it is probable that the transmission of the station will coincide with that of (possibly more than one) other stations, resulting in a collision and the stations' frames being destroyed.

A critical point is the performance of ALOHA. One can see that in order for a packet to reach the destination successfully, it is necessary that:

- no other transmissions begin within one frame time of its start;
- no other transmissions are in progress when the station starts its own transmission; this is because stations in ALOHA are 'deaf', meaning that they do not check for other transmissions before they start their own.

Thus, one can see that the period during which a packet is vulnerable to collisions equals twice the packet transmission size. It can be proven that the throughput $T(G)$ for an offered load of G frames per frame time in an ALOHA system that uses frames of fixed size is given by

$$T(G) = Ge^{-2G} \qquad (2.16)$$

Equation (2.15) gives a peak of $T(G) = 0.184$ at $G = 0.5$. This peak is, of course, very low. A refinement of ALOHA, slotted ALOHA, achieves twice the above performance, by dividing the channel into equal time slots (with duration equaling the packet transmission time) and forcing transmissions to occur only at the beginning of a slot. The vulnerable period for a frame is now lowered to half (the frame's transmission time) which explains the fact that performance is doubled. The throughput $T_s(S)$ for an offered load of G frames per frame time in a slotted ALOHA system that uses frames of fixed size is given by

$$T_s(G) = Ge^{-G} \qquad (2.17)$$

which gives a peak of $T_s(G) = 0.37$ at $G = 1$.

The obvious advantage of ALOHA is its simplicity. However, this simplicity causes low performance of the system. Carrier Sense Multiple Access (CSMA) is more efficient than ALOHA. A CSMA station that has a packet to transmit listens to see if another transmission is in progress. If this is true, the station defers. The behavior at this point defines a number of CSMA variants:

- *P-persistent CSMA*. A CSMA station that has a packet to transmit listens to see if another transmission is in progress. If this is true, the station waits for the current transmission to complete and then starts transmitting with probability p. For $p = 1$ this variant is known as 1-persistent CSMA.
- *Nonpersistent CSMA*. In an effort to be less greedy, stations can defer from monitoring the medium when this is found busy and retry after a random period of time.

Of course, if more than two stations want to transmit at the same time, they will all sense the channel simultaneously. If they find it idle and some decide to transmit, a collision will occur and the corresponding frames will be lost. However, it is obvious that collisions in a

CSMA system will be less than in an ALOHA system, since in CSMA stations ongoing transmissions are not damaged due to the carrier sensing functionality. When a collision between two CSMA nodes occur, these nodes keep collided packets in their buffers and wait until their next attempt, which occurs after a time interval $t = ks$. s is the system's slot time, which for wired networks equals twice the propagation delay of signals within the wire. For wireless networks, s is defined in another way, as described in Chapter 9. The value of k is given by the exponential backoff algorithm, which uniformly selects a number from the interval $[0...,2^{I-1}]$, where $i = \min(c, cw)$. c is the number of consecutive collisions encountered by the frame and cw is a system parameter that governs the maximum random number produced, k.

CSMA has found use in wireless networks, especially WLANs. In wired networks, CSMA is the basis of the IEEE 802.3 protocol, known to most of us as Ethernet. Ethernet is superior to simple CSMA due to the fact that it can detect collisions and abort the corresponding transmissions before they complete, thus preserving bandwidth. However, collision detection cannot be performed in WLANs. This is due to the fact that a WLAN node cannot listen to the wireless channel while sending, because its own transmission would swamp out all other incoming signals. Since collisions cannot be detected, CSMA-based protocols for wireless LANs try to avoid them; thus, CSMA schemes for such networks are known as CSMA-Collision Avoidance (CSMA-CA) schemes. CSMA-CA protocols are the basis of the IEEE 802.11 and HIPERLAN 1 WLAN MAC sublayers which are presented in Chapter 9.

2.6.5 Polling Protocols

Polling protocols are centralized. For a polling protocol to be applied, a central entity (Base Station, BS) is assigned responsibility for polling the stations within the network. If the BS decides that a specific station grants permission to transmit, it polls this station, meaning that it sends to the station a small control frame notifying it that it can transmit one or more frames. After the transmission of this station, the BS proceeds to poll the other stations of the network. If a station is polled but has no traffic to transmit, it notifies the BS of this fact, the procedure continues and the next station is polled.

Polling is an appealing MAC option, however, it demands that the BS possesses knowledge regarding the network topology (the nodes under its coverage) in order for the network nodes to be polled. Such knowledge is difficult to achieve for a wireless network since topology changes occur frequently due to the mobile nature of nodes and the fading wireless links. Several polling protocols tailored to the characteristics of wireless networks have appeared. In this section, we present the Randomly Addressed Polling (RAP) and Group RAP (GRAP) protocols [10–12], which were designed for Wireless LANs (WLANs) as an example. RAP and GRAP also use CDMA in one of their stages.

RAP combats the imprecise knowledge regarding the network topology by working, not with all the nodes under the coverage of the BS, but only with the active ones seeking uplink communication. In RAP, a station is said to be active, if it has a packet to transmit. The RAP protocol assumes an infrastructure cellular topology, which is presented in later paragraphs. Within each cell, multiple mobile nodes exist that compete for access to the medium. The cell's BS initiates a contention period in order for active nodes to inform their intention to transmit packets. The operation of RAP constitutes a number of polling cycles. When a collision between two or more stations occurs, these stations keep the collided packets and compete for access to the medium in the next polling cycle. Newly active stations are usually not allowed to compete with those having collided packets [11]. The collision resolution cycle (CRC) is defined as the period of time that elapses in order for all the active stations at the beginning of the CRC to transmit their packets. In order to keep newly active nodes from entering the competition, the READY message at the beginning of a CRC can have a different form from that at the beginning of a polling cycle commencing inside a CRC. However, the prohibition of newly active stations to compete with those having collided packets is not compulsory [10].

For a RAP WLAN consisting of N active stations, the stages of the protocol are as follows:

- *Contention invitation stage.* Whenever the BS is ready to collect packets from the mobile nodes, it transmits a READY message, which may be piggybacked in a previous downlink transmission.
- *Contention stage.* All active mobile nodes generate a random number R, ranging from 0 to $P - 1$ and transmit it simultaneously to the BS using CDMA transmission. The number transmitted by each station identifies this station during the current cycle and is known as its random address. To combat the medium's fading characteristics a station may transmit its generated random numbers up to q times in a single contention stage. When an error-free transmission is assumed, $q = 1$ suffices. Optionally, the contention stage may be repeated L times. Each time, stations generate and transmit random numbers as described above.
- *Polling stage.* Suppose that at the lth stage ($1 \leq l \leq L$) the BS received the largest number of distinct numbers and these are, in ascending order, $R_1, R_2, \ldots, R_n$. The BS polls the mobile nodes using those numbers. When the BS polls mobile nodes with R_k, nodes that transmitted R_k as their random address at the lth stage transmit packets to the BS. Obviously, if two or more nodes transmitted the same random number at the lth stage a collision will occur. If $n = N$ however, no collision occurs.
- If a BS successfully receives a packet from a mobile node, it sends a positive acknowledgment (PACK). If reception of the packet at the BS is unsuccessful either due to noise or a collision, the BS informs the mobile node by sending a negative acknowledgment (NACK). Acknowledgment packets are transmitted right before polling the next mobile node. If a mobile node receives a PACK, it assumes correct delivery of its packet, otherwise it waits for the current polling cycle to complete and retries during the next one.

Figure 2.34 shows an example of RAP operation with $N = 7$ active stations and $P = 5$ available random addresses. We assume that $L = 2$, thus, at the beginning of the CRC, all seven stations transmit two random addresses to the base station. As we can see, the maximum number of distinct random addresses is received at the second stage, thus, the base station polls according to the received numbers at this stage. Stations C, E and G manage to transmit their packets without a collision, while A, D, B and F proceed to the next polling cycle. At this cycle, the base station polls according to the numbers of the second stage and thus (assuming that no newly active stations are allowed to join re-polling) A and F manage to transmit their packet while B and D collide. During the third polling cycle, B and D transmit their packets. After the completion of the CRC, another one begins and all active stations join the new CRC.

A complete description of RAP and a discussion of implementation issues is provided in Ref. [11]. Numerical results in [11] show that increasing values of L yield better throughput

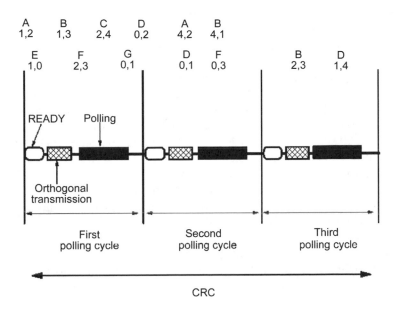

Figure 2.34   Example of RAP operation

results, however, the performance gain with $L > 2$ is very small. As a result, a value of 2 for L seems to be a good choice. Comparisons with CSMA show that although the mean delay in RAP also rises rapidly under heavy load, RAP is characterized by smaller delays for a given throughput value. Moreover, by increasing the value of P, the delay reduces significantly. If the number of active stations, N, is significantly less than P, RAP depicts increased throughput and decreased delay.

A critical point for RAP seems to be the choice of P. Use of large values of P favor performance but also lead to increased circuit complexity. As a result, values of P around 5, which still provide significant performance gains over conventional protocol, are suggested [11].

A modification of RAP, Group RAP is proposed in Ref. [12]. GRAP adopts the superframe structure, consisting of $P + 1$ frames and divides active nodes into groups. At the beginning of each frame only the BS is allowed to transmit. After the BS completes transmission the polling procedure begins. However, GRAP does not allow all active nodes to compete in a single contention period. GRAP states that all nodes that successfully transmitted during the previous polling cycles maintain their random addresses and form the groups from 0 to $P - 1$. A mobile station joins group j if the random address for its previously successful transmission was j. All the new joining stations form the Pth group. Furthermore, all mobile stations that have time bounded packets can join any group for contention. After the $P + 1$ groups have been formed, the polling procedure begins. The members of each group are polled according to the RAP protocol.

An advantage of GRAP over RAP is that it allocates much more bandwidth to the BS. This is because most of the data applications in wireless networks are of client server nature, with the mobile stations being the clients and the downlink traffic from the server demanding an increased portion of bandwidth.

A problem with RAP is that, if the number of active stations N approaches or exceeds P, then the performance of RAP degrades sharply. This is because a small number of available random addresses provides very little space for the contention to be resolved, as for values of N with $P \leq N$, the selection of the same random address by more than one station becomes very likely. As a result, the probability of a successful transmission is lowered, which leads to the decreased throughput and increased delay. TDMA-based RAP (TRAP) [13–15] solves this problem. TRAP employs a variable length TDMA-based contention stage, which lifts the requirement for a fixed number of random addresses. The TDMA-based contention stage comprises a variable number of slots, with each slot corresponding to a random address. However, a mechanism is needed in order for the base station to select the appropriate number of slots (equivalently, random addresses) in the TDMA contention stage. To this end, at the beginning of each polling cycle, all active mobile stations register their intention to transmit via transmission of a short pulse. All active stations' pulses are added at the base station, which uses the aggregate received pulse to estimate the number of active stations. The time slots will obviously be of fixed length, thus, a mobile station that generates a random address p, will transmit its random address at slot p. Based on this approach, the proposed protocol works as follows:

- *Active stations estimation.* At the begging of each polling cycle, the base station sends an ESTIMATE message in order to receive active stations' pulses. After the base station estimates the number of active stations N based on the aggregate received pulse, it schedules the TDMA-based contention stage to comprise an adequate number of random addresses $P = kN$, where k is an integer, for the active stations to compete for medium access.
- *Contention invitation stage.* The base station announces it is ready to collect packets from the mobile nodes, and transmits a READY message, containing the number of random addresses P to be used in this polling cycle.
- *Contention stage.* Each active mobile node generates a random number R, ranging from 0 to $P - 1$. Active nodes transmit their random numbers at the appropriate slot of the TDMA-based contention scheme. As in RAP, stations can generate addresses up to q times in a single contention stage and the contention stage may be repeated L times, with each active station generating a random address for each stage. Obviously, if two or more mobiles select the same random address, their random address transmissions collide and are not received at the base station. Thus, the random addresses received correctly at the base station are always distinct, with each number identifying a single active station.
- *Polling stage.* Suppose that at the lth stage ($1 \leq l \leq L$) the BS received the largest number of distinct numbers and these are, in ascending order, $R_1, R_2, ..., R_n$. The BS polls the mobile nodes using those numbers. When the BS polls mobile nodes with R_k, nodes that transmitted R_k as their random address at the lth stage transmit packets to the BS. Obviously, if two or more nodes transmitted the same random number at the lth stage a collision will occur. If $n = N$ however, no collision occurs.
- If a BS successfully receives a packet from a mobile node, it sends a positive acknowledgment (PACK). If reception of the packet at the BS is unsuccessful either due to noise or

a collision, the BS informs the mobile node by sending a negative acknowledgment (NACK). Acknowledgment packets are transmitted right before polling the next mobile node. If a mobile node receives a PACK, it assumes correct delivery of its packet, otherwise it waits for the current polling cycle to complete and retries during the next one.

Under the assumption of all mobile random address transmissions reaching the base station, the protocol is collision free among data packets. This is because the same random address transmission by two or more stations occurs in the same time slot resulting in a collision of the control packets and the address not being polled. Thus, the data packets do not collide. This is an advantage of TRAP against the original RAP protocol. Due to the orthogonal nature of the contention stage of RAP, when the base station polls a random address that was selected by more than one mobile, the corresponding mobiles' packets will collide. This feature helps preserve bandwidth, since data packets are usually much larger than control packets. Also, it has found use in other WLAN MAC protocols as well, such as IEEE 802.11.

However, the obvious advantage of our proposed protocol is in terms of scalability. Since the number of random addresses can now vary according to the number of active stations, the protocol will not degrade in cases of a large number of competing stations. Simulation results that are presented in Refs. [13–15] reveal that the heuristic estimator $P = kN$ for the number of random addresses is sufficient, since the performance of the protocol at medium and high loads is significantly better than that of RAP. Furthermore, the implementation of TRAP protocol is much simpler than that of CDMA-based versions of RAP, since no extra hardware is needed for the orthogonal reception of the random addresses.

Finally, another interesting polling MAC protocol for a wireless environment is Learning Automata-Based Polling (LEAP) [16]. LEAP is designed for bursty traffic infrastructure WLANs. According to LEAP, the mobile station that grants permission to transmit is selected by the base station by means of a learning automaton [17]. The learning automaton takes into account the network feedback information in order to update the choice probability of each mobile station. The network feedback conveys information both on the network traffic pattern and the base mobile station condition of the wireless links. The learning algorithm asymptotically tends to assign to each station a portion of the bandwidth proportional to the station's needs.

According to LEAP, the BS is equipped with a learning automaton which contains the choice probability $P_k(j)$ for each mobile station k under its coordination. Before polling at polling cycle j those probabilities are normalized in the following way:

$$\prod_k(j) = \frac{P_k(j)}{\sum_{i=1}^{N} P_i(j)} \tag{2.18}$$

Clearly, $\sum_{i=1}^{N} \prod_i(j) = 1$, where N is the number of mobile stations under the coverage of the base station. At the beginning of each polling cycle j, the base station polls according to the normalized probabilities $\prod_i(j)$. Each polling cycle consists of a sequence of packet exchanges between the base station, the selected mobile and a destination mobile station if a packet is to be transmitted by the selected mobile. The protocol uses four control packets, POLL, NO_DATA, BUFF_DATA and ACK whose duration is t_{POLL}, t_{NO_DATA}, t_{BUFF_DATA} and t_{ACK}, respectively. Assuming that the base station polls mobile station k at time position t which marks the beginning of polling cycle j, the propagation delay is t_{PROP_DELAY}, and a station's DATA transmission takes t_{DATA} time to complete, the following events are possible:

1. The poll is received at station k at time $t + t_{POLL} + t_{PROP\_DELAY}$. Then:

  - If station k does not have a buffered packet, it immediately responds to the base station with a NO_DATA packet. If the base station correctly receives the NO_DATA packet, it lowers the choice probability of station k and immediately proceeds to poll the next station. This poll is initiated at time $t + t_{POLL} + 2t_{PROP\_DELAY} + t_{NO\_DATA}$. In the case of no reception at the base station, the choice probability of station k is lowered and the next poll begins at time $t + t_{POLL} + 4t_{PROP\_DELAY} + t_{BUFF\_DATA} + t_{DATA} + t_{ACK}$.
  - If station k has a buffered DATA packet, it responds to the base station with a BUFF_-DATA packet, transmits the DATA packet to its destination and waits for an acknowledgment (ACK) packet. After the poll, the base station monitors the wireless medium for a time interval equal to $t_{BUFF\_DATA} + t_{DATA} + t_{ACK} + 3t_{PROP\_DELAY}$. If it correctly receives one or more of the three packets, it concludes that station k received the poll and has one or more buffered data packets. Thus, it raises station's k choice probability. On the other hand, if the base station does not receive feedback, it concludes that it cannot communicate with station k, lowers the choice probability of k and proceeds with the next poll at time $t + t_{POLL} + 4t_{PROP\_DELAY} + t_{BUFF\_DATA} + t_{DATA} + t_{ACK}$.

2. The poll is not received at station k, k does not respond to the base station and the choice probability of k is decreased. Then the base station proceeds to poll the next station at time $t + t_{POLL} + 4t_{PROP\_DELAY} + t_{BUFF\_DATA} + t_{DATA} + t_{ACK}$.

From the above discussion, it is obvious that the learning algorithm takes into account both the bursty nature of the traffic and the bursty appearance of errors over the wireless medium. Upon conclusion of a polling cycle j, the base station uses the following scheme in order to update the selected station's k choice probability:

$$P_k(j + 1) = P_k(j) + L(1 - P_k(j)),$$

if FEEDBACK$_k(j)$= TRANSMIT

$$P_k(j + 1) = P_k(j) - L_{LEAP}(P_k(j) - a), \qquad (2.19)$$

if FEEDBACK$_k(j)$= IDLE or FEEDBACK$_k(j)$= FAIL

where:

- FEEDBACK$_k(j)$ = TRANSMIT indicates that the base station received feedback indicating that station k, when polled at polling cycle j, transmitted a DATA packet. This means that the base station correctly received one or more of the BUFF_DATA, DATA, and possibly ACK, packets exchanged due to k's transmission.
- FEEDBACK$_k(j)$ = IDLE indicates that the base station received feedback indicating that station k, when polled at polling cycle j, did not transmit a DATA packet. This means that the base station correctly received the NO_DATA packet transmitted by k.
- FEEDBACK$_k(j)$ = FAIL indicates that the base station failed to receive feedback about k's transmission state at cycle j. This is equivalent either to erroneous reception, or to the reception of no packets at all, at the base station for polling cycle j.

For all j, it holds that $L, a \in (0, 1)$ and $P_k(j) \in (a, 1)$. Since the offered traffic is of bursty nature, when the base station realizes that the selected station had a packet to transmit, it is probable that the selected station will also have packets to transmit in the near future. Thus, its choice probability is increased. On the other hand, if the selected station notifies that it does not have buffered packets, its choice probability is decreased, since it is likely to remain in this state in the near future. In general, the background noise and interference at the base station will be the same, if not lower, than that at a mobile station. When the base station fails to receive feedback about the selected mobile's state, the latter is probably experiencing a relatively high level of background noise. In other words, it is 'hearing' the base station over a link with a high BER. Since in wireless communications errors appear in bursts, the link is likely to remain in this state for the near future. Thus, the choice probability of the selected station is lowered in order to reduce the chance of futile polls to this station in the near future.

When the choice probability of a station approaches zero, this station is not selected for a long period of time. During this period, it is probable that the station transits from idle to busy state. The same holds for the status of a high-BER link between the mobile station and the base station. After a period of time, it is probable that the link's state changes to a low BER one. However, since the mobile station does not grant permission to transmit, the automaton is not capable of 'sensing' those transitions. The role of parameter a, is to prevent the choice probabilities of stations from taking values in the neighborhood of zero in order to increase the adaptivity of the protocol.

LEAP updates the choice probabilities of mobile stations according to the network feedback information. The choice probability of each mobile station converges to the probability that this station is ready to transmit, meaning that it has a nonempty queue and it is capable of communicating successfully with the base station. Simulation results in Ref. [16] show superior performance against the RAP-GRAP protocols in wireless environments characterized by bursty packet arrivals.

2.7 Performance Increasing Techniques for Wireless Networks

As mentioned above, the basic problem in wireless networks is the fact that wireless links are relatively unreliable. Thus, a number of schemes that work on the physical layer and try to present a relatively 'clean' medium to higher layers of the network have been considered. Some of them have found use in commercial systems. In our previous discussion, we visited one such technique, spread spectrum, which, as mentioned, effectively combats fading as it splits the transmitted signal over a large bandwidth. Due to the fact that different spectral components inside this large bandwidth fade independently, fading will affect only a part of the transmission. Thus, SS achieves resistance to fading. In this section, we describe some other techniques: diversity, coding, equalization and power control.

2.7.1 Diversity Techniques

In an effort to combat the phenomenon of fading in wireless channels, a family of techniques, known as diversity techniques, is used. Many types of diversity techniques exist, such as time, frequency, antenna (also known as space) and polarization diversity. The principle of diversity systems is to send copies of the same information signal through several different channels. Performance enhancement is achieved due to the fact that these channels fade independently, thus, fading will affect only a part of the transmission. In this section we describe the fundamentals of the most commonly used type of diversity, antenna diversity and its enhancements, smart antennas. Other types of diversity are considered in chapters that present systems that use them.

2.7.1.1 Antenna Diversity

Antenna diversity, also known as space diversity, is commonly used for performance enhancement in wireless systems. It is essentially a method that calls for a set of array elements (also referred to as branches), mostly two, spaced sufficiently apart from each other, with the spacing usually in the order of the wavelength of the used channel. This is due to the fact that multipath fading is considered independent at distances in the order of the channel's wavelength.

Antenna diversity can effectively combat multipath fading in NLOS situations. Nevertheless, the performance gains are lower in LOS cases. Although applicable in both BSs and mobile stations, antenna diversity poses significant challenges for implementation in the mobile stations, due to limitations relating to power consumption and size of the mobile station antenna. It can be used either for transmission (transmit diversity) or reception (receive diversity) of signals. In both cases, the aim is to increase the quality of the signal at the receiver. In receive diversity, the branches of the antenna system pick up a number of differently fading signals and combine them in order to reconstruct the original transmission at the highest possible quality. When applied at the BS of a cellular system, receive antenna diversity obviously enhances the performance of the uplink. Thus, it has found use in the uplink of a number of wireless systems, such as GSM and IS-136. Transmit diversity, on the other hand, calls for transmission of replicas of the signal by each branch of the antenna and when applied to a the BS of a cellular system will favor the reception quality at the downlink. However, it has seen limited support in commercial systems.

Figure 2.35 sketches the way a two-branch receive diversity system can combat interference. A number of algorithms that exploit the signals from the two antenna elements in order to reconstruct the original transmission at the highest possible quality exist. For example, antenna diversity can be used to either select the strongest signal picked up by one of the antenna elements or combine the signals from the two elements in order to reconstruct the original transmission. However, a description of such algorithms is out of the scope of this chapter.

2.7.1.2 Multiantenna Transmission/Reception: Smart Antennas

The term smart (or adaptive) antennas [18] is used to describe antennas that are not fixed but rather change in order to adapt to the conditions of the wireless channel. Smart antennas are considered an enhanced method of antenna diversity. Although applicable to almost all kinds of wireless networks, smart antennas are most commonly considered for use by the BSs of cellular systems. The idea of smart antennas has been around for some years, however, due to the fact that it demands computational power, consideration on its use in commercial systems has started only recently.

Smart antennas combat the deficiencies of conventional omnidirectional antennas. Considering the case of a cellular system, omnidirectional antennas can be regarded as a waste of

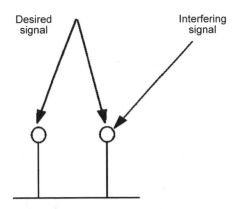

Figure 2.35   A two-branch diversity system

power, due to the fact that they radiate power in all directions while the user being serviced by the antenna is only in a certain direction. Smart antennas surpass this inefficiency since they can (a) focus to the radio transmission of the receiver and (b) focus their own transmission towards the receiver, as seen in Figure 2.36 for a cellular system. This technique is also known as beamforming. According to the principle on which beamforming is based, smart antennas can be categorized into [18]:

- *Switched lobe, or switched beam.* This category of smart antennas is the most simple. It consists of a number of static directive antenna elements and a basic switching function.

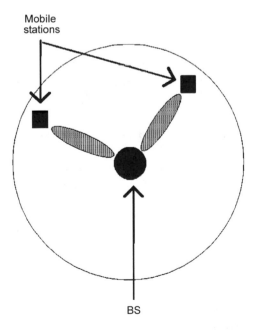

Figure 2.36   Use of smart antennas in cellular systems

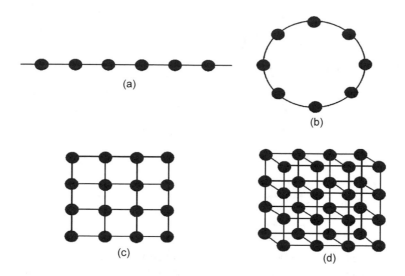

Figure 2.37   Possible placements of array elements in smart antennas

For communication with a receiver, this switching function selects the element that maximizes performance, usually measured by received power. This structure is the most simple but also has the lowest performance gains compared to conventional antennas.

- *Dynamically phased array*. This category utilizes information regarding the direction of arrival of the transmitter's signal. By monitoring this value, antenna elements can be used to track the user as it moves. This category, which can be seen as an enhancement of the switched-beam concept also maximizes performance in terms of received power.
- *Adaptive array*. This category utilizes the direction of arrival value of users nearby the entity, which transmits to the antenna. By using this value the radiation pattern can be adjusted to cancel interference from these users. Furthermore, adaptive arrays can be used to combine the different echoes of a user's transmission and reconstruct the original signal. Adaptive arrays maximize performance by maximizing the received Signal to Interference noise Ratio (SIR).

The above methods describe tracking of the reception signal by the BS which implements antenna diversity. As far as transmission to a mobile is concerned, the BS may utilize the value of direction of arrival of the mobile node transmission at the BS in order to focus its transmission on the mobile receiver. Figure 2.37 shows possible structures of array elements that form smart antennas. The first two structures are used for beamforming on a horizontal plane, which is enough for large cells, typically found in rural areas. For densely populated cells, the third and fourth structures can be used for two-dimensional beamforming. For transmission, the radiation pattern is produced by splitting the signal to be transmitted into a number of other signals. These signals are directed to different array elements and their amplitudes and relative phases of these signals are adjusted in order to maximize performance. The opposite happens in reception: The signals from each element produce a combined signal that enters the decoding circuits of the receiver.

Although realizable today, smart antennas are not likely to start being applicable both in the uplink and downlink of mobile systems. Rather it is envisioned that smart antennas will first be introduced in base stations, thus benefiting uplink (mobile to BS) transmissions. The use of directive antennas to the BSs will occur later on. Finally, when smart antennas find full application in mobile systems, these will be able to accommodate traffic from many users in the same frequency at the same time and separate users by spatial information. This will create a new form of multiple access, Space Division Multiple Access (SDMA), where users will be separated based on the angle of their transmission to the base station.

Due to their capability for directive transmission, smart antennas have the obvious advantage of interference reduction for users nearby the receiver. Furthermore, the capacity of a system is increased since it is possible that the same spectrum can be used at the same time by more than one user. This can be done by exploiting information regarding their position and using smart antennas to direct BS transmissions to these users. This idea gave rise to the concept of geolocation applications which will be used in next generation wireless networks which will be able to extract spatial information of users more precisely than current wireless networks. Geolocation applications are revisited in Chapter 5. Smart antennas are also likely to have an increased range compared to conventional ones and offer an increased level of security, since eavesdropping of a transmission now requires the eavesdropper to be present in the imaginable line formed between the transmitter and the receiver.

However, the use of smart antennas entails some problems too. The first is the increased implementation complexity and the cost of the method. This is incurred by the need for real-time tracking of the receiver position and real-time updating of the corresponding transmission. For a BS this is difficult since BSs in cellular systems are likely to serve many users at the same time. Furthermore, the fact that SDMA separates users based on their angle to the receiver means that BSs will have to switch users to another SDMA channel when angular collisions occur. In densely populated areas, many such collisions are likely to take place thus requiring increased computational efficiency (which means cost) from the BS. However, the advancements of computer technology have driven down costs and systems having the necessary processing power for this task are available. Nevertheless, the cost of a smart antenna system will be larger than a system with a conventional antenna.

2.7.2 Coding

In all kinds of networks, there is a certain possibility that reception of a bit stream is altered by errors. Coding techniques aim to provide resistance to such errors by adding redundant bits to the transmitted bit stream so that the receiver can either detect and ask for a retransmission, or correct the faulty reception. Thus, we have error detection and error correction coding schemes, respectively. The process of adding this redundant information is known as channel coding, or Forward Error Correction (FEC). By recalling the fact that the BER experienced over a wireless channel can be as high as 10^{-3} whereas typical BERs of wired channels are around 10^{-10}, one can easily realize the usefulness of such techniques in wireless systems. A vast number of coding techniques exist in the scientific literature; thus, in this section we present the basic techniques used for FEC.

2.7.2.1 Parity Check

The simplest technique, which is known to almost anyone dealing with computer technology, is the parity bit technique that can detect single-bit errors. According to this technique, the transmitter and the receiver agree whether the number of binary 1s that will be contained in the messages they exchange will be odd or even. Thus, the parity schemes are known as odd parity and even parity, respectively. For the sake of presentation assume that the transmitted message is 10010 and that the agreement is on odd number of binary 1s. After the agreement for odd parity is made, the transmitter adds either a binary 1 or a binary 0 to the end of the original message, so that the number of binary 1s in the resulting bit stream is odd and sends the message. Thus, the actual transmitted message is now 100101. If the bit stream arrives intact at the receiver (100101), then the number of binary 1s will be odd, whereas if a single bit error occurs (e.g. 101101), the receiver will detect that the number of binary 1s is even.

Although very simple to implement, the parity scheme has the disadvantages of (a) not being able to detect a multiple of two bit errors in the same message (e.g. in the above case, it would see the reception of 001101 (two bit errors) as correct) and (b) being able only to detect and not to correct a faulty reception. The problem is that the parity scheme has a Hamming distance of 2. The Hamming distance of a set of binary streams defines the least number of bit inversions that, when applied, can lead from a stream of the set to another stream of the set. By increasing this distance, the scheme can be made more robust. Returning to an example, consider that the receiver and the transmitter agree to exchange only the messages 00000 and 00111, which have a distance of 3. Thus, for the reception of 00000 as 00011 the receiver can detect the double-bit error since the received message is not valid. Furthermore, for the reception of 00000 as 00001, the receiver can correct the single-bit error (as 00001 is closer to 00000 than to 00111) and thus recover the transmitted message.

2.7.2.2 Hamming Code

The above example does not consider adding extra bits to the original message for coding purposes. However, the addition of extra bits for coding should produce a set of valid bit streams that has the maximum possible distance. It holds that if coding leads to a set of valid bit streams of distance d, then it can either detect D errors or correct T errors, where $d \geq D + 1$ and $d \in [2T + 1...2T + 2]$, respectively.

The Hamming code is a very popular error correcting code with distance 3; thus $T = 1$. In the Hamming code, the number of the bits of the coded message, n, the number of the bits of the message to be coded, k, and the number of coding bits, r, are related according to the following equations:

$$n = 2^r - 1, \qquad k = 2^r - 1 - r \qquad (2.20)$$

The Hamming code works by placing check bits in those positions of the coded bit stream that are powers of two (1,2,4,8,...), whereas the remaining positions are filled with data bits. These check bits contribute to the calculation of parity of some, but not all, of the data bits. In order to determine the check bits that are concerned with the integrity of a given data bit, the position of the data bit is rewritten as a sum of numbers that are powers of two and these numbers indicate the coding bits related to the parity of this data bit. This is also the way decoding is done: For every code bit in position p, the receiver checks the parity of the set of bits related to this coding bit. If an error is found, the number of the coding bit is added to a counter p, initialized to 0. At the end of the decoding procedure, if $p \neq 0$, it contains the position of the incorrect bit.

A slight modification of the Hamming code permits it to correct not only one bit error but also a burst error (like the ones appearing in wireless links) of length s in bits. For coded messages of length j this is done by:

- coding the messages to be transmitted according to the Hamming code;
- gathering at least s such messages into the rows of an $s \times j$ matrix A;
- calculation of the $j \times s$ matrix B having the j columns of matrix A as its rows (retrograde matrix);
- transmission of the j messages of length s as those appear in the lines of matrix B.

At the receiver, the inverse function is performed and an $s \times j$ matrix is obtained that contains the results of transmission of the s j-bit messages. Assuming that the medium suffered a long error burst up to s bit times and consequently up to s bit errors arose, this interleaving scheme leads to the reception of up to s messages that suffer a bit error at the same position. Thus, this scheme leads up to s received messages with a single bit error (thus correctable) at the receiver and not to some totally destroyed bit streams.

2.7.2.3 Cyclic Redundancy Check (CRC)

CRC is a widely used error detecting code. For the coding of an m-bit message M with n coding bits, the transmitter and receiver agree on a common $(n + 1)$-bit stream P, with $n < m$. CRC codes this message by appending an n-bit sequence F, known as the Frame Check Sequence (FCS) to the end of the m-bit message. By shifting M n bits left and modulo-2 dividing the result E with P, the FCS F is defined as the remainder of this modulo-2 division. It can be proven that the $n + m$ message produced by appending F to the end of E is exactly divisible by the predetermined number P. After reception of the coded message $T = E + F$, the receiver modulo-2 divides it with P. If there is a nonzero remainder, the message was received with an error, otherwise it is assumed correct. There exists a finite possibility that an error has occurred and the division is still exact, however, this is an unlikely event that can be handled by higher layer protocols.

2.7.2.4 Convolutional Coding

Convolutional codes have found use in several wireless systems, such as the IS-95 cellular standard covered in Chapter 4. Convolutional codes are usually referred to based on the code's rate $r = k/n$ and constraint length K. The code rate of a convolutional code shows the ratio of the number of bits n that are output of the convolutional encoder to the number of bits k that were fed into the encoder. Convolutional coding of a bit stream will produce a larger bit stream. Thus, if the resulting bit stream is to be transmitted to the receiver over the same time period as the source stream, a bandwidth increase is necessary. The constraint length parameter, K, denotes the 'length' of the convolutional encoder; stating how many k-bit stages are available to feed the structure produces the output symbols. In general, the larger the value of K, the less the probability of a bit suffering an error. Specifically, the value of K and this probability are exponentially related.

In order to gain an insight into the operation of a convolutional coder, consider the example of Figure 2.38 which shows a convolutional coder with $K = 4$ and $r = 1/3$. Assume that the bit stream 1001 is to be coded. The first bit of the stream is fed into the first stage of the coder

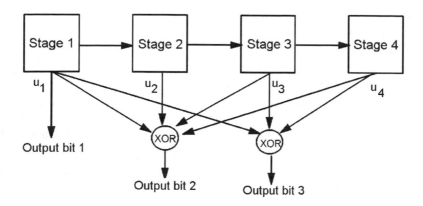

Figure 2.38   A convolutional coder with $r = 1/3$ and $K = 4$

and three bits are produced as a result of coding this bit. This procedure continues until the last bit of the stream reaches the last stage of the coder. Using this structure, the bit stream 1011 would be coded as 111 010 100 110 001 000 011 000. One can notice that the number of bits in the output of the decoder is not 12, as one would expect with what we defined the rate to be, but rather it is higher due to the fact that the operation ends when the last bit of the input sequence exits the encoder. Generally, for an r code and a k-bit input stream, the number of output bits is $(k + K)/r$. Since, in practice, the number of stages K is a very small number compared to the input stream size k, $(k + K)/r \cong k/r$, which is consistent with the definition that $r = k/n$. The operation of a convolutional coder depends on the selection for a value of K, the number of XOR adders and the way these are connected to stage outputs u_i.

   Two categories of convolutional decoding algorithms exist. The first is sequential decoding. It has the advantage of performing very well with convolutional codes of large K, but it has a variable decoding time. The second category, Viterbi decoding, removes this disadvantage by having a fixed decoding time; however, it has an increased computational complexity which is exponentially related to the value of K. The operation of convolutional decoding is out of the scope of this chapter and the interested reader can seek information in the scientific journals.

2.7.3 Equalization

Equalization techniques have found wide use in wireless systems for combating the effect of ISI. The general idea of equalization is to predict the ISI that will be encountered by a transmission and accordingly modify the signal to be transmitted so that the signal reaching the receiver will represent the information the transmitter wants to send. Two categories of equalization techniques exist: linear and nonlinear. Linear equalization techniques are not preferred for wireless communication systems, whereas nonlinear techniques, such as decision feedback equalization (DFE), data directed estimation (DDE) and maximum likelihood sequence estimation (MLSE) are commonly used for wireless systems. Of the nonlinear techniques, the choice for use in wireless systems is usually DFE since MLSE requires an increased computational complexity and knowledge of the channel characteristics.

   DFE employs a set of coefficients that are used for modeling the behavior of the wireless channel. Recent research trends aim to reduce the mobile node's hardware complexity by shifting signal processing tasks from mobiles to BS's thus resulting in asymmetric DFE architectures [19]. The coefficients managed by the BS change their value during a training procedure which is carried out by means of transmission of a fixed-length training sequence by the mobile receiver. Once the coefficients have successfully converged, they are used to "pre-equalize" the channel by canceling the predicted ISI from the transmitted signal. However, the channel estimation, as performed by the converged coefficient, is good only for a small period of time, during which the channel can be assumed identical in both directions. When the behavior of the channel changes, reverse transmission of the training sequence from the mobile station is needed in order to compute a new set of coefficients. Finally, this scheme is obviously useful when the uplink and downlink share the same frequency, as in the opposite case, the ISI due to multipath propagation which is computed from uplink transmission will not be the same as ISI of the downlink.

There are many possible algorithms to compute the coefficients of an equalizer. The most popular are the least mean square (LMS) and the recursive least squares (RLS). In order to choose an algorithm, one must take into account its ability for fast initial convergence during the training phase, good tracking of the channel and low computational complexity. The RLS algorithm satisfies the first two criteria better than the LMS algorithm which, however, is simpler to implement than RLS. However, RLS is used most of the time, especially after the development of its variants, fast RLS and square root RLS, which are mathematically equivalent to RLS. However, a detailed presentation of such algorithms is out of the scope of this chapter.

2.7.4 Power Control

Power Control (PC) schemes try to minimize interference in the system and conserve energy at the mobile nodes by varying transmission power. When increased interference is experienced within a cell, PC schemes try to increase the Signal to Interference noise Ratio (SIR) at the receivers by boosting transmission power at the sending nodes. When the interference experienced is low, sending nodes are allowed to lower their transmitting power in order to preserve energy and lower the interference. Thus, PC has a dual purpose: performance enhancement and energy preservation at the mobile nodes. PC can provide substantial performance increases and, as was mentioned earlier, is useful especially in CDMA systems in order to combat the 'near-far' problem. However, considering the inherent mobility in most wireless systems and the changing nature of wireless links, it is crucial that PC algorithms are fast enough to 'learn' the channel faster than the rate at which the local-mean value of the received signal changes. There are two fundamental types of PC schemes:

- *Open-loop PC.* In this category, for a pair of communicating stations A and B, the transmitter A estimates the channel attenuation on its own, for example, by measuring the strength of the received signal. Based on this estimation, it adjusts the strength of its transmission so as to reach the receiver B with adequate signal strength. For example, if A receives a weak signal from B, it will increase the strength of its own transmission. As a result, attenuation due to distance can be combated, as well as multipath fading provided that the channel between A and B remains in the same condition as that estimated at signal reception at A. The latter statement means that open-loop PC cannot effectively combat

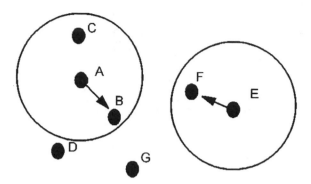

Figure 2.39   Concurrent transmissions in a MAC scheme using PC

multipath fading when transmissions and receptions occur at different channels. This is because, in such a case, multipath fading is not reciprocal, since in general different frequency channels fade independently. However, open-loop PC is easier to implement than the closed-loop scheme described below.

- *Closed-loop PC.* In this category, for a pair of communicating stations A and B, the receiver B measures the quality of the received signal and sends commands to A stating whether the power of A's transmissions to B must be increased or decreased. Thus, closed-loop PC can combat multipath fading; however, this is not always the case since for rapid multipath fading closed-loop PC may lead to inaccurate channel estimation. Furthermore, there is an overhead associated with the sending of PC commands from the receiver to the transmitter.

PC schemes are most commonly considered for uses in cellular networks. However, PC would be beneficial if taken into account during the design of MAC protocols for WLANs. For example, consider the case of Figure 2.39, which shows the topology of a CSMA wireless network. Assuming that all nodes can hear each other, when a data transfer from A to B is in progress, all other nodes are prevented from initiating a transmission (e.g. E to F), since this is likely to collide with the ongoing transmission. However, this wastes bandwidth: If A's transmission could be controlled so as to reach B but not stations further away, a transmission from E to F could take place at the same time. However, although this is an interesting approach, no commercial products based on such a MAC scheme exist and the matter is still under consideration [20].

2.7.5 Multisubcarrier Modulation

Multisubcarrier modulation is another technique that achieves ISI reduction. The channel bandwidth is divided into N subbands. A separate communication link is established over each subband. The data stream is divided into N interleaved substreams, which are used to modulate the carrier of each subband. This results in reduced ISI, since multipath fading does not occur with the same intensity over different frequency channels. An example of multisubcarrier modulation is Orthogonal Frequency Division Multiplexing (OFDM). As OFDM is envisioned to be used for transmission in future generation wireless networks. It is revisited

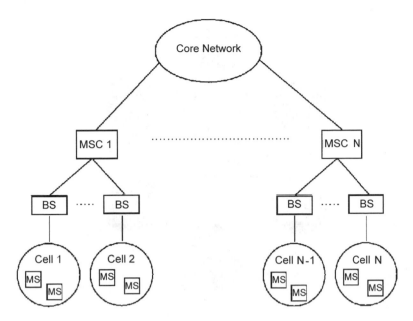

Figure 2.40   Simple cellular architecture

in Chapter 6 which discusses Fourth Generation and beyond wireless networks. OFDM has also found use as an option for transmission in the physical layer of IEEE 802.11 (IEEE 802.11a) and is used in the physical layer of HIPERLAN 2. These are described in Chapters 9 and 10, respectively.

2.8 The Cellular Concept

As already mentioned, one of the basic problems in wireless networks is the fact that the spectrum is a scarce resource. Apart from the techniques presented above which try to increase the capacity over a specific spectrum part, a great increase in efficient spectrum usage has been brought about with the introduction of the cellular concept, which was introduced in early 1970s at Bell Laboratories. The basis of the idea is the concept of the cell, which identifies the users located inside the coverage of the cell's BS. In this discussion, we assume the simple architecture of Figure 2.40, which comprises the following elements found in all cellular systems:

- Mobile terminal, containing at least voice capability.
- Base Station (BS), which manages communications of mobile users within its cell.
- Mobile Switching Center (MSC), which controls a number of BSs and interface the cellular system to/from the core network.
- The Home Location Register (HLR) and Visitor Location Register (VLR) databases that are present in every MSC. The functionality of these databases is explained later.

Moreover, we assume the presence of the following channels, which are also found in all cellular systems:

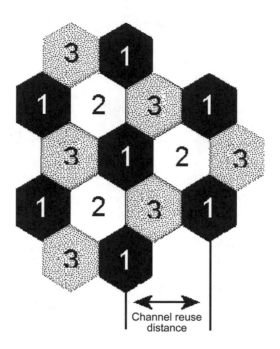

Figure 2.41 Example of frequency reuse in a TDMA/FDMA system having a cluster size of 3

- Broadcast channels, that are used to convey general control information from the BS to all mobile stations within its cell.
- Paging channels, that are used to notify a mobile station of an incoming call.
- Random access channels, which are used by the mobile stations to initiate a call.

The cellular concept enables frequency reuse by stating that instead of using the same set of channels for serving the entire population of a wireless system, the geographical span of the system should be broken into pieces (cells). The available frequency channels are also split into several sets, and the same set of channels is reused by sets of non-neighboring cells. The latter fact reduces interference, as cells that use the same channels are located relatively apart from one another.

The concept of frequency reuse using cells is illustrated in Figure 2.41, where cells are modeled as pentagons. However, this is only for purposes of presentation since most of the times real cells have irregular shapes due to environmental obstacles (such as buildings, hills, etc.). In the figure, we identify three sets of cells, 1, 2 and 3 with cells belonging to the same set using the same channel. Since three different sets of cells are used, the system in Figure 2.41 is said to have a cluster size of three. In real situations, however, larger cluster sizes, seven or even twelve, are used in order to increase the distance between co-channel cells and thus reduce intercell interference.

The frequency reuse scheme that is achieved with the use of cells, results in an increase in overall capacity. Returning to the example of Figure 2.41, if each cell needs to support a channel with bandwidth B, then with frequency reuse a total of $3B$ bandwidth is sufficient to cover the sixteen-cell region. Without frequency reuse, every cell would have to use a different frequency channel, a scheme that would demand a total *16B* of bandwidth. The separation between the channels of neighboring cells depends on the multiple access technique used. Thus, if FDMA is used, neighboring cells use different frequency channels, whereas in the case of TDMA, channels are defined in the time domain and channel separation is performed in this domain. CDMA, on the other hand, has the advantage of using the single-cell clusters, thus simplifying system design. Thus, mobile stations of adjacent cells use the same frequency at the same time and although there is some amount of interference, CDMA manages to cope with it by using proper codes for the transmissions in neighboring cells.

Before proceeding, we explain the concept of sectorization. We consider hexagonal cells divided into three sectors (spaced 120° apart). Sectors within a cell use different frequencies. A cluster is a set of cells. In a cluster, each frequency channel is used only once. For cells having Y sectors each and K available channels, a cluster will comprise K/Y cells and the frequency reuse pattern is referred to as K/KY. For example, for $K = 12$, $Y = 3$, we have a 4/12 reuse pattern. Thus, with an appropriate reuse pattern, frequencies are reused at sectors significantly apart from one another, which enables noninterfering operation of the respective data and control channels.

Using the cellular approach, the effective number of channels per unit area rises, which means that the overall capacity of the system rises as well. For a fixed value for the transmission power of cells' BSs, there is a direct relationship between the frequencies used and the radius of a cell in cellular systems. The use of low frequencies can lead to higher coverage and thus less cells, which means less BSs and consequently less costs. This, together with the fact that people most of the time consider BSs harmful to their health, has led operators of early cellular systems assigned low frequencies (e.g. 400, 900 MHz) to have an advantage over those assigned higher frequencies (e.g. 1.8 GHz). However, from the point of view of spectrum efficiency, it is advantageous to use small cells. Small cell sizes enable better frequency reuse as the number of channels per unit area increases. When the market penetration of cellular systems was such that the need to accommodate an increased number of users became critical, this fact led operators of systems using high frequencies to have an advantage. This is because small cells offer greater overall capacities and as a result are capable for voice systems with high loads and mobile systems serving data applications, which typically require higher capacities than voice systems.

The efficiency of small cells is so useful that it has led to the concept of microcells. These are very small cells that are used to serve increased traffic demands in urban areas, such as streets or buildings. Microcells are the result of splitting larger cells and can exist either on their own or be overlaid on larger cells. The latter situation is known as a multilayer cellular network and its existence is due to the fact that larger cells were built first, whereas microcells were deployed later to serve the increased traffic demand. Picocells, which are even smaller cells than microcells, can be deployed in very small areas such as offices or warehouses.

So far, the discussion has implied that there is a fixed set of channels allocated to each cell. This strategy is also known as Fixed Channel Allocation (FCA). Using FCA, channels are assigned to cells and not to mobiles nodes. The problem with this strategy is that it does not take advantage of user distribution. A cell may contain a few, or no mobiles nodes at all and still use the same amount of bandwidth with a densely populated cell. Therefore, spectrum utilization is suboptimal. The following techniques aim to overcome this problem:

- *Borrowing channel allocation (BCA)*. This is a variation of FCA. In BCA, a heavily loaded cell can ask a lightly loaded neighboring cell to let it use a number of its channels. Although overcoming the aforementioned problem, BCA may introduce intra-cell inter-ference due to the common use of the same channels by the neighboring cells.
- *Dynamic channel allocation (DCA)*. DCA places all available channels in a common pool and MSCs dynamically assign them to cells depending on the cells' current loads. Thus, the system can adapt to varying traffic loads and perform better than an FCA system at the expense of increased computational demands at the MSCs.

2.8.1 Mobility Issues: Location and Handoff

The fundamental advantage of wireless networks is their inherent mobility. In cellular systems, mobility concerns both incoming calls to mobile stations and the management of ongoing calls. The first case is often referred to as the location problem. When a call for a mobile terminal arrives, it is necessary that the network knows the cell where the terminal is located in order to establish the call. However, since in a cellular network users may move from one cell to another (a procedure known as roaming), a mapping of terminals to cells (thus, BSs) is not possible. To solve this problem, the MSCs employ the two databases mentioned above, the HLR and the VLR. Each mobile terminal registers with a specific MSC as its home area and the HLR of an MSC contains the mobile terminals that have registered with that MSC. Furthermore, the VLR contains information on the users that are registered as subscribers to another MSC but happen to be present in this MSC. When users move between cells, these databases are updated in order to reflect changes in the topology. When the terminal moves to another area the HLR for this terminal records this movement along with information regarding the new location of the mobile terminal. Thus, when a call for a specific mobile terminal arrives, its home HLR is checked. If the check result states that the terminal is within the coverage of its home MSC, the call is established, otherwise, the call is redirected to the MSC of the terminal's new location. The new MSC will now check its VLR and if the mobile terminal is found, the new MSC instructs the BSs under its coverage to send a paging message to the terminal in order to notify it of the incoming call. Upon reception of the paging message, the mobile terminal will respond to the BS of the cell it is currently located in and the call establishment can be completed at this BS.

A different procedure takes place when the terminal moving from one cell to another is involved in a call. In this situation, a procedure must be carried out that will change the BS that is involved in the call, from the BS of the old cell to the BS of the new cell. This procedure is known as handoff. Except for the cases of roaming users between cells, handoff also may be initiated for other reasons (the GSM standard identifies more than 40 reasons for a handoff). The basic principle of handoff is the following: for a specific mobile, the quality of connections to more than one BS is observed and whenever a link to a BS with quality exceeding that of the link to the current BS is found, the mobile terminal is 'handed' from the old BS to the new (hence the term handoff) BS. For roaming users crossing cell bound-aries at high speeds, handoff is more difficult to implement as it requires a fast response from the network. In such cases, multilayer cellular networks are beneficial, since they can serve fast users through large cells and reserve microcells for low speed and stationary users.

The decision for a handoff can be made either by the MSCs of the network based on signal measurements made by the BSs, as is the case in first generation cellular systems, or with the cooperation of the mobile terminal, as happens in some second generation TDMA systems. The latter type of handoff is also known as mobile-assisted handoff.

There are generally two types of handoff, soft and hard. In soft handoff, a link is set up to the new BS before the release of the old link. This ensures reliability, as the new BS may be too crowded to support the roaming mobile terminal or the link to the new BS may degrade shortly after establishment. However, the mobile terminal should be able to communicate with two different BSs at the same time. Thus, soft handoff demands increased complexity at the mobile terminals since it demands the capability of supporting two links with different BSs at the same time. Soft handoff is currently used in IS-95 CDMA-based systems. Hard handoff, which is used in most cellular wireless systems, is relatively simpler than soft handoff since the link to the old BS is released before establishment of the link to the BS of the new cell. However, it is somewhat less reliable than soft handoff.

2.9 The Ad Hoc and Semi Ad Hoc Concepts

The concept of ad hoc networking [21–23] is neither new, nor specific to the wireless case. The basic idea behind ad hoc systems stems from the early stages of Internet development during the cold war, where a distributed network of peer nodes capable of operating even when a number of nodes or links are brought down or destroyed was envisioned. The same idea for distributed operation holds for wireless ad hoc systems, which of course have some additional characteristics that stem from the use of wireless transmission. Thus, the term 'wireless ad hoc' stands for a network having no central administration and comprises mobile nodes that use wireless transmission. As seen later, nodes in an ad hoc network can serve as routers as well, by forwarding packets between stations that are out of transmission range of one another. This section aims to introduce the ad hoc concept as this is discussed in detail in Chapter 10. Furthermore, many of the technologies presented in this book, such as the IEEE 802.11 and HIPERLAN WLANs (Chapters 9 and 10) and Bluetooth and HomeRF PANs (Chapter 11) employ ad hoc functionality.

The major characteristics of ad hoc wireless networks are the following:

- *Distributed operation.* The ad hoc concept differs from other wireless systems, such as cellular systems in terms of network operation. An ad hoc network comprises stations that have the same capabilities and responsibilities. No centralized entity that controls the network exists. In an ad hoc network there are no BSs or MSCs and thus all network protocols operate in a distributed manner.
- *Dynamic topology.* In a wireless ad hoc network, nodes are free to move in almost any possible manner. The fact that (a) some mobile stations may be out of range of one another and (b) the wireless medium condition changes rapidly over time results in dynamic network topologies with the nature of topological changes being unknown to the network a priori.
- *Multihop communications.* Due to signal fading and the finite coverage of mobile transmitters, a fully connected topology cannot be assumed for an ad hoc system. Thus, in the case where a station A needs to send data to another station B out of its range, the transmission needs to be relayed through other nodes. Such networks are known as multi-

hop (or store and forward) wireless ad hoc networks; some examples are the HIPERLAN 1 WLAN standard and Bluetooth.

- *Changing link qualities.* This is true for all wireless systems, however, it is more important in the multihop case, since the quality of a multihop path depends on the qualities of all the links that make up the path. Thus, monitoring of link quality is bound to be more difficult in the multihop case.
- *Dependence on battery life.* This applies to most wireless systems, however, in ad hoc systems it is even more important. Consider, for example, the case of cellular systems: BSs are not hindered by finite battery life and overall network performance does not drop when some mobiles switch off due to battery depletion. Rather, it reduces the amount of inter-ference and channel contention and thus increases overall network performance. On the other hand, efficient network operation in ad hoc systems depends on the battery-depen-dent mobile nodes, which are responsible for relaying other nodes' messages when communicating stations are out of range. A fewer number of nodes results in limited support for relaying and this leads to a network with less routing capability.

In recent years, there has been a big interest in wireless ad hoc networks. This is due to the fact that they possess advantages for certain types of applications, such as emergency systems or military communications, that that need quick deployment of a network in cases where a fixed wireless communication infrastructure does not exist or cannot be used due to security, cost, or safety reasons. Since wireless ad hoc networks can be deployed without needing support for a centralized entity, they are very popular in such situations. Similarly, they are useful in applications where increased network reliability is demanded in cases of failing or departing terminals. An example of this is again military applications where wireless ad hoc networks are very efficient due to the fact that the network does not rely on some critical nodes for its organization or control.

The characteristics of dynamic topology and multihop communications make the design and operation of ad hoc systems a challenging task. Such systems need to operate efficiently even in cases of unknown network topologies and absence of direct paths between commu-nicating stations, which leads to multihop connections. It is evident that the performance of ad hoc systems greatly depends on the efficiency of the routing scheme being used. Thus, wireless ad hoc routing algorithms should be efficient for performing their functions; the most common of these are described in the following subsections.

2.9.1 Network Topology Determination

Ad hoc routing protocols must monitor and react to the changing network topologies. Ad hoc systems may employ multihop communications, thus routing protocols must make sure that at least one path exists from any node to any other node. The only case when this is not demanded is, of course, the case of partitioned networks, where the ad hoc network is split into a number of partitions due to the fact that any two nodes belonging to different partitions are not within range of one another. In order to efficiently monitor and adapt to changing network topologies, ad hoc routing protocols must provide all nodes with knowledge regard-ing their neighbors (those nodes of the network with whom they can directly communicate). Due to the distributed nature of ad hoc wireless networks, it is obvious that monitoring of network topology will be done in a distributed manner and information regarding the status of routes should be propagated to all network nodes when topology changes occur.

As an example of network topology determination, we present the case of ad hoc network establishment. Figure 2.42 shows an ad hoc network where nodes join the network one after the other, according to the corresponding numbering. Thus, the network can established when node N_2 comes within the range of node N_1. Both these nodes announce their transmission to form a network by regular beacon transmissions that contain information such as their addresses. Assuming that nodes N_1 and N_2 establish direct communication, an ad hoc network is formed and the routing protocol updates the routing tables in these nodes so as to reflect the change in topology. When a third node, N_3 enters the network, the routing tables are updated to reflect the new topology. If N_3 is within range of both N_1 and N_2, then each node's routing table contains the possible routes from this node to all others. In this case there obviously exist two routes between each pair of nodes. One is direct and the other is relayed through the third node. When N_3 is within range of only N_1 (as shown in the figure) or N2, the routing tables in the nodes are updated correspondingly.

2.9.2 Connectivity Maintenance

After a network's establishment, topological changes are sure to occur either due to node mobility/failure or changing signal propagation characteristics. Thus, routing protocols need to find alternative routes between stations in order to maintain connections. Consider, for example, the case of the ad hoc network in Figure 2.42. If nodes N_3 and N_1 moves so that N_3 goes out of range of N_1 and comes into range of N_2 the topology changes to that of Figure 2.43. In this case N_3 and N_1 can still communicate, although only via node N_2. This fact is first detected by N_2 and N_3, which update their routing tables and is then communicated to all the other nodes of the network.

The performance of a wireless ad hoc system greatly depends on the routing protocol's ability to quickly (a) find loop-free routes between stations when topology is changed and (b) disseminate this information to all the nodes of the network. When network topology changes occur so fast so that the propagation of the previous topology to all nodes update has not yet finished, the performance of the system may degrade significantly. Thus, the application of a specific routing protocol is useful in cases when network topology changes sufficiently

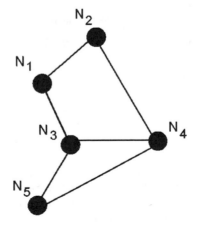

Figure 2.42   An ad hoc wireless network

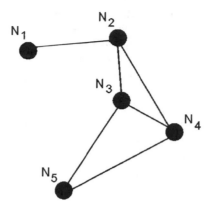

Figure 2.43   Topology change due to mobility

slowly, so as to enable successful propagation of previous topology updates. Wireless ad hoc networks are known as combinatorial stable if and only if they satisfy this constraint.

2.9.3 Packet Routing

As mentioned above, routing schemes are responsible for propagating changes and compute updated routes to a destination when changes in the network topology take place. In order to take into account the characteristics of wireless ad hoc networks, routing protocols for such networks employ a number of additional metrics, apart from the end-to-end throughput and delay metrics that are used in routing protocols for wired systems. The main performance metrics for wireless ad hoc network routing protocols are the following [22]:

- Maximum end-to-end throughput
- Minimum end-to-end delay
- Shortest path between communicating stations
- Minimization of overhead due to control signaling of the routing protocols
- Adaptability to changing topology
- Minimization of total power consumption within the network.

Of course, reaching the optimum values for all of the above constraints cannot be achieved. Rather, routing protocols provide trade-offs between these metrics. For example, the on-demand routing family of protocols, which is examined in Chapter 10, reduce control overhead at the expense of increasing the time needed to calculate new routing information, thus resulting in increased end-to-end delay.

2.9.4 The Semi Ad Hoc Concept

Another wireless networking concept that is related to ad hoc is the semi ad hoc concept. This concept is possible in cases when many radio access networks are available at the same time. In such a case devices can implement a dual mode of functionality, thus having the ability to operate either within a wireless network with centralized control (such as a cellular network) or within a wireless ad hoc network. This approach can increase the robustness of wireless systems. Whenever the entity responsible for the centralized control fails, or users move out of range of the cellular system, devices can set up an ad hoc network of their own. Furthermore, when a few network nodes are still in range of the cellular system, their membership in the ad hoc network will provide coverage extension to the other nodes as well. As can be seen in the corresponding chapters, most commercial ad hoc systems such as 802.11 WLANs, Bluetooth and HomeRF can follow this approach, since they can exploit existing wireless infrastructure to expand the set of services offered.

Ad hoc networks are mainly used by the military whereas most commercial systems are centralized. The integration of the various radio access networks into a combined network with seamless mobility, which is envisioned to be achieved by Fourth Generation (4G) wireless networks makes the semi ad hoc concept a promising approach.

2.10 Wireless Services: Circuit and Data (Packet) Mode

In all communication systems, including wireless systems, transmission of information, either voice or data-related, between a source and a destination station not directly connected to each other, typically employs a number of intermediate nodes. The intermediate nodes are also referred to as switching nodes and the network is known as a switched network [5]. Figure 2.44 shows a simple structure of a switched network, where one can see the user stations (squares) and the switching nodes (circles). Notice that a fully connected topology does not exist, however, at least one route exists between each pair of stations.

2.10.1 Circuit Switching

In a circuit switched network, when a connection is established between two stations, the connection is assigned a dedicated sequence of links between nodes. Thus in Figure 2.44, the data exchanged for a certain connection between stations A and B always follows the same

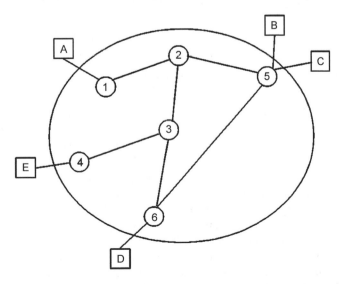

Figure 2.44 A simple structure of a switched network

path (e.g. A, 1, 2, 5, B). Of course, the entire capacity of a physical link is not necessarily dedicated to a single connection but can rather be time or frequency-multiplexed in order to serve more connections. In order for a data transfer to take place in a circuit switched network, the following procedures take place:

- *Circuit establishment.* Before the data transfer takes place, a dedicated sequence of links between nodes that connect the source to the destination must be defined. This is done on a node-to node basis between the source and destination station. Returning to the example of Figure 2.44, for a connection between stations A and B to take place, A sends to node 1 a request for circuit establishment with station B. Node 1 knows that station B is attached to node 5, so it has to find a path to node 5. Based on routing information, which takes into account a number of factors, it sends a circuit establishment request to node 2, which in turn connects to node 5. Thus, the established circuit is defined by nodes 1, 2, 5.
- *Transfer of data.* This is a quite straightforward step that entails the transfer of the data between the communicating nodes.
- *Circuit release.* After the transfer has taken place, the circuit is released which also results in de-allocation of the corresponding resources in the intermediate nodes that serve the circuit.

Circuit switching incurs on overhead for link establishment. However, after link establishment, the delay incurred by switching nodes is insignificant. Thus, circuit switching can support isochronous services such as voice. This is the reason why circuit switching has been widely utilized in cellular systems, which are primarily voice oriented. In such systems, a dedicated path and accompanying network resources are used for the entire duration of a voice call, even if there is a significant time period during which no participant of the call speaks. For data services, this fact makes circuit-switching efficient in transfers of large files and ineffective for bursty applications that transmit small quantities of data at every transmission. The latter is due to the fact that in such situations the circuit will be idle most of the time. Albeit inefficient, some circuit-switched cellular systems employ data transfer functionality.

2.10.2 Packet Switching

The above-mentioned problem of circuit switching for data services is solved by packet switching. Packet switching works by transmitting packets which most of the times are relatively small. If the source station wants to send a large packet, this is broken into a number of smaller packets. Apart from the user's data, each packet carries a control header, which contains information that the network needs to deliver the packet to its destination. The way this delivery is done differs significantly from circuit switching. Instead of defining the same route for all the packets sent from a source station A to a destination B, each packet can follow a different route inside the switched network in order to reach its destination. In each switching node, incoming packets are stored and the node has to pick up one of its neighbors to hand it the packet. This decision entails a number of factors, such as cost, congestion, QoS, etc., and depends on the routing algorithm used.

The benefits of using packet switching for data services are that bandwidth is used more efficiently, since links are not occupied during idle periods. In comparison with circuit switching, packet switching incurs delay at switching nodes, due to the fact that packets are stored there before been handed to the next node. However, this also implies an advantage: In a congested circuit-switched network, a new link establishment will probably fail leading to no communication at all. In a packet-switched one, however, packets will still be accepted, albeit suffering an increased delay due to the fact that they will spend more time stored at the switching nodes. Furthermore, in a packet-switched network, priorities can be used. Packet switching has emerged as an efficient way of handling asynchronous data in cellular systems. Examples of this approach are the CDPD and GPRS standards which are covered later in the book. Furthermore, many other wireless systems (WLANs, PANs, etc.) are packet-switched. The rising significance of data traffic over wireless systems makes the importance of using packet switching in such systems even greater. This can be realized by the fact that the next generations of wireless systems (4G and beyond) are envisioned to be an integrated common packet-switched (possibly IP-based) platform (see Chapter 6).

2.11 Data Delivery Approaches

Data broadcasting has emerged as an efficient way for the dissemination of information over asymmetric wireless environments. Examples of data broadcasting are information retrieval applications, like traffic information systems, weather information and news distribution. In such applications, client needs for data items are usually overlapping. As a result, broadcasting stands to be an efficient solution, since the broadcast of a single information item is likely to satisfy a (possibly large) number of client requests.

Communications asymmetry is due to a number of facts:

- *Equipment asymmetry.* A broadcast server usually has powerful transmitters that are not subject to power limitations, whereas client transmitters are usually hindered due to finite battery life. Moreover, it is desirable to keep the mobile clients' cost low, a fact that sometimes results in lack of client transmission capability.
- *Network asymmetry.* In many cases, the available bandwidth for transmission from the server to the clients (downlink transmission) is much more than that in the opposite direction (uplink transmission). Furthermore, extreme network asymmetry cases exist where the clients have no available uplink channel (backchannel). In the case of backchannel existence, however, the latter may become a bottleneck in the presence of a very large client population.
- *Application asymmetry.* This is due to the pattern of information flow. Information retrieval applications are of client-server nature. This means that the flow of traffic from the server to the clients is usually much higher than that in the opposite direction.

The goal pursued in most proposed data delivery approaches is twofold: (a) determination of an efficient sequence for data item transmissions (broadcast schedule) so that the average time a client waits for an item (mean access time) is minimized and (b) management of the mobile clients' local memory (cache) in a way that efficiently reduces a client's performance degradation when mismatches occur between the client's demands and the server's schedule.

So far, three major approaches have appeared for designing broadcast schedules. These are:

- *The pull-based (also known as on-demand) approach.* In pull-based systems the server broadcasts information after requests made by the mobile clients via the uplink channel.

The server queues up the incoming requests and uses them to estimate the demand probability per data item. This approach has the advantage of being able to adapt to dynamic client demands, since the server possesses knowledge regarding the demands of the clients. However, it is inefficient from the point of view of scalability. When the client population becomes too large, requests will either collide with each other or saturate the server.

- *The push-based approach.* In push-based systems there is no interaction between the server and the mobile clients. The server is assumed to have an a priori estimate of the demand per information item and transmits data according to this estimate. This approach provides high scalability and client hardware simplicity since the latter does not need to include data packet transmission capability. However, it pays the price of being unable to operate efficiently in environments with dynamic client demands.
- *Hybrid approaches.* Hybrid systems employ a combination of push and pull dividing the available downlink bandwidth into two different transmission modes: the periodic broadcast mode, in which the server pushes data periodically to the clients and the on-demand mode which is used to broadcast data explicitly requested by the mobile clients through the uplink channel. Obviously, this approach tries to combine the benefits of pure-push and pure-pull systems.

2.11.1 Pull and Hybrid Systems

In pull-based broadcast systems such as in Ref. [24], adaptivity is trivial to implement. This is because clients submit requests to the server and thus the latter possesses knowledge of client demands. However, pull systems are not easily scalable to large numbers of clients. In such cases, requests carried over the backchannel will either collide with each other or saturate the server. To our knowledge, the only hybrid exception that achieves adaptivity, is proposed in Ref. [25] and its derivation [26]. It uses a periodic probing mechanism to determine whether a particular data item is in demand or not. Nevertheless, hybrid systems have to carefully strike a balance between push and pull and cope with a number of additional issues (determination and dynamic allocation of bandwidth available for push and pull, determination of items to be pushed and those to be pulled, etc.). Furthermore, such systems still impose the need for client packet transmission capability and existence of a backchannel to carry client requests.

2.11.2 Push Systems

Early push systems used the flat approach for broadcasting, which schedules all items with the same frequency. The flat approach was used in the Datacycle project [27,28] and the Boston Community Information System [29]. In an effort to develop more efficient systems, work has led to results which showed that, in order to minimize access time, schedules must be periodic and the variance of spacing between consecutive instances of the same item must be reduced. A popular approach in the area of push systems that satisfied both these constraints was the Broadcast Disks model [30,31]. It proposed a way of superposition of multiple disks spinning at different frequencies on a single broadcast channel. The most popular data are placed on the faster disks and as a result, periodic schedules are produced with the most popular data being broadcast more frequently. This work also proposes some cache management techniques aimed at reducing performance degradation of those clients with demands largely deviating from the overall demands of the client population. This work was augmented later by dealing with issues such as efficient cache management based on prefetching [32], impact of changes at the values of the data items between successive server broadcasts [33] and addition of a backchannel to allow clients to send explicit requests to the server [34]. The latter approach can be considered as a hybrid system.

A drawback of Broadcast Disks is the fact that it is constrained to fixed sized data items and does not present a way of determining either the optimal number of disks to use or their relative frequencies. Those numbers are selected empirically and, as a result, the server may not broadcast data items with optimal frequencies, even in cases of static client demands. Furthermore, the rigid enforcement of the constraint for minimization of the variance of spacing between consecutive instances of the same item leads to schedules with instances of the same item being equally spaced. This fact can lead to schedules that possibly include empty and thus unused periods (holes). Finally, the Broadcast Disks approach is not adaptive, since it is based on the server's knowledge of static client demands resulting in predetermined broadcast schedules.

Push-based systems are also proposed in Ref. [35]. This work proposes broadcast schedules based on the so-called square-root rule. Assuming that the instances of each item are equally spaced in the broadcast, it shows that the access time is minimized when the server broadcasts an item i with frequency directly proportional to the factor $\sqrt{p_i/l_i}$, where p_i is the overall client demand probability for item i and l_i is this item's length.

As stated by its authors, the method in Ref. [35] has the advantage of automatically using the optimal frequencies for item broadcasts, in contrast to Broadcast Disks. Furthermore, the constraint of equally spaced instances of the same item is not rigidly enforced, a fact that leads to elimination of empty periods in the broadcast. Finally, Ref. [35] also works with items of different sizes. This assumption is obviously more realistic compared to that of fixed-length items made in the Broadcast Disks approach. However, the main drawback of the method remains its lack of adaptivity and therefore its inefficiency in environments with dynamic client demands.

2.11.3 The Adaptive Push System

Based on the above discussion, it would be interesting and beneficial in terms both of performance and cost, to reach a method that combines the advantages of the push and pull approaches. The obvious advantage of push and pull systems are their scalability and adaptivity, respectively. Based on the above reasoning, Refs. [36,37] enhance the method in Ref. [35] in order to enable its efficient operation in environments characterized by dynamic client demands. To this end, it incorporates a learning automaton-based adaptation mechanism [17] in the method. This mechanism adapts to overall client demands in order to reflect the overall popularity of each data item. The proposed approach does not increase the computational complexity of the method in Ref. [35]. Simulation studies in Refs. [36,37] show significant performance improvement over the nonadaptive scheme of Ref. [35] in environments with a priori unknown, dynamic client demands.

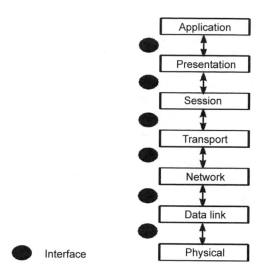

Figure 2.45   The OSI reference model

2.12 Overview of Basic Techniques and Interactions Between the Different Network Layers

All kinds of networks, including wireless networks, are organized in a layering hierarchy. The most widely used layering model is the Open System Interconnection (OSI) model [8]. Although different networks do not implement this model in exactly the same way, we present it here as an introduction to layering architectures. The layers of various wireless networks are explained in the corresponding chapters, along with their functionality.

Figure 2.45 shows the OSI reference model, which comprises seven layers. Each layer describes some functionality, but the exact algorithms that implement this functionality are not defined in the OSI model. For a connection between two hosts, each layer at a host has functionality that enables communication with its peer layer at the other host. Exception to this are the three lower layers, which communicate with the hosts performing routing, in a multihop network.

The stacking of the various layers results in a hierarchy with the most intelligent layers being on the top. Each layer uses services offered by the layer exactly below it. The interfaces shown in Figure 2.45 between the layers define the operations and services that one layer offers to another. The responsibilities of the seven layers of the OSI model, which need not be all implemented in network systems, are briefly summarized below:

- *The physical layer.* This layer is concerned with the transmission of the information over the communications medium. It deals with issues relating to the power of the transmitted signal, the modulation scheme to use, the data rate and a number of other mechanical/ electrical issues that relate to signal transmission.
- *The data link layer.* The data link layer fragments large packets coming from the upper layers into several frames and ensures their correct delivery to their destination. This layer is concerned with presenting to the layers above it an error-free communication medium,

over which the delivery of packets with the proper sequence to the destination is guaranteed. The data link layer thus performs error detection and correction functions and achieves the above goal by using Automatic Repeat Request (ARQ) techniques. Furthermore, it has the responsibility of flow control, regulating the rate the sender sends data in order not to swamp a slower host. Finally, in broadcast networks, this layer has an additional sublayer which is known as the Medium Access Control (MAC) sublayer. The functionality of MAC is to regulate the way the common channel is accessed by the various network hosts. Protocols that achieve such functionality for a MAC sublayer have been presented in Section 2.6.

- *The network layer.* This layer is concerned with governing the operation of the network subnet (formed by the routing hosts), inside a multihop network. The network layer is responsible for routing packets from the source to the destination and also for performing congestion control. The algorithms that perform this operation are known as routing algorithms and their desirable properties were discussed in Section 2.9 for the case of an ad hoc multihop wireless network. In a broadcast network, routing is so simple to implement that this layer is considered insignificant most of the time.

- *The transport layer.* This layer is concerned with managing the connection between two end-stations. It is an end-to-end protocol, since at each station, this layer communicates with its peer at the destination station. It is responsible for accepting data from the upper layers, breaking it into smaller pieces and ensuring that all the pieces arrive error-free and in sequence at the destination station. This layer also regulates the rate the sender sends data, as in the case of the data link layer. However, the difference is that in this case, flow control is between the communicating stations and not between routing hosts, as is the case for the data link layer. This layer is important in systems that use connectionless services, where sequencing and recovery are not performed by the lower layers.

- *The session layer.* This layer is concerned with managing connections between complex application processes. Specifically, it governs the dialogue discipline between such processes. This discipline defines the nature of data flow between the communicating parties, which can be full or half duplex. Furthermore, it provides synchronization services for the case of crashes. This is done by establishing checkpoints for data connections, which enable ongoing operations, such as file transfers, to be reestablished not from the start but from the point defined by the last checkpoint.

- *The presentation layer.* This layer translates and defines the format of data to be exchanged between applications. Furthermore, it performs encryption and compression of data.

- *The application layer.* This layer is the entry point to the OSI model and it is what a user's application sees. It enables file transfers, e-mail management and handling of the various terminal types operated by users.

When an application process wants to send data to a specific destination, the flow of data in the OSI model starts from the top layers. The data exchanged between neighboring layers is organized into units known as Protocol Data Units (PDUs). Each layer accepts data from the one above it and appends some header information that is related to the functions performed by this layer (equivalently, encapsulates the higher layer PDU into a PDU of its own). Then it hands the resulting PDU to the lower layer and the procedure continues until the physical layer transmits the packets to the destination station. There, the reverse procedure is performed: Each layer accepts PDUs and strips off the control information added by its peer layer. Thus, the application layer at the destination station finally gets the information transmitted from its peer layer at the source station.

2.13 Summary

Wireless networks, as the name suggests, utilize wireless transmission for exchange of information, with most commercial systems implementing radio-based transmission. Wireless networks significantly differ from their wired counterparts in a number of issues. This difference is mainly due to the characteristics of wireless transmission. This chapter described the fundamental issues related to wireless transmission systems, which are summarized below:

- The various bands of the electromagnetic spectrum are presented. As spectrum is a scarce resource, it needs to be licensed in order to ensure interference-free operation. Three licensing procedures, comparative bidding, lotteries and auctions, are presented.
- The physical phenomena that govern wireless signal propagation are discussed. Furthermore, the characteristics of signal propagation in street microcells and inside buildings are examined and a scheme used for modeling packet losses in wireless systems is presented.
- Analog and digital transmission are discussed. Analog transmissions have been typically been employed in older generations of wireless systems while the newer generations employ the more efficient digital transmission.
- Several modulation techniques are presented, both for analog and digital systems.
- An overview of techniques that increase the performance of wireless systems by combating the deficiencies of the wireless medium are presented. These include antenna diversity, multiantenna transmission, coding, equalization, power control and multicarrier modulation.
- The cellular, ad hoc and semi ad hoc concepts are discussed.
- The difference between packet mode and circuit mode services is presented.
- Two different approaches for delivering data to mobile clients are presented. These are the push and pull approaches.
- An introductory overview of the basic techniques and interactions at the different network layers is made with the help of the OSI reference model.

WWW Resources

1. *www.palowireless.com*: this is a web site that contains a vast amount of information on wireless networking systems, some relevant to the contents of this chapter.
2. *www.comsoc.org/pubs/surveys*: this is the home page of the IEEE Communications Surveys Magazine, an on-line magazine, free of charge, which publishes articles on wireless systems, including topics related to the contents of this chapter.
3. *www.itu.int*: this is the web site of the ITU, which is responsible for making proposals for worldwide spectrum allocation and use.
4. *www.ero.dk*: this is the web site of the European Radiocommunications Office. It contains information regarding spectrum allocation and usage in Europe.

References

[1] Trabelsi C. and Torun E. Physical Layer Alternatives for high–Speed Outdoor Packet Local Area Networks, *Wireless Personal Communications*, 10, 1999, 189–205.

[2] Neskovic A., Neskovic N. and Paunovic G. Modern Approaches in Modeling of Mobile Radio Systems Propagation Environment, *IEEE Communication Surveys*, Third Quarter, 2000.

[3] Andersen J. B., Rappaport T. S. and Yoshida S. Propagation Measurements and Models for Wireless Communication Channels, *IEEE Communications Magazine*, January, 1995, 42–49.

[4] Gilbert E. Capacity of a Burst Noise Channel, *Bell System Technology Journal*, September, 1960, 1253–1265.

[5] Stallings W. *Data and Computer Communications*, Fifth Edition, Prentice Hall.

[6] Chandra A., Gumalla V. and Limb J. O. Wireless Medium Access Control Protocols, *IEEE Communication Surveys*, Second Quarter, 2000.

[7] Falconer D. D., Adachi F. and Gudmundson B. Time Division Multiple Access Methods for Wireless Personal Communications, *IEEE Communications Magazine*, January, 1995, 50–57.

[8] Tannenbaum A. *Computer Networks*, Third Edition, Prentice Hall.

[9] Kohno R., Meidan R and Milstein L.B. Spread Spectrum Access Methods for Wireless Communications, *IEEE Communications Magazine*, January, 1995, 58 67.

[10] Chen K.-C. Medium Access Control of Wireless LANs for Mobile Computing, *IEEE Network*, September/October, 1994, 50–63.

[11] Chen K.-C. and Lee C.-H. RAP-A Novel Medium Access Control Protocol for Wireless Data Networks, in *Proceedings of IEEE GLOBECOM*, 1993, pp. 1713–1717.

[12] Chen K.-C. and Lee C.-H. Group Randomly Access Polling for Wireless Data Networks, in *Proceedings of IEEE ICC*, 1994, pp. 913–917.

[13] Nicopolitidis P., Papadimitriou G. I., Obaidat M. S. and Pomportsis A. S. TRAP: a High Performance Protocol for Wireless Local Area Networks, *Computer Communications*, July, 25, 2002, 1058–1065.

[14] Nicopolitidis P., Papadimitriou G. I., Obaidat M. S and Pomportsis A. S. A New Protocol for Wireless LANs, in *Proceedings of IEEE International Conference on Communications (ICC)*, 2002.

[15] Nicopolitidis P., Papadimitriou G. I., Obaidat M. S and Pomportsis A. S. Performance Evaluation of a TDMA–based Randomly Addressed Polling Protocol for Wireless LANs, in *Proceedings of IEEE International Conference on Electronics, Circuits and Systems (ICECS)*, 2002.

[16] Nicopolitidis P., Papadimitriou G. I. and Pomportsis A. S. Self-Adaptive Polling Protocols for Wireless LANs: A Learning-Automata-Based Approach, in *Proceedings of IEEE International Conference on Electronics, Circuits and Systems (ICECS)*, 2001, pp. 1309–1312.

[17] Narendra K. S. and Thathachar M. A. L. *Learning Automata: An Introduction*, 1989, Prentice Hall.

[18] Lehne P. H. and Pettersen M. An Overview of Smart Antenna Technology for Mobile Communications Systems, *IEEE Communication Surveys*, Fourth Quarter, 1999.

[19] Badra R. E. and Daneshrad B. Asymmetric Physical Layer Design for High–Speed Wireless Digital Communications, *IEEE Journal on Selected Areas in Communications*, October, 1999, 1712–1724.

[20] Monks J. P., Bharghavan V. and Hwu W. W. Transmission Power Control for Multiple Access Wireless Packet Networks, in *Proceedings of IEEE LCN*, 2000.

[21] Haas Z. J., Gerla M., Johnson D. B., Perkins C. E., Pursley M. B., Steenstrup M. and Toh C.-K. Guest Editorial, Wireless Ad Hoc Networks, *IEEE Journal on Selected Areas in Communications*, August, 1999, 1329–1332.

[22] Toh C.-K. Maximum Battery Life Routing to Support Ubiquitous Mobile Computing in Wireless Ad Hoc Networks, *IEEE Communications Magazine*, June, 2001, 138–147.

[23] Chakrabarti S. and Mishra A. QoS Issues in Ad Hoc Wireless Networks, *IEEE Communications Magazine*, February, 2001, 142–148.

[24] Aksou D. and Franlkin M. Scheduling for Large-Scale On-Demand Data Broadcasting, in *Proceedings of IEEE Infocom*, 1998, pp. 651–659.

[25] Stathatos K., Roussopulos N. and Baras J. S. Adaptive Broadcast in Hybrid Networks, in *Proceedings of VLDB*, 1997, pp. 326–335.

[26] Fernandez J. and Ramamritham K. Adaptive Dissemination of Data in Time-Critical Asymmetric Communication Environments, in *Proceedings of 11th Euromicro Conference on Real-Time Systems*, 1999, pp. 195–203.

[27] Bowen T et al., The Datacycle Architecture, *Communications of the ACM*, December, 1992, 71–81.

[28] Herman G., Gopal G., Lee K. and Weinrib A. The Datacycle Architecture for Very High Throughput Database Systems, in *Proceedings of the ACM SIGMOD Conference*, 1987, pp. 97–103.
[29] Gifford D. Polychannel Systems for Mass Digital Communications, *Communications of the ACM*, February, 1990, 141–151.
[30] Acharya S., Alonso R., Franklin M. and Zdonik S. Broadcast Disks: Data Management for Asymmetric Communication Environments, in *Proceedings of ACM SIGMOD*, 1995.
[31] Acharya S., Franklin M. and Zdonik S. Dissemination–based Data Delivery Using Broadcast Disks, *IEEE Personal Communications*, December, 1995, 50–60.
[32] Acharya S., Franklin M. and Zdonik S. Prefetching from a Broadcast Disk, in *Proceedings of International Conference on Data Engineering*, 1996.
[33] Acharya S., Franklin M. and Zdonik S. Disseminating Updates on Broadcast Disks, in *Proceedings odf VLDB*, 1996, pp. 354–365.
[34] Acharya S., Franklin M. and Zdonik S. Balancing Push and Pull for Data Broadcast, in *Proceedings of ACM SIGMOD*, 1997, pp. 183–194.
[35] Vaidya N. H. and Hameed D. Scheduling Data Broadcast In Asymmetric Communication Environments, *ACM/ Baltzer Wireless Networks*, 5, 1999, 171–182.
[36] Nicopolitidis P., Papadimitriou G. I. and Pomportsis A. S. Using Learning Automata for Adaptive Push-based Broadcasting in Asymmetric Wireless Environment, *IEEE Transactions on Vehicular Technology*, November, 2002.
[37] Nicopolitidis P., Papadimitriou G. I. and Pomportsis A. S. On the Implementation of a Learning Automaton-based Adaptive Wireless Push System, in *Proceedings of SPECTS*, 2001, pp. 484–491.

Further Reading

[1] Taub H. and Schilling D. L. *Principles of Communications Systems*, Second Edition, McGraw-Hill.

3

First Generation (1G) Cellular Systems

3.1 Introduction

As mentioned in Chapter 1, the first public mobile telephone system, known as Mobile Telephone System (MTS), was introduced in 1946. Although it was considered a big technological breakthrough at that time, it suffered many limitations such as (a) the fact that transceivers were very big and could be carried only by vehicles, (b) inefficient way of spectrum usage and (c) manual call switching. IMTS was an improvement on MTS offering more channels and automatic call switching.

However, the era of cellular telephony as we understand it today began with the introduction of the First Generation of cellular systems (1G systems). The major difference between 1G systems and MTS/IMTS was the use of the cellular concept in 1G, which brought about a revolution in the area of mobile telephony. This revolution took a lot of people by surprise, even AT&T who estimated that just 1 million cellular customers would exist by the end of the century, instead of the many hundreds of millions that exist today.

The use of the cellular concept greatly improved spectrum usage, for the reasons mentioned in the previous chapters. However, 1G systems are now considered technologically primitive. Nevertheless, this does not change the fact that a significant number of people still use analog cellular phones and an analog cellular infrastructure is found throughout North America and other parts of the world. The moral lesson from this fact is obvious and has been seen in other areas of technology as well – the market does not entirely follow technological developments. However, the reason why 1G systems are considered primitive is due to the fact that they utilize analog signaling for user traffic. This leads to a number of problems:

- *No use of encryption.* The use of analog signaling does not permit efficient encryption schemes. Therefore, 1G systems do not encrypt traffic. Thus, voice calls through a 1G network are subject to easy eavesdropping. Another problem is the fact that, by listening to control channels, users' identification numbers can be 'stolen' and used to place illegal calls, which are charged to the user.
- *Inferior call qualities.* Analog traffic is easily degraded by interference, which results in

inferior call quality. Contrary to digital traffic, no coding or error correction is applied in order to combat interference.

- *Spectrum inefficiency.* In analog systems, each RF carrier is dedicated to a single user, regardless of whether the user is active (speaking) or not (idle within the call). This is the reason for the inefficient spectrum usage compared to later generations of cellular systems.

3.1.1 Analog Cellular Systems

A number of analog systems have been deployed worldwide [1]. These are briefly described below.

3.1.1.1 United States

The first commercial analog system in the United States, known as Advanced Mobile Phone System (AMPS), went operational in 1982 offering only voice transmission. AMPS has been very successful and even today there are many millions of AMPS subscribers in the United States. Furthermore, AMPS has also been deployed in Canada, Central and South America and Australia. AMPS divides the frequency spectrum into several channels, each 30 kHz wide. These channels are either speech or control channels. Speech channels utilize Frequency Modulation (FM), while control channels can use Binary Frequency Shift Keying (BFSK) at a rates of 10 kb/s. Both data messages and frequency tones are used for AMPS control signaling. In order to combat co-channel interference, AMPS uses either (a) a typical frequency reuse plan with a 12-group frequency cluster with omnidirectional antennas or (b) a 7-group cluster with three sectors per cell. The operating frequency of AMPS consists of $2 \times 25 = 50$ MHz, which are located in the 824–849 MHz and 869–894 MHz bands. In a certain geographical region, two carriers (service providers) can coexist, with each carrier possessing 25 MHz of the spectrum (either the 'A' or 'B' band).

3.1.1.2 Europe

In European countries, several 1G systems similar to AMPS have been deployed. These include:

- Total Access Communications System (TACS) in the United Kingdom, Italy, Spain, Austria and Ireland
- Nordic Mobile Telephone (NMT) in several countries
- C-450 in Germany and Portugal
- Radiocom 2000 in France
- Radio Telephone Mobile System (RTMS) in Italy.

The most popular systems are TACS and NMT, which together accounted for over 50% of analog cellular subscribers in 1995. As in the case of AMPS, all of the above systems employ FM for voice channels and Frequency Shift Keying (FSK) for control channels. Channels are spaced apart and these spacings are as follows: 25 kHz (TACS, NMT-450, RTMS); 10 kHz (C-450); and 12.5 kHz (NMT-900, Radiocom 2000). All these systems base handover decisions on the power received at the Base Station (BS) by the mobile one, except for C-450, which performs handovers based on measurements of round-trip delay.

3.1.1.3 Japan

In Japan, a total of 56 MHz is allocated to analog cellular systems (860–885/915–940 MHz and 843–846/898–901 MHz). The first Japanese analog cellular system was the Nippon Telephone and Telegraph (NTT) system, which began operation in the Tokyo metropolitan area in 1979. The system utilized 600 duplex channels (spaced 25 kHz apart), which were realized via transmission in the 925–940 MHz (uplink) and 870–885 MHz (downlink) bands. Voice channels were again analog and control channels were 300 bps. In 1988, this rate was increased to 2.4 kbps and the number of channels increased to 2400 via use of frequency interleaving (channel spacing of 6.25 kHz). This new improved system allowed backwards compatibility, thus dual mode terminals were built that could access both the old and new system. Currently, NTT DoCoMo provides nationwide coverage in the 870–885/925–940 MHz bands. In 1987, two new operators were introduced:

- 'IDO', which operates the NTT high capacity system discussed above, covering the Kanto-Tokaido areas in the 860–863.5/915–918.5 MHz bands. IDO has also introduced NTACS (a variant of the European TACS system) in the 843–846/898–901 MHz and 863.5–867/918.5–922 MHz bands.
- DDI Cellular Group, which provides coverage outside the metropolitan areas using the JTACS/NTACS systems (a variant of the European TACS system) in the 860–870/915–925 MHz and 843–846/898–901 MHz bands.

IDO and DDI have agreed to provide nationwide service by allowing roaming between their systems.

3.1.2 Scope of the Chapter

The remainder of the chapter examines AMPS and NMT, two representative 1G cellular systems. Although they might seem primitive, they were very successful at the time of their deployment and in some ways have found use as a basis for the development of several 2G systems. An example of this is D-AMPS, which is a 2G system evolving from AMPS; covered in Chapter 4.

3.2 Advanced Mobile Phone System (AMPS)

AMPS [1–3] is a representative 1G mobile wireless system developed by Bell Labs in the late 1970s and early 1980s. As mentioned above, it was designed to offer mobile telephone traffic services via a number of 30 kHz channels between the Mobile Stations (MSs) and the BSs of each cell. These 30 kHz channels are used to carry voice traffic. The latter is a 3 kHz signal that is carried over the AMPS channels via analog transmission.

3.2.1 AMPS Frequency Allocations

The FCC made the first allocation of bandwidth for AMPS in the late 1970 in order to enable the operation of test systems in the Chicago area. The allocated bandwidth was in the 800 MHz part of the spectrum for a number of reasons:

- Limited spectrum was available at lower frequencies, which are primarily occupied either by FM radio or television systems. Lower frequencies are sometimes used by other systems, for example, maritime systems.

- Despite the fact that frequencies above 800 MHz are not very densely used, allocation of frequencies in this bands for AMPS was undesirable due to the fact that signals in those bands (e.g. several GHz) are subject to severe attenuation either due to path loss or fading. Such deterioration of signal qualities could not easily be handled at the time AMPS was developed due to the fact that error correction techniques for an analog system like AMPS were in their infancy.
- The 800 MHz band was a relatively unused band since few systems utilized it.

3.2.2 AMPS Channels

The operating frequency of AMPS consists of $2 \times 25 = 50$ MHz, which are located in the 824–849 MHz and 869–894 MHz bands. In a certain geographical region two carriers (service providers) can coexist, with each carrier possessing 25 MHz of the spectrum (either the 'A' or 'B' band). The transmit and receive channels of each BS are separated by 45 MHz. Both traffic channels for carrying analog voice signals and control channels exist. In a certain geographical area, two operators can exist and a different set of channels is assigned to each operator. The two channel sets, A and B, comprise channels from 1 to 333 and from 334 to 666, respectively. Channels from 313 to 333 and from 334 to 354 are the control channels of bands 'A' and 'B', respectively. Thus, each operator has 312 voice channels and 21 control channels at its disposal. Each control channel can be associated with a group of voice channels, thus each set of voice channels (either of bands 'A' or 'B') can be split into groups of 16 channels, each group controlled by a different control channel.

As mentioned above, traffic channels (TCs) are 30-kHz analog FM channels used to serve voice traffic. The main traffic channels are the Forward Voice Channel (FVC) and the Reverse Voice Channel (RVC) carrying voice traffic from the BS to the MS and from the MS to the BS, respectively. The network assigns them to the MS upon establishment of termination of a call.

Control channels (CCs) carry digital signaling and are used to coordinate medium access of Mobile Stations (MSs). Specifically, each MS that is not involved in a call (idle MS) is locked onto the strongest CC in order to receive control information. The CCs of AMPS are summarized below:

- *The Forward Control Channel (FOCC)*. This is a dedicated continuous data stream that is sent from the BS to the MS at 10 kbps. FOCC is a time division multiplexed channel comprising three data streams: (a) streams A and B, which are identified via the least significant bit of the MS's Mobile Identity Number (described later), with bit 0 identifying stream A and bit 1 identifying stream B and (b) the busy-idle stream, which is used to indicate the status of the RECC (described below). The use of the busy-idle stream reduces the possibilities of collisions on the RECC, as this might be used by more than one MSs. The FOCC is also used by the BS to inform a MS which RVC to use for a newly established call.
- *The Reverse Control Channel (RECC)*. This is a dedicated continuous data stream that is sent from the MS to the BS at 10 kbps.

SAT frequency (KHz)	SCC value
5.97	00
6	01
6.03	10

Figure 3.1   Mapping of SATS to SCC codes.

AMPS used both data messages and frequency tones for control signaling. The Supervisory Audio Tone (SAT) and the Signaling Tone (ST) are described below.

3.2.2.1 The Supervisory Audio Tone (SAT)

SAT is sent on the voice channels and is used in order to ensure link continuity and enable MSs and BSs to possess information on the quality of the link that connects them. Both the BS and the MS send this tone on the FVC and RCC, respectively, and the tone is added prior to the modulation of the voice signal. When a MS is switched on or has roamed under the coverage of a new BS, it tunes to the FOCC and reads a 2-bit field known as the SAT color code (SCC). The value of the SCC informs the MS which SAT to expect. SAT codes are shown in Figure 3.1. SAT determination is performed every 250 ms and the three defined SATs are at the following frequencies: 5.97 kHz, 6 kHz and 6.03 kHz.

3.2.2.2 The Signaling Tone (ST)

The ST is used to send four signals:

- The 'request to send' signal, which is used to allow the user to enter more data on the keypad while engaged in an ongoing conversation, T;
- The 'alert' signal, which, once the MS has been alerted, is continuously sent on the RVC until the user of the MS answers the call;
- The 'disconnect' signal, which is sent by the MS over the RVC in order to indicate call termination;
- The 'handoff confirmation' signal, which is sent by the MS in response to the network's request for handoff of this MS to another BS.

3.2.3 Network Operations

Prior to describing some basic network operations in AMPS, we describe the three identifier numbers used in AMPS:

- *The Electronic Serial Number (ESN).* The ESN is a 32-bit binary string that uniquely identifies an AMPS MS. This number is set up by the MS manufacturer and is burned into a Read Only Memory (ROM) in an effort to prevent unauthorized changes of this number. The fact that this number is stored in a ROM means that the MS will become inoperable if someone tries to rewrite the ESN. The format off an ESN is shown in Figure 3.2. It comprises three fields: (a) part 1, comprising bits from 24 to 31; this 8-bit field is the

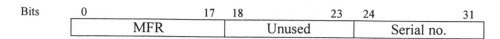

Figure 3.2   Structure of the 32-bit ESN.

manufacturers code (MFR), which uniquely identifies each manufacturer; (b) part 2 which comprises bits from 18 to 23 and has remained unusable; and (c) part 3, which comprises bits 0–17, which are assigned by the manufacturer to the MS. These bits are essentially the MSs serial number. When a manufacturer has produced so many MSs that 18 bits are no longer able to provide additional serial numbers for its MSs, it can apply to the FCC for an additional MFR. Thus, it can continue to produce MSs and MSs will be identified by a different MFR/serial number combination.

- *The System Identification Numbers (SIDs).* These are 15-bit binary strings that are assigned to AMPS systems and uniquely identify each AMPS operator. SIDs are (a) transmitted by BSs to indicate the AMPS network they belong to and (b) used by MSs to indicate either the AMPS network they belong to (in cases of two collocated AMPS networks), or to determine roaming situations.
- *The Mobile Identification Number (MIN).* This is a 34-bit string that is derived from the MSs 10-digit telephone number.

3.2.3.1 Initialization

Once an AMPS MS is powered up, a sequence of events takes place. This sequence is briefly described below:

- *Event 1.* The MS receives systems parameters in order to conFigure 3.itself to use one of the two AMPS networks.
- *Event 2.* The MS scans the 21 control channels of the selected AMPS network to receive control messages. If a control channel with an acceptable quality is found, this is selected.
- *Event 3.* The MS receives a message on the control channel containing system parameters.
- *Event 4.* The message received in Event 3 provides the MS with information that is needed in order to update information that was received in possible previous initializations. Furthermore, the MS reads the SID of the AMPS network in this message, compares it to the SID of the network it belongs to and when the MS is in the service area of another network, the MS can prepare for roaming operations.
- *Event 5.* The MS identifies itself to the network by sending its MIN, ESN and SIDS via the RECC.
- *Event 6.* The AMPS network examines the parameters transmitted by the MS in Event 5 in order to determine whether this MS is a roaming one or not.
- *Event 7.* The BS verifies initialization parameters by sending a control message to the MS.
- *Event 8.* The MS enters idle state and waits for a call establishment request. During idle mode, the MS must perform operations to (a) ensure synchronization with the BS, (b) make the network aware of the MS's location.

3.2.3.2 Call Setup from a MS

The procedure of placing a call from an MS can be described via a number of events. These events are summarized below:

- *Event 1.*The MS sends to the BS a message containing the MS's MIN, ESN and the phone number dialed.
- *Event 2.*The BS passes the information sent by the MS to the network for processing.
- *Event 3.*The BS indicates to the MS the channel number that will be used for the voice call. Furthermore, information related to the SAT frequency to be used is relayed to the MS.
- *Event 4.* Both MS and BS switch to the voice channels.
- *Event 5.*The BS sends a control message on the FVC via the SAT signal.
- *Event 6.*The MS confirms link continuity via the SAT on the RVC.
- *Event 7.*The call is established.

3.2.3.3 Call Setup to an MS

The procedure of placing a call to an MS can be described via a number of events. These events are summarized below:

- *Event 1.*The identification of the MS is passed to the BS.
- *Event 2.*Control information, including the channel number to be used, is conveyed to the MS.
- *Event 3.*The MS responds by sending its MIN, ESN and other control-related information.
- *Event 4.*Information related to the SAT frequency to be used is relayed to the MS.
- *Event 5.*Both MS and BS switch to the voice channels.
- *Event 6.*The BS sends a control message on the FVC via the SAT signal.
- *Event 7.*The MS confirms link continuity via the SAT on the RVC.
- *Event 8.*The call is established.

3.2.3.4 Call Handoff

The procedure of handoff in AMPS can be described via a number of events. These events are summarized below:

- *Event 1.*The BS serving the MS notices a decrease in the MS's transmission power.
- *Event 2.*The BS sends a handoff measurement request to its MSC.
- *Event 3.*The MSC instructs BSs in the neighborhood of the current BS to perform measurements of the MS's signal strength.
- *Event 4.*The MSC selects the best choice for a BS to serve the MS.
- *Event 5.* The MSC allocates a traffic channel to the selected BS.
- *Event 6.*The selected BS acknowledges the traffic channel allocation.
- *Event 7.* The MSC sends a handoff message to the current BS.
- *Event 8.*The current BS sends the handoff message to the MS. This message informs the MS which traffic channel to use and the power level of its transmission under the new BS.
- *Event 9.*The MS confirms the current BS's message and switches to the traffic channel.
- *Event 10.*The MS starts scanning and eventually receives the new BS's SAT.

- *Event 11.*The MS confirms link continuity to the new BS via the SAT on the RVC.
- *Event 12.*The new BS confirms the handoff to the MSC.

3.3 Nordic Mobile Telephony (NMT)

NMT [4] has been deployed in several European countries. There are two versions of the system: the first operates in the area around 450 MHz and the second operates in the area around 900 MHz. These variants, are known as NMT 450 and NMT 900, respectively.

3.3.1 NMT Architecture

An NMT system is made up of four basic parts:

- Mobile Telephone Exchange (MTX)
- Home Location Register (HLR), integrated in MTX or as a separate node
- Base Station (BS)
- Mobile Station (MS)

The MTX and HLR control the system and include the interface to the Public Switched Telephone Network (PSTN). This interface can be made at local or international gateway levels. BSs are permanently connected to the MTX and are used to handle radio communication with the mobile stations. BSs also supervise radio link quality via supervision tones. The set of BSs that are connected to the same MTX form an MTX service area, which in turn can be divided into subareas called Traffic Areas (TAs). The maximum number of BSs stations in a TA can be as high as 256.

MSs can be vehicle-mounted, transportable or hand-portable. In order to set up a call to a mobile, a paging signal must be sent out in parallel from all BSs in the TA in which the mobile station resides, instead of being sent out on all BSs in the service area. The aim of this approach is to reduce call set-up time and system load.

A number of network elements may also exist. These are:

- Combined NMT/GSM Gateway (CGW)
- Mobile Intelligent Network (MIN)
- Authentication Register (AR).

CGW is a gateway that can interrogate an NMT HLR and a GSM HLR. This is an optional feature for GSM MSCs that demands no new hardware. The HLR is used to store data about every subscriber, its services and location. In large networks where subscriber numbers are high, HLRs are preferably utilized as separate nodes, whereas in small networks, HLRs can be integrated with MTXs. The signaling protocol between MTSs and HLRs is according to CCITT Number 7 standard. Finally, The MIN adds intelligence to the network in order to enable introduction of new, customized services.

The radio network consists of cells, each having a Calling Channel (CC) and a set of Traffic Channels (TC). In order to enable frequency reuse, adjacent BSs obviously employ different operating frequencies. The frequency reuse schemes that are typically employed divide the available frequencies among groups of 7, 12, or 21 cells. The reuse plan is then built up by repeating these groups by trying to optimize the distance between BSs that employ the same frequency. In order to adjust to variable traffic intensities, cell size may change correspondingly. Radio coverage is provided in the cells by placing BSs either at (a) the center of the cell or (b) at a corner of the cell (omni cells or sector cells). The latter option gives the advantage of using one BS for several cells, thus reducing the number of BSs used and obviously deployment costs. The coverage of a BS ranges from 15 to 60 km for NMT 450 and from 2 to 30 km for NMT 900, depending on the BS placement height and the actual environment.

3.3.2 NMT Frequency Allocations

Connections between BSs and MSs are utilized via full-duplex radio channels (ether in the 450 or 900 MHz band as mentioned before), which allow information to be exchanged simultaneously in both directions. These full duplex channels are utilized via a pair of uplink and downlink channels with BSs transmissions occurring in higher frequency bands than the transmissions of MSs. In NMT 450, 180 channels exist, separated via 25 kHz of spectrum. An optional extension band exists that can offer 20 more channels. With interleaved channels the system can use a total of 359 channels, which become 399 if the extended band is used.

3.3.3 NMT Channels

There are four channel types in NMT. These are (a) the Calling Channel (CC), (b) the Traffic Channel (TC), (c) the Combined Calling and Traffic Channels (CC/TC) and (d) the Data Channel (DC).

- *Calling Channel (CC).* Each NMT BS uses one channel as the calling channel. The CC is used by the BS for transmission of a continuous signal that identifies this BS to the mobiles. MSs within the cell of a BS lock onto the BSs CC. The CC is also used by the BS to page MSs under its coverage. Upon response of the MS, an additional channel, known as a TC, is allocated to the mobile. Finally, the CC may also be used for priority calls, meaning that messages over a CC can cause a user to terminate his call in order to receive one of a higher priority.
- *Traffic Channel (TC).* The purpose of the TC is to carry the voice traffic. A TC can be in three different states: (a) 'free marking' state, in which the TC is mainly used for setting up calls from mobile stations; (b) 'busy' state, in which the TC is occupied by a voice call and (c) 'idle' state, in which the TC is not occupied.
- *Combined Calling and Traffic Channel (CC/TC).* The CC of the BS can also operate as a combined calling and traffic channel. This is useful in cases where all traffic channels are occupied. In such cases, an MS can use the calling channel to set up a call. In such an event, the BS will completely lack a calling channel for some time. When a traffic channel becomes free, it functions as a combined calling and traffic channel. Thus, the BS's CC will be used only when no other traffic channels are available.
- *Data Channel (DC).* The DC is used to make signal strength measurements on mobile stations that are involved in a voice call on order from the MTX. The results of these measurements are used by the MTX at handover decisions.

Every BS should have one CC, or some free TCs and one DC. Nevertheless, it is possible that a BS uses up to four DCs. This results in improved capacity for signal strength measurements, which is beneficial in situations characterized by increased traffic density or small cell sizes.

3.3.4 Network Operations: Mobility Management

3.3.4.1 Paging

Paging is used to determine the position of a MS. The service area of an MTX can be divided into a number of traffic areas. Paging involves sending over all CCs in the traffic area where the subscriber is expected to be (the area where the last registration of this MS was made) a page with the number of the paged MS number. Paging only the traffic area that is known to contain the MS helps reduce paging load on the system. However, if the MS is not found, then paging will be reinitiated and performed on all traffic areas of the MTX rather than only in that where the MS is expected to be. Upon reception of the page message, the MS will respond to the BS of the cell where the MS is currently located. If a certain time period elapses without a reply from the paged MS, then the page is considered unsuccessful. If the paging is unsuccessful, it is repeated once more.

3.3.4.2 Handover

In order for handover to be performed, the radio connection quality is measured during the call. When the quality of the connection lowers, the BS that is currently serving the call signals the MTX. The purpose of this procedure is to investigate whether a BS with a better link quality to the mobile unit can be found. If such a BS is found and it has an available channel to serve the call, then a handover of the call to the new BS is initiated. If the handoff is to be performed, the MTX indicates to the mobile station that it must change its operating frequency to that of the new traffic channel selected in the new BS. The switch is made in the MTX at the same time as the mobile station changes its frequency. After a successful handover, the old channel is released. If, however, a BS with a better link quality to the mobile unit is not found, then the call continues with the current BS on the current channel and periodical signal measurements will be made in order to enable a successful handoff later. Normally 20–30 s periods are used between successive attempts. If the handoff never takes place and the link quality continues to worsen (probably due to the subscriber moving far away from the BS) then the connection serving the call is dropped. A handover includes (a) seizing of the most suitable channel in the new BS, (b) supervision of the quality of the new channel, and (c) switching of the speech path towards the new channel.

3.3.4.3 Signal Strength Supervision

The MTX also performs continuous supervision of channel quality through signal strength measurements. This operation improves call quality, as handovers will be performed at an earlier stage.

3.3.4.4 Intra-cell Handover

This handover type involves moving a MS from a TC that experiences interference to another TC in the same BS. This procedure obviously improves call quality.

3.3.4.5 Handover Queue

In cases of highly loaded systems, handovers may be burdened with channel congestion, which imposes a difficulty when performing handovers. The handover queue tries to solve this problem. The MTX performs signal strength measurements on BSs surrounding the mobile and stores the first and the second best BS alternatives. If a handover is required and the BS with the best value does not have a TC available, the second best BS is chosen. If the second BS also does not have an available TC, then the handover is delayed and will be retried with the best BS alternative after a predetermined time period. This time period varies between 0 and 10 s and is adjustable. If no traffic channel becomes available during the waiting time period, the handover attempt is terminated and the call continues on the old channel.

3.3.4.6 Traffic Levelling

This feature increases the capacity and improves the success rate for call setups during peak time. The handover parameters are changed dynamically for the BS carrying high traffic load. Thus, handovers of calls occur to less loaded BSs.

3.3.4.7 Location Updating

This function keeps continuous track of the MS in the network. It comprises two parts: (a) automatic location updating call from a mobile station and (b) updating of location data in the MTX. The location data indicates the current traffic area where calls to the mobile station can be directed.

3.3.4.8 Roaming Updating

Each MS subscriber is registered permanently in its Home Location Register (HLR) where all information relating to a mobile station is stored. Whenever an MS roams to the service area of another MTX which is controlled by an MTXV, then updating information is exchanged between the MTXs and the HLR. The HLR is updated to reflect the new subscriber location and the MTXV receives a copy of the MS subscriber's categories, etc. Upon movement of an MS to another MTX area, the corresponding MTXV is ordered by the HLR to erase data that concerns that MS subscriber.

3.3.4.9 Inter-exchange Handover

This is an extension of the handover function that allows switching of calls in progress, even to BSs controlled by other MTXs. In conjunction with roaming, this feature makes the MS independent of the service areas of MTXs. The necessary signaling to perform this procedure is based on CCITT Number 7 and a specific Handover User Part (HUP). In inter-exchange handover more than one exchange is involved. These are (a) the 'anchor exchange', which controls the service area where the MS was at the original call set-up, (b) the 'serving exchange', which has radio contact with the MS and (c) 'target exchange' which involves the exchange to the BS which was identified by the BS as the most suitable BS for the handover. Furthermore, there are two different handover categories: (a) the basic handover, which is done from an anchor exchange to a target one and (b) the subsequent handover, which is done either from a serving nonanchor exchange back to an anchor exchange, or from a nonanchor serving exchange to a third exchange. The inter-exchange handover is controlled by the anchor exchange.

3.3.4.10 Subscription Areas

This feature enables operators to define mobility limits for MS subscribers. This procedure involves the definition of a restricted geographical area inside which the MS may place and/or receive calls. Representative examples of the usefulness of this feature are; (a) the subscription area is only one BS; this could be useful in cases where fixed telephony is more expensive than mobile telephony; (b) the subscription area is the coverage area of the whole system excluding large city areas; this can be used in the case of rural subscribers that enjoy a special tariff.

3.3.5 Network Operations

In NMT, subscribers are able to receive and originate calls both in their home and visited MTX. When a MS moves from one cell to another during a call, a handover will take place, enabling the call to continue.

3.3.5.1 Searching for a CC

A channel is selected randomly and the search starts from this channel. Then, additional channels are tested. The first time the search is performed, the searching MS operates at reduced sensitivity, in order to prevent itself from locking onto a channel with a weak signal. If, however, the MS detects no channel during this search, it reinitiates the search, this time at an increased level of sensitivity. If the MS still detects no channel, it scans for a third time, operating at full sensitivity. When the MS has found a calling channel, the traffic area information is detected. The MS makes a comparison with the information stored in its memory. If the memory is empty or contains other information, the MS makes an updating call to the MTX on a traffic channel. If the MS is locked to the CC but experiences low quality for the CC, then it will start a search for a new CC as described above. However, the MS may have to check all frequencies in the CC band in order to find a high quality CC within the traffic area where the MS is registered for the moment. When a CC cannot be found, the MS locks itself to a calling channel in some other traffic area.

3.3.5.2 Searching for a Free TC or AC

This search operation uses the sensitivity reduction procedure mentioned above, however, with a maximum total number of 15 scans instead of three.

3.3.5.3 Transmission Quality Supervision

This function aims to ensure the best possible transmission quality of a call in progress, irrespective of a subscriber's movement within the service area. This is made possible by selecting the most appropriate BS to serve the MS calls. This selection is based on signal strength measurement performed at the current and all the neighboring BSs. Supervision of transmission quality is made by BSs in two ways: (a) measurement of the signal strength of the carrier from the MS; (b) measurement of the signal to noise ratio of a special supervision signal, which is transmitted by the BS and returned from the MS via the TC. The supervision signal is a tone above the speech band. It is also used as an identification signal to secure that handovers are being performed between the right BSs. Four different analog signals separated by 30 Hz are used: (a) signal number 1 at 3955 Hz; (b) signal number 2 at 3985 Hz; (c) signal number 3 at 4015 Hz; and (d) signal number 4 at 4045 Hz. Furthermore, it is possible to use an additional set of 35 digital supervisory signals. When transmission quality drops below a certain limit, the BS informs the exchange. In such a case, the MTX will request the RF carrier signal strength measurements of neighboring BSs. These results are then evaluated and ranked by the MTX. The result of this operation is a possible increase or decrease in the MS transmission power, or a handover to a new channel in the same or a different BS.

3.3.5.4 Blocking of Disturbed Channels

Idle traffic channels, which experience interference, either due to systems other than the network or the network itself (such as inter-channel interference) are automatically blocked for the duration of the disturbance. Therefore, they are not used for traffic, new calls, or handovers.

3.3.5.5 Discontinuous Reception

The purpose of this feature is to save battery energy in MSs. Battery saving is achieved by switching off the MS receiver most of the time, with only a clock function active during the low-power mode. Calls received during this time are buffered in the exchange. Once the MS has exited the power-saving mode, they are paged with information relating to buffered calls. After the paging, MSs can re-enter low-power mode.

3.3.6 NMT Security

As in all cases of networking, security is a major issue of concern in NMT. The NMT features that aim to provide security are summarized below.

3.3.6.1 Mobile Station Identity Check

Unauthorized use of a MS can be prevented via use of a password that is attached to the identity of each MS. This password, which is stored in the MS and in the HLR, is a three-digit part of the mobile station identity, also known as the security code. Password validity is checked on calls to and from mobile subscribers and also on roaming updating messages. When an incorrect password is detected, the call is disconnected and roaming is not performed. In order to prevent repeated call attempts with illegal passwords, thorough supervision and logging of the identity check procedures are available.

3.3.6.2 Subscriber Identity Security (SIS)

This feature improves the security of the subscriber identity beyond the level achieved by the three-digit password mentioned above. This feature protects subscribers from illicit use of their identities through an authentication mechanism based on a challenge-response method between the MTX and the MS including encryption of the dialed B-number. A Secret Authentication Key (SAK) is installed in the MS and the Authentication Register (AR), which is an external database that provides the HLR with authentication data. This data is generated in the form of a triplet comprising three values: (a) Key for B-number enciphering (B-KEY); (b) Random Number (RAND); and (c) Signed Response (SRES). This triplet is transferred on request to the HLR over a C7 signaling link. The HLR stores one or more triplets for every subscriber of the SIS system. The identity is checked every time a call is made from the mobile station. The check is performed by the MTX sending the random number to the mobile station, which computes an answer by using its SAK. The answer is sent to the MTX, which compares the answer to the result received from AR; when they correspond, access is allowed. If they do not correspond, the call is rejected. The MTX can handle MSs with and without SIS. Thus, SIS activation is optional. The operator can control the permission to roam for mobile stations with and without SIS. This flexibility is especially useful in international roaming in order to prevent illegal access to the network. SIS authentication can also be made for MSs that receive calls. This prevents unauthorized users receiving calls. The authentication is then done immediately after the call is set up and when the authentication indicates an unauthorized user the call is dropped. As in the case of the MS identity check, thorough supervision and logging of failed SIS authentication are available, in an effort to prevent repeated call attempts with illegal passwords and thus improve fraud prevention.

3.3.6.3 Location Dependent Call Barring

Selective call barring according to the location of the MS is also possible. The idea of this approach is to neglect location updating for MSs that are in a certain MTX service area or traffic area. Restrictions could then be put on outgoing calls and on roaming situations. Placing restrictions on roaming could be useful in several cases, such as avoid expensive charging for forwarded incoming calls for MS roaming abroad.

3.3.6.4 PIN code

NMT provides an additional security method through use of a secret code at the MS known as the Personal Identification Number (PIN). PIN codes can be used to control roaming. For instance, when a subscriber is visiting a foreign MTX, he or she will only become fully updated in the visited MTX only upon dialing the PIN code. It is up to the operators to define whether it is necessary to dial the PIN code every time a call is made or only the first time a call is placed for a user that has entered a certain MTX service area or traffic area. The PIN code can also be used to control barring of outgoing calls, such as local or international calls. The subscriber may control the type of barring by dialing a special service code including the PIN code.

3.4 Summary

The era of cellular telephony as we understand it today began with the introduction of the first generation of cellular systems (1G systems). Such systems served mobile telephone calls via analog transmission of voice traffic. Despite the fact that 1G systems are considered technologically primitive today, the fact remains that a significant number of people still use analog cellular phones and analog cellular infrastructure is found throughout North America and other parts of the world. Furthermore, they have found use as a basis for the development of several second generation systems. An example of this is D-AMPS, which is a 2G system evolving from AMPS. This chapter described the Advanced Mobile Phone System (AMPS) and Nordic Mobile Telephony (NMTS) 1G cellular systems. AMPS divides the frequency spectrum into several channels, each 30 kHz wide. These channels are either speech or control channels. Speech channels utilize Frequency Modulation (FM), while control channels can use Binary Frequency Shift Keying (BFSK) at a rate of 10 kb/s. Both data messages and frequency tones are used for AMPS control signaling and two operators can be collocated in the same geographical area. There are two versions of NMT. The first operates in the area around 450 MHz and the second operates in the area around 900 MHz. These variants are known as NMT 450 and NMT 900, respectively.

WWW Resources

1. *www.telecomwriting.com*: this web site contains information on early mobile telephony systems, including some information on 1G cellular systems.

References

[1] Padgett J. E., Gunther C. G. and Hattori T. Overview of Wireless Personal Communications, *IEEE Communications Magazine*, January, 1995, 28–41.
[2] Black U. *Second Generation Mobile and Wireless Networks*, Prentice Hall.
[3] Hubbel Y. C. A Comparison of the Iridium and AMPS Systems, *IEEE Network*, March/April, 1997, 52–59.
[4] NMT System Description, Ericsson Document.

4

Second Generation (2G) Cellular Systems

4.1 Introduction

As was mentioned in the previous chapter, the era of mobile telephony began with the development and operation of the First Generation (1G) of cellular systems in the late 1970s. Although these systems have found widespread use and are still used nowadays, the evolution of technology has enabled the industry to move to Second Generation (2G) systems, the successors of 1G systems. 2G systems overcome many of the deficiencies of 1G systems mentioned in the previous chapter. Their increased capabilities stem from the fact that, contrary to 1G systems, 2G systems are completely digital. Compared to analog, digital technology has a number of advantages:

- *Encryption.* Digitized traffic can be easily encrypted in order to provide privacy and security. Encrypted signals cannot be intercepted and overheard by unauthorized parties (at least not without very powerful equipment). On the other hand, powerful encryption is not possible in analog systems, which most of the time transmit data without any protection. Thus, digital systems provide an increased potential for securing the user's traffic and preventing unauthorized network access.
- *Use of error correction.* In digital systems, it is possible to apply error detection and error correction techniques to the user traffic. Using these techniques the receiver can detect and correct bit errors, thus enhancing transmission reliability. This obviously leads to signals with little or no corruption, which of course translates into (a) better voice call qualities, (b) higher speeds for data applications, and (c) efficient spectrum use, since fewer retransmissions are bound to occur when error correction and error detection techniques are used. Furthermore, digital data can be compressed, which increases the efficiency of spectrum use even more. It is actually this increased efficiency that enables 2G systems to support more users per base station per MHz of spectrum than 1G systems, thus allowing operators to provide service in high-density areas more economically.
- In analog systems, each RF carrier is dedicated to a single user, regardless of whether the user is active (speaking) or not (idle within the call). In digital systems each RF carrier is shared by more than one user, either by using different time slots or different codes per user. Slots or codes are assigned to users only when they have traffic (either voice or data) to send.

The movement from analog to digital systems was made possible due to the development of techniques for low-rate digital speech coding and the continuous increase in the device density of integrated circuits. Contrary to 1G systems, which employ FDMA for user separation, 2G systems allow the use of Time Division Multiple Access (TDMA) and Code Division Multiple Access (CDMA) as well. Since the standards that will be discussed in this chapter employ either TDMA or CDMA (sometimes with a combination with FDMA), we briefly revisit the three approaches.

In order to accommodate various nodes inside the same cellular network, FDMA divides the available spectrum into subbands each of which are used by one or more users. Each user is allocated a dedicated channel (subband), different in frequency from the channels allocated to other users. When the number of users is small relative to the number of channels, this allocation can be static, however, for many users dynamic channel allocation schemes are necessary. In cellular systems, channel allocations typically occur in pairs. Thus, for each active mobile user, two channels are allocated, one for the traffic from the user to the Base Station (BS) and one for the traffic from the BS to the user. The frequency of the first channel is known as the uplink (or reverse link) and that of the second channel is known as the downlink (or forward link). For an uplink/downlink pair, uplink channels typically operate on a lower frequency than the downlink one in an effort to preserve energy at the mobile nodes. This is because higher frequencies suffer greater attenuation than lower frequencies and consequently demand increased transmission power to compensate for the loss. By using low frequency channels for the uplink, mobile nodes can operate at lower power levels and thus preserve energy. Due to the fact that pairs of uplink/downlink channels are allocated by regulation agencies, most of the time they are of the same bandwidth. This fact makes FDMA relatively inefficient since in most systems the traffic on the downlink is much more heavier than that in the uplink. Thus, the bandwidth of the uplink channel is not fully used.

TDMA is the technology of choice for a wide range of second generation cellular systems such as GSM, IS-54 and DECT. TDMA divides a band into several time slots and the resulting structure is known as the TDMA frame. In this, each active node is assigned one (or more) slots for transmission of its traffic. Nodes are notified of the slot number that has been assigned to them, so they know how much to wait within the TDMA frame before transmission. Uplink and downlink channels in TDMA can either occur in different frequency bands (FDD-TDMA) or time-multiplexed in the same band (TDD-TDMA). The latter technique obviously has the advantage of easy trading uplink to downlink bandwidth for supporting asymmetrical traffic patterns.

TDMA is essentially a half-duplex technique, since for a pair of communicating nodes, at a specific time, only one of the nodes can transmit. Nevertheless, slot duration is so small that the illusion of two-way communication is created. The short slot duration, however, imposes strict synchronization problems in TDMA systems. This is due to the fact that if nodes are far from one another, the propagation delay can cause a node to miss its turn. In order to protect inter-slot interference due to different propagation paths to mobiles being assigned adjacent slots, TDMA systems use guard intervals in the time domain to ensure proper operation.

Instead of sharing the available bandwidth either in frequency or time, CDMA places all nodes in the same bandwidth at the same time. The transmissions of various users are separated through a unique code that has been assigned to each user.

All nodes are assigned a specific n-bit code. The value of parameter n is known as the system's chip rate. The various codes assigned to nodes are orthogonal to one another, meaning that the normalized inner product of the vector representations of any pair of codes equals zero. Furthermore, the normalized inner product of the vector representation of any code with itself and the 1s complement of itself equals 1 and -1, respectively. Nodes can transmit simultaneously using their code and this code is used to extract the user's traffic at the receiver. Obviously, the receiver knows the codes of each user in order to perform the decoding.

The use of TDMA or CDMA in cellular systems offers a number of advantages:

- Natural integration with the evolving digital wireline network.
- Flexibility for mixed voice/data communication and the support of new services.
- Potential for further capacity increases as reduced rate speech coders are introduced.
- Reduced RF transmit power (which obviously translates into increasing battery life in handsets).
- Reduced system complexity (mobile-assisted handoffs, fewer radio transceivers).

4.1.1 Scope of the Chapter

The remainder of this chapter describes several 2G standards. D-AMPS, the 2G TDMA system that is used in North America and descends from the 1G AMPS is described in Section 4.2. CdmaOne, which is the only 2G system based on CDMA is discussed in Section 4.3. The widely used Global system for Mobile Communications (GSM) is described in Section 4.4. Section 4.5 describes IS-41, which is actually not a 2G standard but rather a protocol that operates on the network side of North American cellular networks. Section 4.6 is devoted to data transmission over 2G systems and discusses a number of approaches, including GRPS, HSCSD, cdmaTwo, etc. Furthermore, Section 4.6 discusses the problems faced by TCP in a wireless environment, mobileIP, an extension of the Internet Protocol (IP) that supports terminal mobility and the Wireless Access Protocol (WAP). Section 4.7 discusses Cordless Telephony (CT) including the Digital European Cordless Telecommunications Standard (DECT) and Personal Handyphone System (PHS) standards. The chapter ends with a brief summary in Section 4.8.

4.2 D-AMPS

In an effort to increase the performance of AMPS a standard known as D-AMPS (standard name is IS-54) was developed. D-AMPS maintains the 30-kHz channel spacing of AMPS and is actually an overlay of digital channels over AMPS. D-AMPS was designed in a way that enables manufacturing of dual-mode (AMPS and D-AMPS) terminals. Thus, the development of D-AMPS has led to a hybrid standard. This is necessary to accommodate roaming subscribers, given the large embedded base of AMPS equipment.

The main difference between AMPS and D-AMPS is that the latter overlays digital channels over the 30 kHz carriers of AMPS. Each such digital channel can support three times the users that are supported by AMPS with the same carrier. Thus, D-AMPS can be seen as an overlay on AMPS that 'steals' some carriers and changes them to carry digital traffic. Obviously, this does not affect the underlying AMPS network, which can continue to serve regular AMPS users. In fact, each D-AMPS MS initially accesses the network via the traditional AMPS analog control channels. Then the MS can make a request to be assigned a digital channel and if such a channel is available, it is allocated to the D-AMPS MS; otherwise the MS will operate in AMPS mode.

Finally, as far as handoffs are concerned, D-AMPS supports Mobile Assisted Handoff (MAHO). MSs make measurements of the signal strength from various neighboring BSs and report these measurements to the network, which uses this information to decide whether a handoff will be performed, and to which BS. The difference with AMPS is that in AMPS, MSs do not perform signal strength measurements. Rather these measurements are made by the BSs as can be seen in Chapter 2 from the sequence of events that describes a handoff in AMPS.

Both D-AMPS and its successor IS-136 support voice as well as data services. Supported speeds for data services are up to 9.6 kbps.

4.2.1 Speech Coding

D-AMPS utilizes Vector-Sum Excited Linear Predictive Coding (VSELP). This method breaks the PCM digitized voice bit-stream into parts corresponding to 20 ms speech intervals. Each such bitstream forms the input to a codebook whose output replaces the input bitstream with the codeword that is closest to the actual value of the input bitstream. This codeword is what will be transmitted over the wireless link. Each codeword will be later provided with protection against the fading wireless environment. This protection comprises: (a) a CRC operation on the most significant bits of each speech coder output; (b) convolutional coding to protect the most vulnerable bits of the speech coder output; and (c) interleaving the contents of each coder output over two time slots. Each digital channel provides a raw bit rate of 48.6 kbps, achieved using π/4 DQPSK.

4.2.2 Radio Transmission Characteristics

D-AMPS operates at the same frequency band with AMPS. Uplink digital channels occur in the 824–849 band and downlink ones in the 869–894 band. Each digital channel is organized into 40 ms frames and each frame comprises six 6.67 ms time slots. Each user can use either 2 slots (either 1 and 4, 2 and 5 or 3 and 6) or 1 slot within each frame. The first configuration is used with the full-rate voice codec, producing transmission of actual voice information up to 7.95 kbps (5.05 kbps with Forward Error Correction (FEC)). The second configuration is used with the half-rate voice codec producing transmission of actual voice information up to 3.73 kbps (2.37 kbps with FEC). The corresponding values for data speeds are 9.6 without FEC and 3.4 kbps with FEC.

The overall access method is shown in Figure 4.1. It can be seen that the uplink and downlink slots have a slightly different internal arrangement. The slot parts are described below:

- *The training part.* This part has enables the MS and BS to 'learn' the channel. This is because a signal is bound to arrive at the receiver over a number of paths due to reflections from objects in the environment. Thus, equalization is used to extract the desired signal from the unwanted reflections. The IS-54 standard also provides for an adaptive equalizer to mitigate the intersymbol interference caused by large delay spreads, but due to the relatively low channel rate (24.3 kbaud), the equalizer will be unnecessary in many situations.

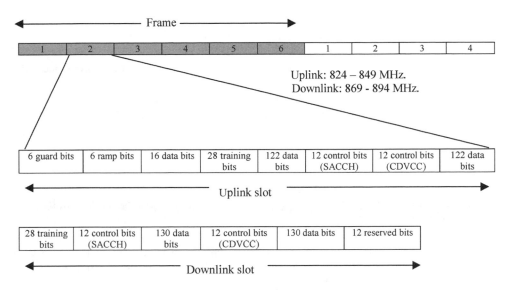

Figure 4.1   Structure of IS-54 slot and frame

- *The traffic (data) parts.* These parts carry user traffic, either voice or data-related. As the channels are digital, user traffic can be encoded or encrypted, thus the whole traffic part is not always entirely dedicated to the transfer of user data but also contains encryption/coding overhead.
- *The guard part.* This provides guard intervals in the time domain in order to separate a slot from the previous slot and the next slot. The need for these parts is due to propagation delay, which can cause a node to miss its slot when nodes are very far from one another.
- *The ramp bits.* These are used to ramp up and down the signal during periods where the signal is in transition.
- *The control parts.* These carry control signaling via the channel shown in parentheses.

Uplink and downlink frames are offset in time by 8.518 ms. As the uplink and downlink occur in different carriers, this offset allows an MS to operate at half-duplex mode since with this arrangement MSs never transmit and receive at the same time.

4.2.3 Channels

D-AMPS reuses the AMPS channels described in Chapter 2. However, it also introduces some new digital channels. The channel definitions for AMPS are as follows:

- *Forward Control Channel (FOCC).* Same as AMPS.
- *Forward Voice Channel (FVC).* Same as AMPS. The analog channel carrying voice traffic from the BS to the MS.
- *Forward Digital Traffic Channel (FDTC).* This is a BS to MS channel carrying digital traffic (both user data and control data). It consists of the Fast Associated Control Channel (FACCH) and Slow Associated Control Channel (SACCH). FACCH is a blank-and-burst operation, meaning that the traffic channel is pre-empted by control signaling. SACCH is a

continuous channel also associated with control signaling. However, it differs from FACCH in that a certain amount of bandwidth is allocated a priori to SACCH.

- *Reverse Control Channel (RECC).* Same as AMPS.
- *Forward Voice Channel (RVC).* Same as AMPS. The analog channel carrying voice traffic from the MS to the BS.
- *Reverse Digital Traffic Channel (RDTC).* This is an MS to BS channel carrying digital traffic (both user data and control data). It consists of a FACCH and SACCH.

4.2.4 IS-136

IS-136 is an upgrade of AMPS that also operates in the 800 MHz bands. However, there are planned upgrades to the 1900 band. While D-AMPS is a digital overlay over AMPS, IS-136 is a fully digital standard. IS-136 has much in common with GSM (such as convolutional coding, interleaving, etc.). However, their air interfaces are incompatible. Due to the similarities between GSM and IS-136, we do not make a detailed presentation of the former. Rather, we present the organization of the air interface of IS-136, which as can be seen from Figure 4.2 builds on top of that of D-AMPS.

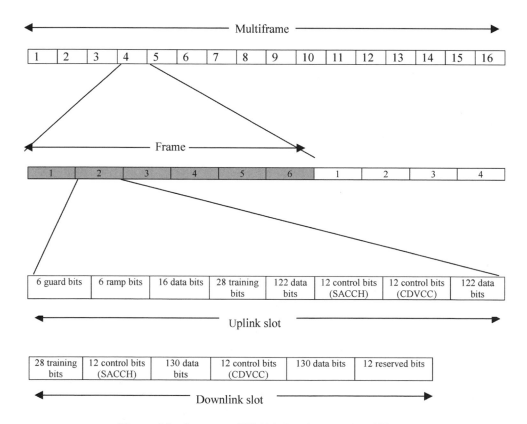

Figure 4.2   Structure of IS-136 slot, frame and multiframe

4.3 cdmaOne (IS-95)

In 1993 cdmaOne, a 2G system also known as IS-95, has been standardized and the first commercial systems were deployed in South Korea and Hong Kong in 1995, followed by deployment in the United States in 1996. cdmaOne utilizes Code Division Multiple Access (CDMA). In cdmaOne, multiple mobiles in a cell, whose signals are distinguished by spreading them with different codes, simultaneously use a frequency channel. Thus, neighboring cells can use the same frequencies, unlike all other standards discussed so far. cdmaOne is incompatible with IS-136 and its deployment in the United States started in 1995. Both IS-136 and cdmaOne operate in the same bands with AMPS. cdmaOne is designed to support dual-mode terminals that can operate either under an cdmaOne network or an AMPS network. cdmaOne supports data traffic at rates of 4.8 and 14.4 kbps.

4.3.1 cdmaOne Protocol Architecture

Figure 4.3 shows the protocol architecture of the lower two layers of cdmaOne and its correspondence to the layers of the OSI model. Layer 1 obviously deals with the actual radio transmission, frequency use, etc. These issues will be discussed briefly in the next subsection. Layer 2 offers a best effort delivery of voice and data packets. The MAC sublayer of this layer also performs channel management. This sublayer maintains a finite-state

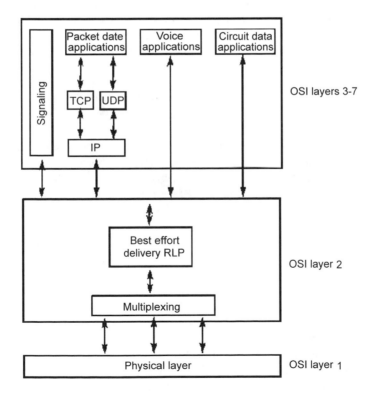

Figure 4.3   cdmaOne protocol architecture

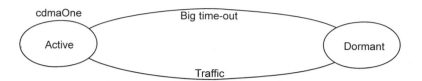

Figure 4.4 cdmaOne MAC states

machine with the two states shown in Figure 4.4. Reflecting the status of packet or circuit data transmissions, a different machine is maintained for each transmission. cdmaOne mobiles maintain all their channels and go to the dormant state after a 'big' timeout (big period during which the MS is idle). In this state, mobiles do not maintain any channels. Thus, there exists no mechanism for sending user data while in the dormant state; rather the mobile must request channel assignment, thus incurring an overhead for infrequent data bursts. Upon having traffic to send, they return to the active state where channels are assigned to the mobile. Finally, data originating from different sources are multiplexed and handed for transmission to the physical layer.

4.3.2 Network Architecture-Radio Transmission

As mentioned above, cdmaOne reuses the AMPS spectrum in the 800 MHz band. cdmaOne uses a channel width of 1.228 MHz both on the uplink and downlink. Therefore, 41 30 kHz AMPS channels are grouped together for cdmaOne operation. A significant difference between cdmaOne and the other cellular standards stems from the fact that in cdmaOne, the same frequency is reused in all cells of the system. This leads to a frequency reuse factor of 1 and is due to the fact that cdmaOne identifies the transmissions of different mobiles via the different spreading codes that identify each mobile. Both cdmaOne BSs and MSs utilize antennas that have more than one element (RAKE receivers) in order to combat the fading wireless medium via space diversity.

The use of CDMA for user separation imposes the need for precise synchronization between BSs in order to avoid too much interference. This synchronization problem is solved via the use of the Global Positioning System (GPS) receivers at each BS. GPS receivers provide very accurate system timing. Once the BSs are synchronized, it is their responsibility to provide timing information to the MSs as well. This is achieved by conveying from the BSs to the MSs a parameter identifying the system time, offset by the one way or round-trip delay of the transmission. In this way, it is ensured that BSs and MSs remain synchronized.

Finally, as far as the network side is concerned, cdmaOne utilizes the IS-41 network protocol that is described in a later section.

4.3.3 Channels

4.3.3.1 Downlink Channels
Downlink channels are those carrying traffic from the BS to the MSs. The cdmaOne downlink is composed of 64 channels. These logical channels are distinguished from each other by using different CDMA spreading codes, W0 to W63. The spreading code is an orthogonal code, or called Walsh function. The cdmaOne downlink comprises

common control and dedicated traffic channels, the most important of which are summarized below.

- *Pilot channel.* This channel provides the timing information to the MS regarding the downlink and signal strength comparisons between BSs. The actual content of the pilot channel is a continuous stream of 0s at a rate of 19.2 kbps.
- *Sync channel.* This optional channel is used to transmit synchronization messages to MSs. The sync channel is usually present, but may be omitted in very small cells. In that case, a mobile will get synchronization information from a neighboring cell. The channel operates at a rate of 1200 bps.
- *Paging channel.* This is an optional channel. There are up to seven paging channels on the downlink which can carry four major types of messages: overhead, paging, order, and channel assignment. This channel operates at one of the following data rates: 2400, 4800, or 9600 bps.
- *Traffic channels.* Traffic channels carry user data, at 1200, 2400, 4800, or 9600 bps. All traffic channels are spread by a long code (PN code), which provides discrimination between mobile stations.

Except for the pilot channel, all channels on the downlink are coded and interleaved. The vocoder uses the Code Excited Linear Predictive (CELP) algorithm. The vocoder is sensitive to the amount of speech activity present on its input, and its output will appear at one of four available rates. The bit rate of the vocoder changes in proportion to how active the speech input may be at any time. The rate may vary every 20 ms. The output of the vocoder is first encoded by the convolutional encoder into a constant 19.2 ksps (1000 symbols/second) binary stream, each data bit is represented by two symbols, with one redundancy bit inserted (rate 1/2). The output of the convolutional coder is input to a repetition function, which is used to repeat the data pattern of reduced rates (1200, 2400, or 4800 bps) to form a constant output rate of 19.2 ksps. The encoded binary stream is then interleaved randomly by the interleaver (at an interval of 20 ms) into frames (frame interleaving). The purpose of using interleaving is to combat the multipath fading environment, which causes burst errors on the radio channel. The output of the interleaver is then modulo-2-added to a 19.2 kcps (1000 chips/second) scrambling code from a 1/64 decimator. The decimator selects every 64th bit from a 'long code' generator running at 1.2288 Mcps. The 'long code' generator creates a very long codes ($2^{42} - 1$ bits) based on the user-specific information, such as the Mobile Identity Number (MIN) or the user's Electronic Serial Number (ESN). Long codes provide a very high level of security, because of the long length. This information is also made available to the network when the MS sends its handshaking information to the BS. After modulated by a long code, the resulting 19.2 ksps data stream is spread by a Walsh function running at a rate of 1.2288 Mcps. Walsh spreading provides every channel with a unique identification number. Finally, the spread 1.2288 Mcps signal is spread one more time by a short code running at 1.2288 Mcps. Short code is also a Pseudonoise (PN) code, and is $2^{15} - 1$ bits in length. All base stations use the same short code, but with different offsets. There exist 512 different offsets, thus this scheme can uniquely identify 512 different cdmaOne BSs. A mobile can easily distinguish transmissions from two different base stations by their short-code offsets. The resulting signal is transmitted over the wireless medium via Quadrature Phase Shift Keying (QPSK) modulation.

4.3.3.2 Uplink Channels

There are two types of uplink channels, access and traffic. There can be up to 32 access channels on the uplink, each of which operates at 4800 bps. These channels are used by MSs to initiate calls and respond to paging messages. An access channel contains information that the BS needs to properly log the mobile into service. There can be up to 62 traffic channels on the uplink. These are used to carry user data. The payload of a traffic channel comes from a variable rate vocoder with four possible output rates: 9600, 4800, 2400 and 1200 bps.

The data from the vocoder is convolutionally encoded by a 1/3 rate encoder, which adds two redundancy bits to each data bit, thus multiplying the data rate by three, resulting in a binary stream of rate 28.8 ksps. The encoded data is interleaved randomly before entering the block encoder, which examines the content of the input data stream in a 6-bit segment and replaces the 6-bit segment with the corresponding 64-bit Walsh function.

After leaving the block encoder, the data stream is spread by the long code and short codes, respectively. The resulting spread data stream has a rate of 1.2288 Mcps and is transmitted over the wireless medium via Offset Quadrature Phase Shift Keying (OQPSK) modulation. OQPSK provides more Forward Error Correction (FEC) than QPSK since MSs cannot coordinate their transmissions as efficiently as BSs.

4.3.4 Network Operations

4.3.4.1 Handoff

There are four handoff categories in cdmaOne, soft, softer, hard and idle handoff. A handoff occurs when a MS detects a pilot channel of higher quality than that of the BS currently serving the MS. In soft handoff, a link is set up to the new BSs before the release of the old link. This ensures reliability, as the new BS may be too crowded to support the roaming mobile terminal or the link to the new BS may degrade shortly after establishment. However, the mobile terminal should be able to communicate with two different BSs at the same time. Thus, soft handoff causes increased complexity at the mobile terminals since it demands the capability of supporting two links with different BSs at the same time. When a soft handoff takes place between sectors inside the same cell, it is also known as softer handoff. Hard handoff is relatively simpler than soft handoff since the link to the old BS is released before establishment of the link to the BS of the new cell. However, it is somewhat less reliable than soft handoff. Finally, the cdmaOne specification defines the idle handoff. The main difference of idle handoff with the previous handoff types is that in the previous types the MS being handed off is involved in an active call. However, in an idle handoff the MS is in idle mode.

4.3.4.2 Power Control

Power control is critical in cdmaOne due to the fact that the use of CDMA imposes the need for all MS transmissions to reach the BS with strength difference of no more than 1 dB. If the signal received from a near user is stronger than that from a far user, the former signal will be swamped out by the latter. This is known as the 'near-far' problem. Another reason for implementing power control is to increase capacity. Power control is implemented on both the uplink and downlink.

On the uplink, both open-loop and closed-loop power control is used (the principle of which has been described in Chapter 2). On the downlink, a scheme known as slow power control is employed. According to this scheme, the BS periodically reduces its transmitted power to the MS. The latter makes periodic measurements on the frame error ratio (FER). When the FER exceeds a predefined limit, typically 1%, the MS requests a boost in the transmission power of the BS. This adjustment occurs every 15–20 ms. The dynamic range of the downlink power control is around six times less than that of the composite open-loop and closed-loop power control scheme employed on the uplink.

4.4 GSM

The origins of the Global System for Mobile Communications (GSM) can be found in Europe in the early 1980s. At that time, Europe was experiencing a spectacular growth of analog cellular systems, mainly with NMT in Scandinavia and TACS in Great Britain, Italy, Spain and Ireland. Moreover, other European countries had deployed other 1G systems, such as C-450 in Germany and Portugal, Radiocom 2000 in France and RTMS in Italy. These systems were generally not compatible with each other so the European market suffered from a divergence of standards. This was an undesirable situation, because (a) mobile equipment operation was limited within national boundaries, which was obviously bad when taking into account the European Community (EC, nowadays European Union, EU) aim of a unified Europe and (b) limited the market for each type of equipment, so economies of scale and the subsequent savings could not be realized.

Acknowledging this problem, in 1992 the EC formed a study group called the Groupe Special Mobile (later renamed to Global System for Mobile Communications). GSM [1], which comes from the initials of the group's name, had the task of studying and developing a pan-European public land mobile system. The proposed system had to meet certain criteria:

- Good subjective speech quality;
- Low terminal and service cost;
- Support for international roaming;
- Ability to support handheld terminals;
- Support for range of new services and facilities;
- Spectral efficiency;
- ISDN compatibility.

In 1989, GSM responsibility was transferred to the European Telecommunication Standards Institute (ETSI), and phase I of the GSM specifications was published in 1990. Commercial deployment of GSM systems started in 1991, and by 1993 there were 36 GSM networks in 22 countries around Europe. GSM is nowadays the most popular 2G technology; by 1999 it had 1 million new subscribers every week. This popularity is not only due to its performance, but also due to the fact that it is the only 2G standard in Europe. This existence of one standard boosted the cellular industry in Europe, contrary to the situation in the United States, where several different 2G systems have been deployed thus leading to a fragmented market.

Despite the fact that GSM was standardized in Europe, it has been deployed in a large number of countries worldwide (approximately 110). Overall, there are four versions of the GSM system, depending on the operating frequency. These systems are shown in Figure 4.5. The system that operates at 900 MHz was the first to be used. The operating frequency was

GSM variant	Upling frequency (MHz)	Downlink frequency (MHz)
GSM 900	890 - 915	935 - 960
GSM 1800 (DCN)	1710 - 1785	1805 - 1880
GSM 1900 (PCS)	1850 - 1910	1930 - 1990
GSM 450	450.4 - 457.6 or 478 - 486	460.4 - 467.6 or 488.8 - 496

Figure 4.5   GSM variants

chosen at 900 MHz in order to reuse the spectrum used by European TACS systems. The next GSM variants to appear were those operating at 1800 MHz in Europe and 1900 MHz in America. These variants are known as Digital Communications Network (DCN) and Personal Communications System (PCS), respectively, but they are essentially GSM operating at another frequency. The fourth variant operates at 450 MHz in order to provide a migration path from the 1G NMT standard that uses this band to 2G GSM systems.

The primary service supported by GSM is voice telephony. Speech is digitally encoded and transmitted through the GSM network as a binary bitstream. For emergency situations, an emergency service is supported by dialing a certain three-digit number (usually 112).

GSM also offers a variety of data services. It allows users to send and receive data, at rates up to 9600 bps. Data can be exchanged using a variety of access methods and protocols, such as X.25. A modem is not required between the user and GSM network due to the fact that GSM is a digital network. Other data services include Group 3 facsimile. GSM also supports the Short Message Service (SMS) and Cell Broadcast Service (CBS). Finally, GSM supports a number of additional services, such as call forward (call forwarding when the mobile subscriber is unreachable by the network), call barring of outgoing or incoming calls, caller identification, call waiting, multiparty conversations, etc.

4.4.1 Network Architecture

A GSM network comprises several functional entities, whose functions and interfaces are specified. Figure 4.6 shows the layout of a GSM network.. The GSM network can be divided into the three broad parts described below. As can be seen from the figure, the MS and the BSS communicate across the Um interface, also known as the air interface or radio link. The BSS communicates with the MSC across the A interface.

4.4.1.1 Mobile Station (MS)

The MS consists of the terminal (TE) and a smart card called the Subscriber Identity Module (SIM). The SIM provides personal mobility, so that the user can have access to subscribed services irrespective of a specific terminal. Furthermore, the SIM card is the actual place where the GSM network finds the telephone number of the user. Thus, by inserting the SIM

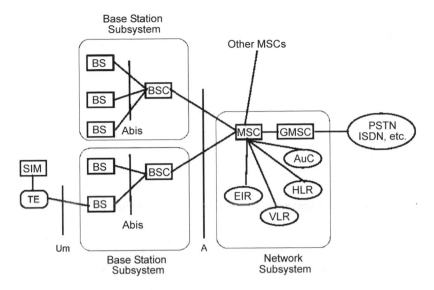

Figure 4.6   GSM network architecture

card into another GSM terminal, the user is able to use the new terminal to receive calls, make calls and user other subscribed services while using the same telephone number.

The actual GSM terminal is uniquely identified by the International Mobile Equipment Identity (IMEI). The SIM card contains the International Mobile Subscriber Identity (IMSI) used to identify the subscriber to the system, a secret key for authentication, and other information. The IMEI and the IMSI are independent, thereby allowing personal mobility. Furthermore, the SIM card may be protected against unauthorized use by a password or personal identity number.

The structures of the IMEI and the IMSI are shown in Figures 4.7 and 4.8, respectively. The IMEI can be up to 15 digits and comprises the following parts:

- A 3-digit Type Approval Code (TAC). This is given to the unit after it passes conformance tests.
- A 1 or 2-digit Final Assembly Code (FAC). This identifies the place of final manufacture or assembly of the MS unit.
- The MS unit serial number.
- 1 spare digit reserved for future assignment.

TAC (3 digits)	FAC (1 or 2 digits)	Serial number (up to 11 digits)	1 spare digit

Figure 4.7   IMEI structure

MCC (3 digits)	MNC (2 digits)	MSIC (up to 10 digits)

Figure 4.8   IMSI structure

The IMSI is also up to 15 digits and comprises the following parts:

- *A 3-digit Mobile Country Code (MCC).* This identifies the country where the GSM system operates.
- *A 2-digit Mobile Network Code (MNC).* This uniquely identifies each cellular provider.
- *The Mobile Subscriber Identification Code (MSIC).* This uniquely identifies each customer of the provider.

4.4.1.2 Base Station Subsystem (BSS)

The BSS contains the necessary hardware and software to enable and control the radio links with the MSs. It comprises two parts, the Base Station (BS) and the Base Station Controller (BSC). These communicate across the standardized Abis interface, allowing (as in the rest of the system) operation between components made by different suppliers. The BS contains the radio transceivers that define a cell and handles the radio-link protocols with the MS. In a large urban area, there will potentially be a large number of BSs deployed, thus the BSC typically manages the radio resources for one or more cells. BSs are responsible for frequency administrations and handovers. The BSC is the connection between the mobile station and the Mobile service Switching Center (MSC). BSCs are quite intelligent and perform many of the necessary functions to enable the link between the BSs and the MSs. Finally, we mention that BSs and BSCs may be collocated. Another option is for the BSC and the Mobile Switching Center (MSC) to be collocated.

4.4.1.3 Network Subsystem

The central component of the network subsystem is the Mobile Switching Center (MSC). The MSC performs switching of user calls and provides the necessary functionality to handle mobile subscribers. This functionality includes support for registration, authentication, location updating, handovers, and call routing to a roaming subscriber. Furthermore, the MSC interfaces the GSM network to fixed networks. Such an MSC is known as a Gateway MSC (GMSC) and performs the necessary interworking functions (IWF) to interface the GSM network to a fixed network such as the Public Switched Telephone Network (PSTN) or ISDN. Signaling between functional entities in the network subsystem uses Signaling System Number 7 (SS7), which is widely used in public networks.

   The MSC contains no information about particular mobile stations. Rather, this information is stored in the two location registers of GSM. These are the Home Location Register (HLR) and the Visitor Location Register (VLR). These two registers together with the MSC provide the call-routing and roaming capabilities of GSM. The HLR contains all the administrative information for the subscribers. This information includes the current locations of the MSs (that is the VLR of the subscriber, which is described later). There exists one HLR per GSM network, although it may be implemented as a distributed database.

   The Visitor Location Register (VLR) contains selected administrative information from the HLR, necessary for call control and provision of the subscribed services, for each mobile roaming in the area controlled by the VLR. VLR is implemented together with the MSC, so that the geographical area controlled by the MSC corresponds to that controlled by the VLR in order to simplify signaling.

There exist two additional registers, which are used for authentication and security purposes. These are the Equipment Identity Register (EIR) and the Authentication Center (AuC). The EIR is a database that contains a list of all valid MSs on the network, each uniquely identified by its IMEI as mentioned above. Invalid MSs are those that have either been stolen or their operation has been prohibited due to other reasons. Invalid MSs are identified by marking their IMEI as invalid. The actual markings that can be used for an MS's IMEI are:

- *White-listed:* This marking means that the MS is allowed to connect to the network.
- *Grey-listed:* This marking means that the terminal is under observation from the network for possible problems.
- *Black-listed:* This marking means that the terminal has either been reported as stolen, or is prohibited from using the network for some other reason.

The Authentication Center (AuC) is a protected database that stores a copy of the secret key stored in each subscriber's SIM card, which is used for authentication and encryption over the radio channel.

4.4.2 Speech Coding

Voice needs to be converted from its analog form to a digital form that will be transmitted over the digital GSM wireless network. However, PCM, which is used in ISDN is not applicable to the case of wireless networks due to its high capacity demands (64 kbps). The GSM group studied several speech coding algorithms on the basis of subjective speech quality and complexity (which is related to cost, processing delay, and power consumption once implemented) before arriving at the choice of a Regular Pulse Excited-Linear Predictive Coder (RPE-LPC) with a long term predictor loop. Basically, information from previous samples, which does not change very quickly, is used to predict the current sample. Speech is divided into 20 ms samples, each of which is encoded as 260 bits, giving a total bit rate of 13 kbps. This is the so-called full-rate speech coding. Recently, an Enhanced Full-Rate (EFR) speech coding algorithm has been implemented by some North American GSM1900 operators. This is said to provide improved speech quality using the existing 13 kbps bit rate. Furthermore, a half-rate codec has been made possible due to the advances of microelectronics. This codec halves the bandwidth needed per call with only a slight degradation in quality.

4.4.3 Radio Transmission Characteristics

In this section we discuss the air interface of GSM (the Um interface), which actually defines the way information is transmitted over the air. As with every other wireless network, GSM encodes data into waves in order to send it over the wireless medium. The actual modulation scheme that is used is Gaussian Minimum Shift Keying (GMSK), which achieves 270.8 kbps over each of the 200-kHz wide GSM channels. The available bandwidth in GSM is split into 124 carriers, each 200 kHz wide. GSM uses a combination of Time and Frequency Division Multiple Access (TDMA/FDMA) for user separation. One or more carrier frequencies are assigned to each BS of the GSM network and each of those carriers is divided in the time

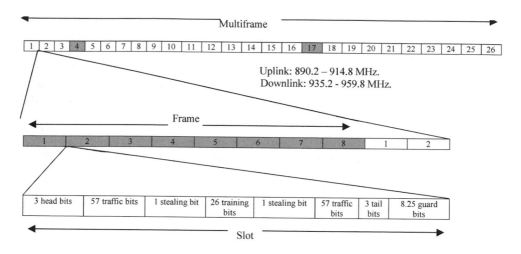

Figure 4.9   Structure of GSM slot, frame and 26-frame multiframe

domain. Each time period is called a slot and lasts 0.577 ms. A slot comprises the following parts, which are also shown in Figure 4.9:

- *The head and tail parts.* These parts are 3 bits each and are used to ramp up and down the signal during periods where the signal is in transition.
- *The training sequence part.* This part comprises a fixed sequence of 26 bits. Its purpose is to enable the MS and BS to 'learn' the channel. This is because a signal is bound to arrive at the receiver over a number of paths due to reflections from objects in the environment. Thus, equalization is used to extract the desired signal from the unwanted reflections. As mentioned in Chapter 2, equalization works by finding out how a known transmitted signal is modified by multipath fading, and constructing an inverse filter to extract the rest of the desired signal. The 26-bit training sequence constitutes a signal known to both the BS and the MS. The receiver will compare the incoming signal corresponding to the 26 bit training sequence to the original one and will use it to 'equalize' the channel. The actual implementation of the equalizer is not specified in the GSM specifications.
- *The stealing bits parts.* These bits are used to identify whether the lot carries data or control information.
- *The traffic part.* This part is 57 bits long and carries user traffic, either voice or data-related. User traffic can be encoded or encrypted, thus the whole traffic part is not always entirely dedicated to the transfer of user data.
- *The guard interval.* This is 8.25 bits long. It is essentially empty space whose purpose is to provide guard intervals in the time domain in order to separate a slot from the previous slot and the next slot. The need for this is due to propagation delay, which can cause a node to miss its slot when nodes are very far from one another. In order to protect inter-slot interference due to different propagation paths to mobiles being assigned adjacent slots, GSM systems use the guard interval to ensure proper operation. Using this interval, the effects of propagation delay are negated for distances up to 35 km from the GSM antenna of the BS. For MS–BS distances that exceed 35 km the propagation delay becomes large relative to the slot duration, thus resulting in the GSM phone losing its slot. Therefore, in such a case a GSM phone cannot operate even in the presence of a signal of good quality.

Eight slots make up a GSM frame with duration of 4.615 ms. An actual channel assigned to an MS is served via a certain slot within the GSM frame. The fact that each MS is assigned only one slot within each frame limits the maximum speeds offered by GSM for data services to 33.9 kbps; 1/8 of the 270.8 kbps capacity of a 200 kHz GSM carrier. Due to FEC and encryption overhead, the actual speeds are much lower and are typically around 9.6 kbps.

As will be seen later, channels are divided into dedicated channels, which are allocated to an active mobile station and common channels, which can be used by all mobile stations in idle mode. Users cannot use all frames; rather, every 26 GSM frames, one is 'stolen' and used by the network for signaling purposes, while a second one is reserved for other traffic types such as Caller Line Identification (CLI), etc. A multiframe comprises 26 GSM frames and is shown in Figure 4.9, which also shows the frequencies allocated for the downlink and the uplink for the 900 MHz GSM variant. In this figure, the shaded frames are those that are stolen by the network for control signaling. However, stolen frames are not always the same; rather, stolen frames move on by one frame for every multiframe. This fact helps with timing.

For the control channels, there is a different multiframe structure that comprises 51 GSM frames. This structure is shown in Figure 4.10. In this figure, one can also see that there are four different possibilities for the actual content of each frame of the 51-frame multiframe. All these comprise two tail parts, 3 bits each, and an 8.25 bit guard interval unless stated otherwise. The different contents are summarized below:

- *The frequency correction slot.* This contains a sequence of 142 bits each having a value of 0. Its purpose is to synchronize the MS with the system master frequency.

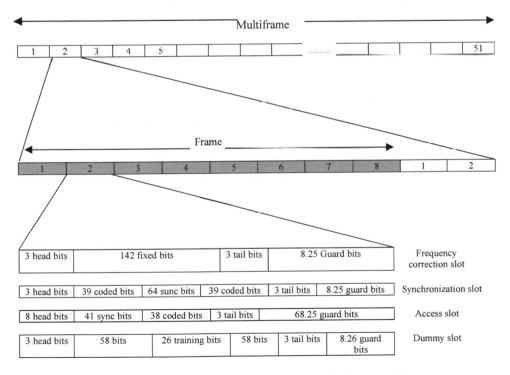

Figure 4.10   Structure of GSM slot, frame and 51-frame multiframe

- *The synchronization slot.* This aims to synchronize in time the MS and the BS. It comprises two 39-bit pairs of coded bits separated by 64 synchronization bits. The coded bits contain information that enables the MS to know the position and identity of all slots in the TDMA transmissions and receptions. Furthermore, they contain information relating to the code of the BS, the national code, etc. The synchronization bits play the same role as those found in the slot structure shown in Figure 4.9, that is, to provide for BS-MS synchronization.
- *The access slot.* This is used to enable the random access channel (this is explained later) that is used by MS to request slot assignment. The 41 synchronization bits are used for BS–MS synchronization and the coded bits contain information relating to the success of the MSs random attempt. The longer 68.25 guard period of this slot ensures that the slot can be used at MS-BS distances up to 75.5 km.
- *The dummy slot.* This is used to fill empty slots.

The overall GSM framing structure combines the 26 and 51 multiframes into a higher-level structuring comprising superframes and hyperframes. Multiframes are grouped into superframes, with each superframe comprising 1326 frames and lasting 6.12 s. Each superframe comprises 1326 frames, because this is the least common multiple of 26 and 51. Thus, this configuration leads to no empty slots at a superframe. The hyperfame is the largest set and comprises 2048 hyperframes and lasts 3 h, 28 min, 53 s and 760 ms. Obviously, these definitions are cyclic which means that after a frame, multiframe, superframe, or hyperframe have elapsed, a new corresponding structure is issued by the system.

GSM uses convolutional encoding and block interleaving to protect transmitted data. The exact algorithms used differ for speech and for different data rates. The method used for speech blocks is described below. Recall that the speech codec produces a 260 bit block for every 20 ms speech sample. From subjective testing, it was found that some bits of this block were more important for perceived speech quality than others. The bits are thus divided into three classes:

- *Class Ia.* These are the 50 bits that are considered to be most sensitive to bit errors.
- *Class Ib.* These are the 132 bits that are considered to be moderately sensitive to bit errors.
- *Class II.* These are the 78 bits that are considered to be least sensitive to bit errors.

Class Ia bits have a 3-bit Cyclic Redundancy Code (CRC) added for error detection. These 53 bits, together with the 132 Class Ib bits and a 4 bit tail sequence (a total of 189 bits), are input into a 1/2 rate convolutional encoder of constraint length 4. Each input bit is encoded as two output bits, based on a combination of the previous 4 input bits. The convolutional encoder thus outputs 378 bits, to which are added the 78 remaining Class II bits, which are unprotected. Thus, every 20 ms speech sample is encoded as 456 bits, giving a bit rate of 22.8 kbps.

To further protect against the burst errors common to the radio interface, each sample is interleaved. The 456 bits output by the convolutional encoder are divided into 8 blocks of 57 bits, and these blocks are transmitted in eight consecutive slots. Since each slot can carry two 57-bit blocks, each burst carries traffic from two different speech samples. This provides diversity and enhances the resistance of GSM to interference.

4.4.4 Channels

4.4.4.1 Traffic Channels

A traffic channel (TCH) is used to carry speech and data traffic. Traffic channels are defined using the GSM multiframe structure. TCHs for the uplink and downlink are separated in time by 3 slots so that the mobile station does not have to transmit and receive simultaneously, thus simplifying the electronics. In addition to these full-rate TCHs, there are also half-rate TCHs defined to work with the half-rate speech codec. Eighth-rate TCHs are also specified, and are used for signaling. They are called Stand-alone Dedicated Control Channels (SDCCH).

4.4.4.2 Control Channels

Control channels can be accessed both by idle and active mobiles. These are common channels and are used by idle mode mobiles to exchange the signaling information required to change to dedicated mode. Mobiles already in dedicated mode monitor the surrounding base stations for handover and other information. The control channels are defined within the 51-frame GSM multiframe, so that active mobiles using the 26-frame multiframe TCH structure can still monitor control channels. The control channels are summarized below:

- *Broadcast Control Channel (BCCH)*. Continually broadcasts, on the downlink, information including BS identity, frequency allocations, and frequency-hopping sequences.
- *Frequency Correction Channel (FCCH) and Synchronization Channel (SCH)*. These are used to synchronize the mobile to the time slot structure of a cell by defining the boundaries of time slots and the time slot numbering. Every cell in a GSM network broadcasts exactly one FCCH and one SCH, which are by definition on time slot number 0 (within a TDMA frame).
- *Random Access Channel (RACH)*. This is a used by the mobile to request access to the network. Mobiles compete for access to this channel using slotted Aloha.
- *Paging Channel (PCH)*. This channel is used to alert the mobile station to an incoming call.
- *Access Grant Channel (AGCH)*. This channel is used to allocate an SDCCH to a mobile for signaling following a request on the RACH.

4.4.5 Network Operations

A GSM MS can seamlessly roam nationally and internationally. This requires that registration, authentication, call routing and location updating functions exist and are standardized in GSM networks. These functions along with handover are performed by the network subsystem, mainly using the Mobile Application Part (MAP) built on top of the Signaling System No. 7 protocol.

The signaling protocol in GSM is structured into three general layers. Layer 1 is the physical layer, which uses the channel structures discussed above over the air interface. Layer 2 is the data link layer. Across the Um interface, the data link layer is a modified version of the LAPD protocol used in ISDN, called LAPDm. Across the A interface, the Message Transfer Part layer 2 of Signaling System Number 7 is used. Layer 3 is divided into the 3 sublayers described below. Following this description, handover and power control in GSM are discussed.

4.4.5.1 Radio Resources Management

The radio resources (RR) management layer oversees the establishment of a link, both radio and fixed, between the MS and the MSC. An RR session is always initiated by the MS side either for an outgoing call or in response to a paging message. The RR layer handles among other things radio features, such as power control, discontinuous transmission and reception, frequency hopping and management of channel changes during handovers between cells.

4.4.5.2 Mobility Management

The Mobility Management (MM) layer is built on top of the RR layer and works with the HLR and VLRs. It is concerned with handling issues arising due to the mobility of the MS (such as location management and handoff), as well as authentication and security aspects. Location management is concerned with the procedures that enable the system to know the current location of a powered-on mobile station so that incoming call routing can be completed. The actual location updating mechanism in GSM organizes cells into groups called location areas. MSs send update messages to the network whenever the MS moves into a different location area. This approach can be thought of as a compromise between two extremes: (a) for every incoming call, page every cell in the network in order to find the desired MS; (b) the MS notifies the network whenever it changes a cell. Location update messages are conveyed via the Location Update Identifier (LAI), shown in Figure 4.11. The first two fields of this structure have been explained earlier. The third field, the Location Area Code (LAC) identifies a group of cells. Whenever the MS roams into a cell having a different LAC than the previous one, a LAI is sent to the network, which records the new location of the mobile and then makes the appropriate updating at the HLR and the MSC/VLR covering the area where the MS is located. If the subscriber is allowed to use the requested service, the HLR sends a subset of the subscriber information, needed for call control to the new MSC/VLR. Then the HLR sends a message to the old MSC/VLR to cancel the old registration. For reliability reasons, GSM also has a periodic location updating procedure. In the case of a HLR or MSC/VLR failure, these databases are updated not from scratch but rather as subsequent location updating events occur. Both the enabling of periodic updating and the time period between periodic updates, are controlled by the operator and constitute a trade-off between signaling overhead and speed of recovery. Finally, the detach procedure relates to location updating. A detach procedure lets the network know that the MS is unreachable, in order to avoid futile channel allocations and pages to the MS. Similarly, there is an attach procedure, which informs the network that the mobile is reachable again.

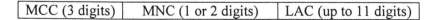

| MCC (3 digits) | MNC (1 or 2 digits) | LAC (up to 11 digits) |

Figure 4.11   LAI structure

4.4.5.3 Communication Management

The Communication Management (CM) layer is responsible for setting up and tearing down calls, supplementary service management and Short Message Service (SMS) management. Each of these may be considered as a separate sublayer within the CM layer. Other functions performed by the CC sublayer are call establishment and call release.

4.4.5.4 Handover

Handover is performed by the RR layer. There are four different types of handover in the GSM system:

- Handover of a call between channels (slots) in the same cell.
- Handover of a call between cells under the control of the same BSC.
- Handover of a call between cells under the control of different BSCs, which belong to the same MSC.
- Handover of a call between cells under the control of different MSCs.

The first two types of handover are called internal handovers, involve only one BSC and are managed by the BSC alone without intervention of the MSC. The last two types of handover are called external handovers and are handled by the MSCs involved. In external handovers, the original MSC remains responsible for most call-related functions, with the exception of subsequent inter-BSC handovers under the control of the new MSC.

Handovers can be initiated by either the mobile or the MSC. The latter option provides a procedure for the network to perform traffic load balancing. During its idle time slots, the mobile scans the BCCH of up to 16 neighboring cells, and forms a list of the six best candidates for possible handover, based on the received signal strength. This information is passed to the BSC and MSC, at least once per second.

There are two basic algorithms used in order to determine when to perform a handoff, both closely tied in with power control. This is because the BSC usually does not know whether the poor signal quality is due to multipath fading or to the mobile having moved to another cell. The first algorithm gives precedence to power control over handover, so that when the signal degrades beyond a certain point, the power level of the mobile is increased. If further power increases do not improve the signal, then a handover is considered. The second algorithm uses handover to try to maintain or improve a certain level of signal quality at the same or lower power level. Therefore it gives precedence to handover over power control.

4.4.5.5 Power Control

There are five classes of MS defined (three in North American GSM standards), according to their peak transmitter power, rated at 20, 8, 5, 2, and 0.8 W. To minimize co-channel interference and to conserve power, both the mobiles and the base transceiver stations operate at the lowest power level that will maintain an acceptable signal quality. The mobile station measures either the signal strength or the signal quality (obviously the Bit Error Rate (BER) of the received signal) and passes this information to the BSC, which decides whether the power level of transmission should be changed.

4.4.5.6 Initialization

Once a GSM MS is powered up, a sequence of events takes place. This sequence is briefly described below:

- *Event 1.* The MS locks onto the strongest frequency channel and then finds the FCCH.
- *Event 2.* The MS locates the SCH and obtains synchronization and timing information.
- *Event 3.* The MS locates the BCCH and reads system values such as LAC.
- *Event 4.* The MS uses the RACH to request a SDCCH. The BS grants his request via the AGCH.
- *Event 5.* The MS possibly initiates a location update procedure in order to inform the network of its new position. The MS knows is previous position by storing the previous LAC in its memory.
- *Event 6.* The authentication procedure for the MS starts.
- *Event 7.* After successful authentication, the network informs the MS that traffic will be encrypted.
- *Event 8.* The HLR and VLRs are updated and the MS is ready to receive a call.

4.4.5.7 Call Setup to an MS

The procedure of placing a call to a MS can be described via a number of events. These events are summarized below:

- *Event 1.* The BS notifies the MS of the incoming call via a page on the PCH.
- *Event 2.* The MS uses the RACH to request a SDCCH. The BS grants this request via the AGCH.
- *Event 3.* The MS responds at the page via the assigned SDCCH.
- *Event 4.* The authentication procedure for the MS starts.
- *Event 5.* After successful authentication, the network informs the MS that traffic will be encrypted.
- *Event 6.* Establishment of a Temporary Mobile Station Identifier (TMSI), which is good only for the duration of the call.
- *Event 7.* The MS is assigned a TCH for the call.

4.4.6 GSM Authentication and Security

Authentication involves two entities, the SIM card in the MS and the Authentication Center (AuC). Each subscriber is given a secret key. Copies of this key are stored both in the SIM card of the subscriber and in the AuC. During authentication, the AuC generates a random number and sends it to the MS. Based on this number and the subscriber's secret key, both the MS and the AuC use a ciphering algorithm called A3 to generate a signed response (SRES). The MS then sends the calculated SRES to the AuC. If the number sent by the mobile is the same as the one calculated by the AuC, the subscriber is authenticated.

The same initial random number and subscriber key are also used to perform encryption of traffic. Based on these numbers, a ciphering key is produced by using an algorithm called A8. This ciphering key, together with the TDMA frame number, use the A5 algorithm to create a

114-bit sequence. This sequence is then that is XORed with every 114 user data bits and the resulting bitstreams are sent over the two 57 bit parts of every GSM slot.

4.5 IS-41

IS-41 [2] is the protocol standard that operates on the network side of North American cellular networks. This is the reason that, contrary to GSM, a description on the network side elements has not been made for D-AMPS and cdmaOne. As mentioned above, the wireless cellular market in North America is fragmented into a number of incompatible standards. IS-41 provides for interworking between such incompatible standards. IS-634 is a successor to IS-41 that defines the operations between BSs and MSCs. This section covers several issues including the way IS-41 handles handoff and the feature of automatic roaming.

4.5.1 Network Architecture

The topology of IS-41 is quite similar to that of the network side of GSM. It is defined by a number of functional entities. The way two functional entities communicate and exchange information is defined by the corresponding interface. Most of these entities and interfaces also appear in the case of GSM and thus they have the same functionality. The topology of IS-41 is shown in Figure 4.12 [3]. The entities shown in this figure are described briefly below:

- AC, Access Control
- BS, Base Station
- CSS, Cellular Subscriber Station

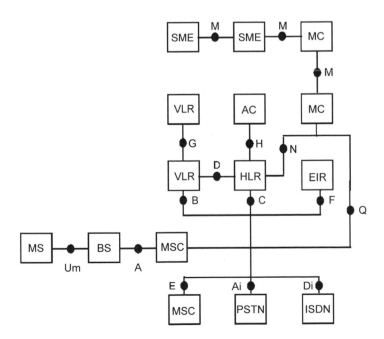

Figure 4.12 IS-41 network topology

- EIR, Equipment identity Register
- HLR, Home Location Register
- ISDN, Integrated Services Digital Network
- MC, Message Center
- MSC, Mobile Switching Center
- PSTN, Public Switched Telephone Network
- SME, Short Message Entity
- VLR, Visitor Location Register

4.5.2 Inter-system Handoff

IS-41 can manage handoffs between different systems. Two types of handoff exist, handoff-forward and handoff-back. These are described below followed by a description of the path-minimization procedure that IS-41 employs during handoffs.

- *Handoff-forward.* This handoff type entails a MS X involved in an active call that moves from the serving area of a specific MSC (e.g. A) to that of another MSC (e.g. B). As shown in Figure 4.13, the handoff from the service area of MSC A to that of MSC B requires setting up a circuit from MSC A to MSC B. Through this circuit, the call to X continues to be served.
- *Handoff-back.* This handoff type entails a MS X involved in an active call. X is located in the service area of a specific MSC (e.g. B). X is moving towards the service area of another MSC (e.g. A). Furthermore, while in the service area of MSC B, the call of X passes through another MSC A. As shown in Figure 4.14, the handoff from the service area of MSC B to that of MSC A causes release of the circuit from MSC A to MSC B. Through this circuit, the call to X continues to be served.
- *Handoff with path minimization.* This property of IS-41 is shown in Figure 4.15. In this example, MS X which is involved in an ongoing call (a) is located at the service area of MSC B, (b) the call to X passes via MSC A. Upon movement of X to the service area of MSC C, IS-41 checks to see whether an intersystem circuit can be established between

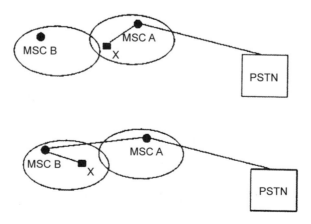

Figure 4.13   Handoff-forward

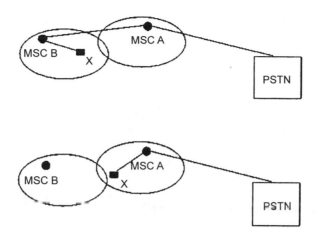

Figure 4.14   Handoff-back

MSCs A and C in order to drop the circuit from MSC A to B. If such a circuit can be established, then the call is handed off to MSC C and continues to be served via the link from MSC A to MSC C, as can be seen from the bottom part of Figure 4.15. If the circuit's path cannot be minimized, then the A-B-C MSC link will serve the call.

4.5.3 Automatic Roaming

Automatic roaming allows a roaming user to originate a call inside the visited system after credit worthiness has been validated, to invoke in the visited systems the subscribed features

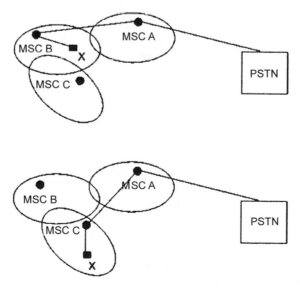

Figure 4.15   Handoff with path minimization

and to receive calls. The processes of registration, call origination and call delivery are described below.

4.5.3.1 Registration

A roaming user can notify the visited system of his/her presence either via autonomous registration or call origination. Once the MSC of the service area where the MS is located identifies the MS, the associated VLR is notified. The VLR then notifies the MSs HLR of the MSs current location and requests the MSs, and requests from this HLR, information relating to the credit worthiness of the specific user along with a service profile of the user. When the MS has previously registered elsewhere, the corresponding VLR of the previous MSC service area will be notified that the MS has left its system. Upon receipt of this message, this VLR erases all information related to the roaming MS.

4.5.3.2 Call Origination

When the roaming MS places a call establishment request, the MSC requests from the associated VLR information relating to the credit worthiness of the MS. If the VLR possesses such information it passes it to the MSC, otherwise it contacts the HLR of the MS in order to obtain the information. Once the serving MSC gets the desired information, it can decide whether or not the MS is allowed to place the call.

4.5.3.3 Call Delivery

When a call request is made with the roaming MS as the destination, the home MSC requests from the MS's HLR routing information in order to be able to establish a connection to the roaming MS. The HLR requests this information from the VLR of the visited system, which in turn relays the request to the MSC of the visited system. Then, the requested information is returned to the home MSC, which finally routes the call to the MSC of the visited system. When the MSC of the visited system receives the call destined for the MS, then the BSs of this MSC page the specific MS and the call is established.

4.6 Data Operations

4.6.1 CDPD

Cellular Digital Packet Data (CDPD) [4,5] is an extension that offers the ability to send data. This is actually a packet switching overlay to both AMPS and D-AMPS. It is the only way to offer data transfer support in an analog AMPS network.

CDPD operates using the idle voice channels of AMPS. These idle channels are used to transmit short data messages and establish a packet-switching service. In order to utilize idle AMPS channels, MSs hop around the several AMPS channels in order to select an idle one. CDPD station operation is completely transparent to the AMPS network. Thus, for an AMPS voice call, the network may decide to use channels that are already used by CDPD stations. In such a case, a power ramp-up indicating the initiation of voice traffic is detected, which triggers a channel hopping procedure. The MDBS (described later) sends a special signal that

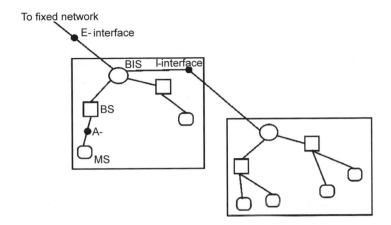

Figure 4.16   The CDPD network architecture

closes down the channel. The CDPD station leaves this channel and searches for a new idle AMPS channel. Alternatively, the BS can relay to the CDPD MS the new AMPS channel to be used. Thus, CDPD can occupy any idle capacity in a cell, without interfering with the voice system. However, the existence of dedicated CDPD channels is also possible.

The air interface of CDPD operates at a raw data rate of 19.2 kbps using GMSK modulation. It provides FEC to combat the interference and fading of the wireless environment. However, this protection poses an overhead for the system and the final data transfer speeds the user sees are around 9.6 kbps. Uplink and downlink channels are realized over different frequency carriers. Thus, transmissions occur in full-duplex mode.

Figure 4.16 shows a CDPD network. It is comprised of three kinds of stations and three different interfaces. These are briefly summarized below:

- *Mobile end systems.* These are actually CDPD MSs.
- *Mobile data base systems.* These are CDPD BSs.
- *Mobile data intermediate systems.* These are CDPD Base Interface Stations (BIS). Such a station interfaces all the BSs in a CDPD network to a fixed router in order to provide connectivity to backbone networks (such as the Internet).
- *The E interface.* This connects a CDPD area to a fixed network.
- *The I-interface.* This connects two CDPD areas. It allows roaming of CDPD MSs.
- *The A-interface.*This is the air interface between the BSs and the MSs.

Data is sent over the CDPD network into blocks of 420 bits. These blocks are produced by (a) wrapping up 274 compressed and encrypted user data bits into 378-bit blocks using a Reed–Solomon error correcting code and (b) adding 7 flag words, each being 6 bits long, to the output of the Reed–Solomon coder. This procedure produces a 420-bit block referred to as microblock. Microblocks are what is actually sent over the uplink and downlink channels.

The operation of the CDPD downlink is a relatively trivial application. This is due to the fact that the BS is the only sender in the downlink and thus channel access conflicts do not occur. The BS just broadcasts packets on the downlink and it is up to the intended receiver to receive the packet destined for itself. Obviously CDPD supports broadcast.

However, the operation of the uplink is more complicated. This is due to the fact that in this case, there is more than one potential sender. These are the CDPD MSs. Thus, in order to coordinate uplink channel access, CDPD utilizes the Digital Sense Multiple Access (DSMA) protocol, a variation of CSMA. DSMA is a slotted protocol, which is very similar to the CSMA/CA protocol that was described in Chapter 2. CSMA/CA will be revisited in Chapter 9. In DSMA, when a CDPD MS wants to send a frame, it senses the channel. Channel sensing is of virtual nature, meaning that the CDPD MS is informed on the status of the uplink channel via examination of a certain flag of the downlink channel. This flag reflects the status of the uplink channel. If the uplink channel is found idle, the MS will perform a microblock transmission. Otherwise, it executes a binary exponential backoff algorithm (this was described in Chapter 2) and retries.

Nevertheless, it is obvious that collisions can still appear due to the fact that more than one CDPD MSs may detect a busy uplink and calculate the same backoff period. Thus, there must be a mechanism that enables CDPD stations to acquire knowledge regarding the success or failure of their microblock transmission. This is realized via setting the value in a certain flag of each downlink microblock. The flag regarding transmission of uplink microblock n reaches the MS just before microblock $n + 2$ is transmitted. Thus, if a CDPD station has more than one microblock to send, it sends them without trying to re-acquire the channel. If the collision/success notification states that microblock n did not suffer a collision, the MS continues to transmit its microblocks sequentially; otherwise it stops its transmissions.

4.6.2 HCSD

HSCSD is a very simple upgrade to GSM. Contrary to GSM, it gives more than one time slot per frame to a user; hence the increased data rates. HSCD allows a phone to use two, three or four slots per frame to achieve rates of 57.6, 43.2 and 28.8 kbps, respectively. Support for asymmetric links is also provided, meaning that the downlink rate can be different from that of the uplink. A problem with HSCSD is the fact that it decreases battery life, due to the fact that increased slot use makes terminals spend more time in transmission and reception modes. However, due to the fact that reception requires significantly less consumption than transmission, HSCSD can be efficient for web browsing, which entails much more downloading than uploading.

4.6.3 GPRS

GPRS [6–8] operation is based on the same principle as that of HSCSD: allocation of more slots within a frame. However, the difference is that GPRS is packet-switched, whereas GSM and HSCSD are circuit-switched. This means that a GSM or HSCSD terminal that browses the Internet at 14.4 kbps occupies a 14.4 kbps GSM/HSCSD circuit for the entire duration of the connection, despite the fact that most of the time is spent reading (thus downloading) Web pages than sending (thus uploading) information. Therefore, significant system capacity is lost. GPRS uses bandwidth on demand (in the case of the above example, only when the user downloads a new page). Is GPRS, a single 14.4 kbps link can be shared by more than one user, provided of course that users do not simultaneously try to use the link at this speed; rather, each user is assigned a very low rate connection which for short periods can use additional capacity to deliver Web pages. GPRS terminals support a variety of rates, ranging from 14.4 to 115.2 kbps, both in symmetric and asymmetric configurations.

The two new network elements that are introduced with GPRS to the GSM architecture are:

- *The Serving GPRS Support Node (SGSN).* This provides authentication and mobility management. At a high level, the SGSN provides similar functionality to the packet data network that the MSC/VLR provides to the circuit-switched network.
- *The Gateway GPRS Support Node (GGSN).* This provides the interface between the mobile and the backbone IP or X.25 network. The GGSN tunnels packets from the packet data network using the GPRS tunneling protocol. When a mobile wants to send data, it must set up what is referred to as a packet data protocol (PDP) context between the SGSN and the GGSN, which is more or less equivalent (at least in the context of IP) to obtaining an IP address. After setting up a PDP context, the mobile can then begin using GPRS point-to-point or point-to-multipoint services.

GPRS maintains the modulation scheme (GMSK) and time-slot structure employed in GSM. GPRS defines a new burst mode of operation for packet data transfer in which bursts consist of 456 bits of coded information interleaved across the equivalent of four time slots (4 slots with 114 bits each). GPRS defines four different levels of coding, CS-1 to CS-4, each of which offers a different level of protection.

GPRS cells may dedicate or share one or more physical channels, called Packet Data Channels (PDCH), for packet-switched services within the cell. The PDCH consists of three logical channels. GPRS bursts are carried over the logical packet data traffic channel (PDTCH). In addition to the logical PDTCH, GPRS defines a logical packet broadcast control channel (PBCCH) as well as a Packet Common Control Channel (PCCCH). The PBCCH is used to convey system information to the mobile. The PCCCH serves multiple functions. On this channel, a mobile listens or 'camps' awaiting the arrival of signaling messages – for example, a paging message indicating the start of a packet transfer. Not all cells will necessarily have a PCCCH. In this instance, a mobile will listen to the standard GSM common control channel instead. A mobile also uses the PCCCH to respond to a paging message as well as signal the base station when it wishes to initiate a packet transfer. Finally, the PCCCH is also used by the base station to indicate uplink resource allocations that have been made to a particular mobile for packet data transfer. The reservations are ultimately transferred on the PDCH using the uplink.

GPRS defines three classes of terminals: A, B, and C. A class A terminal supports simultaneous circuit-switched and packet-switched traffic. Thus a user of such a terminal can simultaneously talk and browse the Internet. A class B terminal can be attached to the network as both a circuit-switched and packet-switched client but can only support traffic from one service at a time. Thus, when a user of such a terminal receives a call, his Internet connection is suspended. Finally, a class C terminal uses only packet-switched services. Thus, when a user of such a terminal receives a call, his Internet connection is dropped.

4.6.4 D-AMPS+

Similar to HSCSD and GRPS in GSM, an enhancement of D-AMPS for data, D-AMPS+ offers increased rates, ranging from 9.6 to 19.2 kbps. These are obviously smaller than those supported by GSM extensions.

D-AMPS+ will be based on the GPRS architecture. To implement GPRS-136, a new physical packet data channel (PDCH) is defined, as in the GPRS/GSM case. During operation, the MS continually listens to the PDCH. The PDCH consists of two logical channels on the uplink: a random access channel and a logical payload channel. On the downlink, the packet data channel consists of four logical channels: broadcast, paging, payload, and feedback. The broadcast channel is used for conveying system information, and the paging channel is used for tunneling IS-41 signaling messages to the mobile over the PDCH. The downlink payload channel is self-explanatory and the packet channel feedback (PCF) is used to acknowledge previously sent packets as well as allocate additional slots to a particular mobile on the uplink. The PCF field occupies 24 bits in one downlink time slot. The first 12 bits are used to either acknowledge (ACK) or negatively acknowledge (NACK) the data sent in the previous uplink slot. The last 12 bits are used for reservation of the uplink. The PCF can either indicate that the next available slot is reserved for a given user or indicate that the slot is not reserved, thereby indicating that a mobile may attempt a random access on the PDCH.

Another key enhancement used to provide higher data rates over the PDCH is the use of adaptive modulation. Depending on channel conditions, the data within a given time slot will be encoded using either $\pi/4$-DQPSK or 8-PSK. When 8-PSK is used, not all fields are encoded using 8-PSK; some fields, such as those containing synchronization information, are still modulated using $\pi/4$-DQPSK, in order to provide backward compatibility with handsets that only support $\pi/4$-DQPSK. In addition, allowances have been made to incorporate both fixed coding and incremental redundancy schemes with both types of modulation, 8-PSK, when used in conjunction with incremental redundancy achieves the highest data rates since more bits are encoded per modulated symbol and redundant bits are only transmitted on an as needed basis.

4.6.5 cdmaTwo (IS-95b)

An extension of cdmaOne, known as cdmaOneb or cdmaTwo [9,10], offers support for 115.2 kbps by letting each phone use eight different codes to perform eight simultaneous transmissions. Thus, cdmaOneb actually allows more than one simultaneous CDMA transmission by the same station.

On the uplink, an MS is initially assigned a fundamental channel code. When the MS has data to transmit, it uses the fundamental channel code to set up a channel in order to request additional codes from the BS. The BS informs the MSC and it is up to the MSC to coordinate access among other active mobiles. If the MSC decides to grant additional codes to the MS, it relays a supplemental channel assignment message (SCAM) to the MS via the BS. The SCAM indicates up to seven supplemental channel codes (in addition to the fundamental channel code) that will be used by the MS. Each of the assigned supplemental channel codes is based on a shift of the fundamental channel code.

On the downlink, the MSC informs the MS that it should prepare to receive a data burst by transmitting a SCAM message. In this message, the MSC indicates the number of channel codes (up to eight) as well as the actual Walsh codes to be used for each channel.

4.6.6 TCP/IP on Wireless-Mobile IP

Wireless networks pose an extremely interesting field for the Internet protocols, especially for TCP applicability [11,12]. One approach to supporting the wireless environment is to terminate the use of TCP at some point in the wireless network infrastructure and use a different protocol over the wireless link. Thus, the transport protocol used within the wireless environment is a transport protocol that has been specifically adapted to the wireless environment. The most common implementation of this approach is to extend a Web client into the mobile wireless device, using some form of proxy server at the boundary of the wireless network and the Internet. This is actually the approach followed by the Wireless Access Protocol, which is described in the next section.

Alternatively, TCP may be used end-to-end. This allows mobile wireless devices to function as any other Internet-connected device. However this approach faces some difficulties due to the fact that TCP was originally developed for wired networks. Thus, it makes certain assumptions, which do not apply in a wireless environment. Below we list some of these assumptions along with a brief comment on their truth in a wireless environment:

- *Packet loss is the result of network congestion.* This is hardly true in a wireless network, where packet losses are mainly due to the increased Bit Error Rate (BER) of the wireless medium. In such an error-prone environment, the TCP sender initially attempts fast retransmit of the missing segments; when this does not correct the condition, the sender will experience an ACK time-out. This will cause the sender to collapse its sending window and continue to retransmit from the point of packet loss via the slow-start mode. Obviously, the problem here is that TCP mistakes the packet losses for congestion and thus will lower the amount of data it 'sends' to the wireless host. Sadly, this is turn lowers throughput even more.
- *Round-trip times (RTT) have some level of stability.* Due to the increased occurrence of frame reception errors in a wireless network, the link-level protocol uses a stop and wait protocol. This protocol automatically retransmits corrupted data, resulting in this wireless segment latency which has high variability.
- *Link bandwidth is constant.* As many wireless networks utilize adaptive coding and modulation techniques that adapt to the wireless link BER, both coding overhead and bits per baud in transmission are variable, resulting in nonconstant link bandwidths.
- *Session durations will justify the initial TCP handshake overhead.* This is not true in wireless environments, which typically support short-duration sessions.

Several schemes have been proposed to improve performance of TCP over wireless links. These can be divided into two classes. In the first, the TCP sender is unaware of the losses due to wireless link so the TCP at the sender need not be changed. In the second class, the sender is aware of the existence of the wireless link in the network and attempts to distinguish the losses due to the wireless link from that due to congestion. So the sender does not invoke congestion control algorithms when the data loss is due to the wireless link.

4.6.6.1 MobileIP

In conventional fixed networks, the IP address of each host identifies the point of attachment to the network. This poses a problem for applying IP to the mobile environment, since it is undesirable that a change of location results in a change of IP address. Equivalently, we want a scheme where a host maintains its IP address even after changes at its point of attachment to the network.

The above problem is solved by an extension of IP, MobileIP. Terminal mobility in Mobile IP includes both roaming and handoff. Roaming takes place when a MS powers on, and registers, in a new network. Handoff takes place when the MS moves between the coverage areas of different BSs while maintaining connections. An important aspect of Mobile IP is that an MS can communicate with terminals not implementing the mobility extension to IP as well as with other MSs.

In MobileIP, the routing of datagrams to and from an MS away from its home network (where its original point of attachment, which also gives it its IP address, is located) takes place following a successful registration with the Home Agent (HA). The HA is typically a router that is responsible for sending traffic to the MS when the latter is not in the home network. Furthermore, the HA is responsible for maintaining location information regarding the MSs of its network. Each network that allows its users to roam in another area has to create a HA.

Each 'foreign' network that allows users to roam in its area has to create a Foreign Agent (FA). Whenever an MS roams into a 'foreign' network, it contacts the FA and registers with it. The FA then contacts the MS's HA and gives it a 'Care-of Address'. This is the address of the FA. In order for datagrams to be sent to a station that has roamed into a foreign network, the following events take place:

- A datagram to the MS arrives on the home network via standard IP routing.
- The datagram is intercepted by the HA and is tunneled to the 'Care-of Address'.
- The datagram is de-tunneled and delivered to the MS.
- Datagrams sent by the MS are routed via standard IP routing to their destination.

When MobileIP is used for micromobility support, it results in high control overhead due to frequent notifications to the HA. Also, in the case of a Quality of Service (QoS) enabled MS, acquiring a new care-of address on every handoff would trigger the establishment of new QoS reservations from the HA to the FA even though most of the path remains unchanged. Thus, while Mobile-IP should be the basis for mobility management in wide-area wireless data networks, it has several limitations when applied to wide-area wireless networks with high mobility users that may require QoS.

4.6.7 WAP

The Wireless Application Protocol (WAP) [13–15] is a data-oriented service that targets easy access to Internet services via cellular phones. Nokia, Ericsson, Motorola and Unwired Planet started WAP with the formation of the WAP Forum. Actually what preceded the formation of the Forum was a competition initiated by Omnipoint to develop a mobile web service. These four companies initially submitted their own proprietary standards. Such a divergence, however, was not accepted by Omnipoint, which asked the four companies to work together on a single standard. Thus, WAP emerged.

The WAP protocol stack follows the OSI model, although not exactly. Most of the first WAP products ran over GSM connections. However, this was inefficient both in terms of speed and billing. The latter is due to the circuit-switched connection of GSM which charges the user based on the duration of the connection. The most promising technology to 'carry' the WAP traffic is currently the packet-switched GPRS, although WAP can also function over CDPD.

The following entities are defined in the WAP:

- *Micro-browser*. This can be compared to a standard Internet browser such as Netscape Navigator. However, the micro-browser has fewer capabilities than conventional browsers.
- *Wireless Markup Language (WML)*. This is a scripting language, similar to JavaScript, which defines the way information appears on the micro-browser.
- *Wireless Telephony Application (WTA) interface*. This is an interface that allows WAP to access several phone features, such as the telephone directory.
- *Content formats*. Several predefined content formats.
- *A layered telecommunication stack*. This provides for transport, security encryption, etc.

4.7 Cordless Telephony (CT)

Cordless telephony systems differ significantly from cellular systems. The main difference is the fact that while CTs are optimized for low complexity equipment and high-quality speech in a relatively confined static environment (regarding user speeds), cellular systems target maximization of bandwidth efficiency and frequency reuse in a macrocellular, high-speed fading environment. The more aggressive requirements of cellular systems are obviously reflected in an increase in the complexity and cost of cellular BSs and MSs over CT ones.

4.7.1 Analog CT

Cordless telephones (CTs) first appeared in the 1970s and since then have experienced a significant growth. They employed analog voice transmission and were originally designed to provide mobility within small coverage areas, such as homes and offices. Cordless telephones comprise a portable handset, which communicates with a BS connected to the Public Switched Telephone Network (PSTN). Thus, cordless telephones primarily aim to replace the cord of conventional telephones with a wireless link.

Since 1984 analog cordless telephones in the United States have operated on ten frequency pairs in the 46.6–47.0 MHz band (BS transmit) and 49.6–50.0 MHz band (handset transmit). Prior to 1984, five of the 49 MHz frequencies were paired with five frequencies near 1.6 MHz. However, this configuration provided inferior performance due to the different performance of the two links. These analog telephones used Frequency Modulation (FM) for carrying the analog voice signals. Due to the widespread use of these devices, FCC licensed 15 additional frequency pairs in 1992 in order to serve high-density areas. Despite the emergence of digital CTs, these analog telephones working in the 49 MHz band are still popular in the United States due to their low cost.

A standard similar to that in the United States appeared in Great Britain. This standard, also known as CT0 allowed eight channel pairs near 1.7 MHz (base unit transmit) and 47.5 MHz (handset transmit). Most CT0 units could access only one or two channel pairs. A similar standardization approach was adopted in France. In the rest of the Europe, the analog cordless standard known as CEPT/CT1 has been adopted. This allows forty duplex channels, each 25 kHz wide and the system works in the 914–915 and 959–960 MHz bands.

4.7.2 Digital CT

Analog CTs suffered poor call qualities, since handsets were subject to interference. This situation changed with the introduction of the first generation of digital cordless telephones, which offer voice quality equal to that of wired phones. Such a standard was the CT2/Common Air Interface (CAI), the most important features of which are digital transmission of voice. CT2 operates in the frequency area between 864 and 868 MHz and uses 40 100 kHz channels. Voice is digitized via a 32 kbps Adaptive Differential Pulse Code Modulation (ADPCM) encoder. The resulting bitstream is then compressed and transmitted along with control data over the air at a rate of 72 kbps rate via Gaussian Frequency Shift Keying (GFSK). The control data bits are protected against errors. BS to MS and MS to BS traffic is separated via using a Time Division Duplex (TDD) access scheme. The system can utilize both transmission and reception antenna diversity in order to combat the fading wireless environment. Finally, CT2 support data transmission up to 32 kbps. However, the standard does not provide for mobility. An effort to solve this problem is made by the CT2+ standard which dedicates 5 of its 40 channels for signaling. These signaling channels support location registration, location updating and paging,

Furthermore, several digital CTs offered additional features such as the ability for a handset to be used outside of a home or office. These systems are also known as telepoint systems and allowed users to use their cordless handsets in places such as train stations, busy streets, etc. The advantages of telepoint over cellular phones were significant in areas where cellular BSs could not be reached (such as subway stations). If a number of appropriate telepoint BSs were installed in these places, a cordless phone in range of such a BS could register with the telepoint service provider and be used to make a call. However, the telepoint system was not without problems. One such problem was the fact that telepoint users could only place and not receive calls. A second problem was that roaming between telepoint BSs was not supported and consequently users needed to remain in range of a single telepoint BS until their call was complete. Telepoint systems were deployed in the United Kingdom where they failed commercially. Nevertheless, in the mid-1990s, they did better in Asian countries due to the fact that in these deployments, telepoint could also be used for other services (such as dial-up in Japan). However, due to the rising competition by the more advanced cellular systems, telepoint is now a declining business.

4.7.3 Digital Enhanced Cordless Telecommunications Standard (DECT)

The evolution of digital cordless phones led to the DECT system [16]. This is a European cordless standard for transfer of both voice and data that provides support for mobility. DECT is not only intended for CT but also for applications like Wireless Local Loop (WLL), telepoint, etc. A building can be equipped with multiple DECT BSs that connected to a Private Brach Exchange (PBX). In such an environment, a user carrying a DECT cordless handset can roam from the coverage area of one BS to that of another BS without call disruption. This is possible as DECT provides support for handing off calls between BSs. In this sense, DECT can be thought of as a cellular system. DECT, which has so far found widespread use only in Europe, also supports telepoint services. In the following subsections a number of features of DECT are discussed, including the air interface, handover and security.

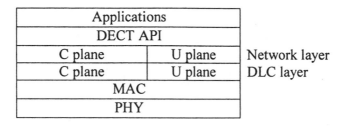

Applications		
DECT API		
C plane	U plane	Network layer
C plane	U plane	DLC layer
MAC		
PHY		

Figure 4.17   DECT protocol architecture

4.7.3.1 DECT Protocol Architecture

The protocol architecture of DECT is shown in Figure 4.17. It is organized around the lower layers of the OSI model. The physical layer is concerned with the actual radio transmission and operates via a TDD method. The physical layer is examined in the next subsection. The Media Access Control (MAC) layer is responsible for managing the connections between DECT MSs and BSs, multiplexing and demultiplexing information over the air interface and to provide (a) broadcast services, (b) connection-oriented services and (c) connectionless services. The Data Link Control Layer (DLC) has the functionality of the corresponding layer of OSI. It is divided in the C and U planes, which provide functionality specific to applications and common to all applications, respectively. The network layer corresponds to the third layer of the OSI model and is responsible for the establishment and release of connections. The C-plane of the network layer offers the following services (among others): (a) management of calls; (b) connection-oriented and connectionless services; and (c) mobility management.

4.7.3.2 Radio Transmission Characteristics

These are defined by the physical layer specification of DECT. According to this, DECT operates in the 1880–1900 MHz band. It digitally encodes speech at a rate of 32 kbps using ADPCM and transmits the resulting bitstream information over the air via GFSK modulation. The DECT air interface utilizes a Multicarrier, Time Division Multiple Access, Time Division Duplex (MC/TDMA/TDD) medium access method. The multicarrier part stems from the use of 10 different frequency channels. Each frame comprises 24 slots with each slot carrying 320 bits of useful data. The first 12 slots of each frame carry traffic from the BS to MSs and the remaining 12, carry traffic from the MSs to the BS. The TDD part stems from this alternation in the direction of traffic within each frame. This MC/TDMA/TDD structure allows up to 12 simultaneous basic DECT (full duplex) voice connections per transceiver. The DECT MC/TDMA/TDD medium access method is shown in Figure 4.18.

Like other systems, DECT uses FEC to protect data, synchronization fields in the DECT frame to provide for BS-MS synchronization and guard intervals to protect successive slots from multipath fading. For data transmission purposes rates up to a maximum of 552 kbps can be achieved. Moreover, DECT BSs can be equipped with antenna diversity. Due to the fact that both the uplink and downlink operate at the same frequency, antenna diversity not only improves the uplink quality but also the downlink quality, at slow speed.

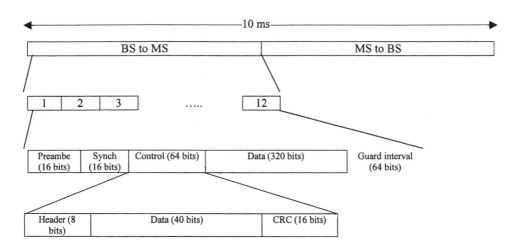

Figure 4.18   The DECT medium access method

4.7.3.3 Handover

DECT MSs can initiate either intracell (to another radio channel within the same cell) or intercell (between BSs of different cells). The two radio links are temporarily maintained in parallel with identical speech information being carried across while the quality of the links is being analyzed. After some time, the BS determines which radio link has the best quality and releases the other link. The decision to perform the handover is initiated by the DECT MS and performed at the time when the signal strength of a new BS becomes higher than that of the current BS.

4.7.3.4 Security

In order to provide for security, DECT includes both subscription and authentication methods. Furthermore, the standard includes an encryption method for user traffic. The subscription process is the process by which the network opens its service to a particular portable. The network operator or service provider provides the portable user with a secret subscription key (PIN code) that will be entered into both the BS and the user's MS before the procedure starts. In order for a DECT MS to initiate subscription, the MS should also know the identity of the fixed part to subscribe to (for security reasons the subscription area could even be limited to a single designated, low power base station of the system). The time to execute the procedure is usually limited and the subscription key can only be used once in an effort to further minimize the risk of misuse. A radio link is set up and both ends verify that they use the same subscription key. Then both the BS and the MS (each on their own) calculate a shared secret authentication key to be used for authentication at every call setup. The authentication procedure that follows is a challenge–response procedure. The BS sends a random number to the MS, which, based on the authentication key and a random sequence, calculates a response message and sends it back to the BS. Upon receipt of this message, the BS compares it with the message that would be expected from a station that shares the same authentication key with the BS. If the BS confirms that the MS possesses the correct key, the MS is authenticated.

During authentication both sides also calculate a cipher key. This key is used to encrypt the transmitted data. At the receiving side the same key is used to decrypt the information. Encryption implementation is not mandatory for DECT units.

4.7.4 The Personal Handyphone System (PHS)

A standard similar to DECT is being used in Japan. This is known as the Personal Handy-phone System (PHS). The objectives of PHS are to be efficient for both home/office use and have public access capability. The potential subscriber base for PHS is estimated to be 5.5 million in 1998 and 39 million in 2010.

PHS operates in the 1895–1918.1 MHz band and supports 77 channels, each 300 kHz wide. The 1906.1–1918.1 MHz band (40 frequencies) is designated for public systems, and the 1895–1906.1 MHz band (37 frequencies) is used for home/office applications. Like DECT, the PHS standard uses a MC/TDMA/TDD medium access method with the difference that each frequency carrier can carry up to four duplex 32 kbps voice channels rather than 12 and the frame duration is 5 ms instead of 10 ms. Channels are selected based on their signal strength. At the physical layer, information is transmitted using $\pi/4$ DQPSK.

PHS uses a 32 kb/s ADPCM voice coder. To combat medium errors, support for Cyclic Redundancy Check (CRC) is provided, but no error correction code is used. Due to channel reciprocity, which is attributed to the use of TDD and low-speed mobility assumption, transmission diversity is provided on the forward link. Reception diversity at the base station can be used on the reverse link. PHS supports (optionally) handoffs although it is confined to walking speed.

4.8 Summary

The era of mobile telephony as we understand it today, is dominated by second generation cellular standards. This chapter describes several 2G standards:

- D-AMPS, the 2G TDMA system that is used in North America and descends from the 1G AMPS is described in Section 4.2. D-AMPS operates in the 800 MHz band and is an overlay over the analog AMPS network. It maintains the 30-kHz channel spacing of AMPS and uses AMPS carriers to deploy digital channels. Each such digital channel can support three times the users that are be supported by AMPS with the same carrier. Digital channels are organized into frames, with each frame comprising six slots. The actual channel a user sees comprises of one or two slots within each frame. D-AMPS can be seen as an overlay on AMPS that 'steals' some carriers and changes them to carry digital traffic. Obviously, this does not affect the underlying AMPS network, since an AMPS MS can continue to operate. IS-136 is a descendant of D-AMPS that also operates in the 800 MHz bands. However, there are planned upgrades to the 1900 band. While D-AMPS maintains the analog channels of AMPS, IS-136 is a fully digital standard. Both D-AMPS and its successor IS-136 support voice as well as data services. Supported speeds for data services are up to 9.6 kbps.
- cdmaOne, which is the only 2G system based on CDMA is discussed in Section 4.3. It is a fully digital standard that operates in the 800 MHz band, like AMPS. In cdmaOne, multiple mobiles in a cell, whose signals are distinguished by spreading them with different

codes, simultaneously use a frequency channel. Thus, neighboring cells can use the same frequencies, unlike all other standards discussed so far. cdmaOne supports data traffic at rates of 4.8 and 14.4 kbps.

- The widely used Global System for Mobile Communications (GSM) is described Section 4.4. Four variants of GSM exist, operating at 900 MHz, 1800 MHz, 1900 MHZ and 450 MHz. It is a fully digital standard and manages channel access via a TDD mechanism that splits the available bandwidth in the time domain. The resulting access method is actually a hierarchy of slots, frames, multiframes and hyperframes. Apart from voice services, GSM also offers data transfer services. The data speeds a user sees are typically round 9.6 kbps.
- Section 4.5 describes IS-41, which is actually not a 2G standard but rather a protocol that operates on the network side of North American cellular networks.
- Section 4.6 is devoted to data transmission over 2G systems and discusses a number of approaches, including GRPS, HSCSD, cdmaTwo and D-AMPS+. HSCSD is a simple upgrade to GSM that allows each MS to be allocated up to four slots within each frame thus resulting in maximum speeds up to 57.6 kbps. The problem with HSCSD stems from its circuit-switched nature. GPRS also allocates up to eight slots to each MS thus resulting in maximum speeds up to 115.2 kbps. However, it has the advantage of being packet-switched. CdmaTwo is an upgrade of cdmaOne that lets a MS use up to eight spreading codes. This is equivalent to performing more than one CDMA transmission, thus resulting in speeds up to 115.2 kbps. Section 4.6 also discusses the problems faced by TCP in a wireless environment, mobileIP, an extension of the Internet Protocol (IP) that supports terminal mobility and the Wireless Access Protocol (WAP).
- Section 4.7 discusses Cordless Telephony (CT) including analog and digital CT, the Digital European Cordless Telecommunications Standard (DECT) and Personal Handy-phone System (PHS) standards.

WWW Resources

1. *www.gsmworld.com*: this is the web site of the GSM association. It contains information on GSM, GPRS and HSCSD.
2. *www.telecomwriting.com*: this web site contains, among others, information on Second Generation Cellular Networks.
3. *www.cdg.org:* this is the web site of the CDMA development Group. It contains CDMA-system related information, such as information on cdmaOne.
4. *www.etsi.org*: this is the web site of the European Telecommunications Standards Institute (ETSI), a non-profit organization that produces European standards in the telecommunication industry. Information on the ETSI DECT standard can be found here.
5. *www.wapforum.org:* this is the web site of the WAP Forum. The WAP specification can be found here.

References

[1] Rahnema M. Overview of the GSM System and Protocol Architecture, *IEEE Communications Magazine*, April, 1993, 92–100.

[2] Yu J. I. IS-41 for Mobility Management, in *Proceedings of IEEE ICUPC 92*, 158–161.

[3] Black U. *Second Generation Mobile and Wireless Networks*, Prentice Hall.

[4] Salkintzis A. K. Packet Data Over Cellular Networks: the CDPD Approach, *IEEE Communications Magazine*, June, 1999, 152–159.

[5] Badr N. G. Cellular Digital Packet Data CDPD, in *Proceedings of the IEEE 14th Annual International Phoenix Conference*, 1995, pp. 659–665.

[6] Akesson S. *GPRS, General Packet Radio Service*, pp. 640–643.

[7] Ferrer C. and Oliver M. Overview and Capacity of the GPRS (General Packet Radio Service), in *Proceedings of PIMRC 98*, 1998, pp. 106–110.

[8] Granbohm H. and Wiklund J. GPRS-General packet radio service, *Ericsson Review*, 2, 1999, 82–88.

[9] Knisely D., Li Q. and Ramesh N. S. Cdma2000: a Third Generation Radio Transmission Technology, *Bell Labs Technical Journal*, July–September, 1998, 63–78.

[10] Knisely D., Kumar S. and Nanda S. Evolution of Wireless Data Services: cdmaOne to cdma2000, *IEEE Communications Magazine*, October, 1998, 140–149.

[11] Huston G. TCP in a Wireless World, *IEEE Internet Computing*, March/April, 2001, 82–84.

[12] Xylomenos G., Polyzos G. C., Mahonen P. and Saaranen M. TCP Performance Issues over Wireless Links, *IEEE Communications Magazine*, April, 2001, 52–58.

[13] WAP Forum, The WAP Architecture, Version 12 - July, 2001.

[14] Erlandson C. and Ocklind P. WAP - The Wireless Application Protocol, *Ericsson Review*, 4, 1998, 150–153.

[15] Pehrson S. WAP - The catalyst of the mobile Internet, *Ericsson Review*, 1, 2000, 14–19.

[16] DECT Forum. DECT, The Standard Explained, February, 1997.

5

Third Generation (3G) Cellular Systems

5.1 Introduction

The big success of first (1G) and second-generation (2G) wireless cellular systems can be attributed to the user need for voice communication services, a need that follows the 3A paradigm: Anywhere, Anytime, with Anyone. By dialing a friend or colleague's mobile phone number, one is able to contact him/her in a variety of geographical locations, thus overcoming the disadvantage of fixed telephony. For more than a decade, the 2G systems presented in the previous chapters (GSM, IS-136, IS-95) have performed very well as far voice communication is concerned. This has led to 400 million 2G mobile subscribers for the year 2000 with estimates bringing this number up to 1.8 billion for the year 2010. At the same time, the market penetration of 1G systems is following a decreasing path. Figure 5.1 shows the increasing number of worldwide cellular subscribers.

Despite their great success and market acceptance, 2G systems are limited in terms of maximum data rate. While this fact is not a limiting factor for the voice quality offered, it makes 2G systems practically useless for the increased requirements of future mobile applications. It is expected that the increased popularity of both multimedia applications and Internet services will have a significant impact on the world of mobile networks in a foreseeable time period. According to a survey, in the year 2010 about 60% of mobile traffic will concern multimedia applications [1]. People will want to be able to use their mobile platforms for a variety of services, ranging from simple voice calls, web browsing and reading e-mail to video conferencing, real-time and bursty-traffic applications. To realize the inefficiency of 2G systems for such applications, consider a simple transfer of a 2 MB presentation. Such a transfer would take approximately 28 min employing the 9.6 kbps GSM data transmission. It can be clearly seen that future services cannot be realized over the present 2G systems.

Third generation (3G) mobile and wireless networks aim to fulfill the demands of future services. 3G systems will offer global mobile multimedia communication capabilities in a seamless and efficient manner. Regardless of their location, users will be able to use a single device in order to enjoy a wide variety of applications. The term 3G is usually accompanied by some vagueness, as sometimes different people mean different things when they refer to it. 3G was originally defined to characterize any mobile standard that offered performance

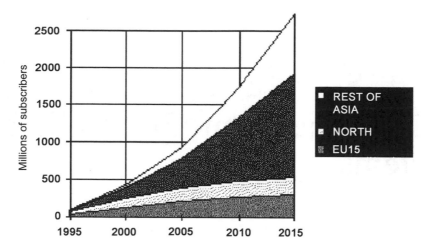

Figure 5.1 Number of cellular subscribers worldwide (Source: UMTS Forum)

quality at least equal to that of ISDN (144 kbps). 3G systems will provide at least 144 kbps for full mobility applications in all cases, 384 kbps for limited mobility applications in macro- and microcellular environments and 2 Mbps for low mobility applications particularly in the micro- and picocellular environments. Those speeds are enough for the support of future mobile multimedia applications. Returning to the example of the previous paragraph, the presentation transfer would take only 8 s over the 2 Mbps link of a 3G system, which results in a significant performance improvement over 2G. It should be noted that speeds similar to those of ISDN are offered by some of the 2.5G standards presented in earlier chapters (GPRS, IS-95B). However, these speeds occur under ideal channel conditions and only match the lower speeds of 3G systems. Some key characteristics of 3G systems are [2]:

- Support for both symmetric and asymmetric traffic.
- Packet-switched and circuit-switched services support, such as Internet (IP) traffic and high performance voice services.
- Support for running several services over the same terminal simultaneously.
- Backward compatibility and system interoperability.
- Support for roaming.
- Ability to create a personalized set of services per user, which is maintained when the user moves between networks belonging to different providers. This concept is known as the Virtual Home Environment (VHE) [3]

.Standardization for 3G systems was initiated by the International Telecommunication Union (ITU) in 1992. The outcome of the standardization effort, called International Mobile Tele-communications 2000 (IMT-2000), comprises a number of different 3G standards. Each of these standards was submitted by one or more national Standards Developing Organizations (SDO). The plurality of standards aims to achieve smooth introduction of 3G systems so that backward compatibility with existing 2G standards is maintained. In order to facilitate the development of a smaller set of compatible 3G standards, several international projects were created, such as the Third Generation Partnership Proposal (3GPP), and 3GPP2. According to the country of deployment, a suitable radio access standard (also known as the air interface) has to be selected, in an effort to provide backward compatibility with existing legacy 2G systems and conform to the country's spectrum regulation issues. As explained in a later section, spectrum assignment for 3G networks is a troublesome activity due to the fact that spectrum is not identically regulated in every country.

The aim of 3G networks is towards convergence. 3G services will combine telephony, the Internet and multimedia services into a single device. It is interesting to note that when the first recommendations for 3G networks were made back in 1992, the Internet was still a tool for the academic and technical society and multimedia applications were much simpler than those of the present day. As a result, the need to support Internet and multimedia was not directly identified in those days. However, this has changed over the years and the present 3G standards will provide efficient support for advanced Internet services like web-browsing and high performance multimedia applications.

5.1.1 3G Concerns

In order to enable the market penetration of 3G data services, pricing schemes that are flexible and appealing to the consumer should be adopted. This, however, poses a problem for the service providers. Data applications, especially multimedia ones, are bandwidth hungry. As the bandwidth is a scarce resource, offering spectrum-demanding data applications will impose a significant cost for the service providers. Thus, pricing schemes that are appealing to both the user and operator communities need to be identified.

As far as battery technology is concerned, it is desirable to have long-life batteries. This results in less maintenance activities (such as recharging) for the user. For 3G data services, the need for increased battery life is even more significant since call durations will be much higher for data than for voice services. However, battery technology improvements occur in small steps. On the contrary, the energy efficiency of new electronics and software shows a significant increase. As a result, the development of more energy efficient electronics and software is desired in order to extend 3G terminal operating times between recharges.

The standardization of APIs for 3G applications will offer the ability to efficiently create 3G applications. Such APIs will allow the abstraction of both the terminal and network, providing a generic way for applications to access 3G services. 3G APIs will enable the rapid use of 3G services, allowing the same application to be used on a wide variety of terminal types.

3G data services will need the development of intelligent new protocols. Most of the protocols used today over wireless links are the same as those used over the wire. However, such protocols do not perform optimally in the wireless environment. Middle-ware protocols try to combat the defects of the wireless link, removing this burden from applications and thus reducing application complexity. The development of efficient middle-ware protocols will significantly improve application performance over 3G systems.

However, applications still need to contain added intelligence. When moving to a location with bad connection quality, the link offered to the user will be inferior in terms of capacity. Applications should posses the intelligence to adapt to such situations by lowering quality or shutting down certain features. For example, a teleconferencing application could compensate for the reduced capacity offered by either initiating a compression increase or shutting off the video feeds.

Another issue regarding intelligence is the ability to create a personalized set of services for each user, which is available at all times. This concept is known as the Virtual Home Environment (VHE). The VHE allows a user to personalize the set of services he has subscribed to and tries to support these services when the user roams between networks of different providers. If the user is using an application but is roaming to a network that does not support it, the VHE will service him by using the application closest to his needs. For example, consider a user that usually exchanges voice mail with his colleagues. The user is on a business trip, which triggers roaming between two networks of different providers. It would be of great benefit to him if the provider of the network he just roamed to offers support for voice mail. However, if this is not possible, the VHE will convert the voice mail messages to text messages and vice versa in an effort to provide support for the voice mail service.

Furthermore, it would be desirable to develop intelligence for the transfer of application states between different terminals. Consider a user of a videoconferencing application. Suppose that the user decides to leave his office in the middle of an ongoing conference, still wanting, however, to participate in the conference while driving to home. It would be nice for him to have the ability of seamless transfer of the videoconferencing application from his computer screen to his mobile phone upon leaving his office. This transfer could take place either directly between the two devices, or through built-in network intelligence.

3G multimedia applications will comprise several video and audio feeds. Returning to the example of the previous paragraph, the ability to seamlessly transport the multimedia feeds of the videoconferencing application among various types of networks (LAN at the office, 3G while driving) will be of significant importance since the properties of various networks may have an impact on the content. A single multimedia session should be served efficiently using a combination of different networks, such as 3G, Ethernet, ATM and X.25.

5.1.2 Scope of the Chapter

The remainder of this chapter provides an overview of the 3G area. In Section 5.2, spectrum regulation issues are examined and the need for additional spectrum is identified. The several candidate extension bands are presented followed by a number of technologies that can alleviate problems attributed to nonuniform worldwide spectrum regulation and spectrum shortage. Section 5.3 begins with a brief explanation of the difference between services and applications and presents the main service classes that will be offered by 3G networks from a capacity point of view. This section also presents some representative 3G applications. Standardization projects and issues, the three 3G air interface standards and the use of ATM and IP technology in the fixed network are discussed in detail in Section 5.4. The chapter ends with a brief summary in Section 5.5.

5.2 3G Spectrum Allocation

5.2.1 Spectrum Requirements

ITU plays an important role in spectrum regulation. ITU licensed a guideline for worldwide IMT-2000 spectrum usage in parts around the 2 GHz band. It would be ideal if every country in the world would follow the ITU guideline. All 3G systems would operate in the same frequency band, a fact that would greatly ease global roaming, especially among operators

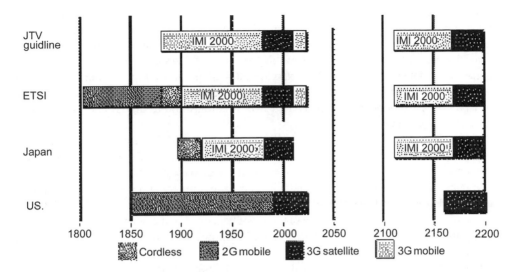

Figure 5.2 Spectrum allocation

following the same IMT-2000 standard. However, the national communications organizations are not bound to follow the ITU guideline exactly. A globally unique spectrum allocation is impossible since most countries have populated the frequency spectrum in different ways according to their needs.

The only country that has exactly followed the ITU guideline is China. Europe and Japan did not fully adapt to the guideline, as part of the IMT-2000 spectrum was already being used for cordless devices and GSM. To make things worse, the entire range of the IMT-2000 spectrum is already in use in America by Personal Communication Services (PCS) and cordless devices. Figure 5.2 shows the current state of spectrum allocation for some of the most economically advanced countries in the world, which have not adapted to the ITU spectrum regulation guideline.

Although in many cases the spectrum proposed by ITU for IMT-2000 is already in use, it is still possible to offer 3G services over the spectrum bands now designated for 2G networks. Some of the 3G standards that are covered later were developed with this approach in mind. In general, we can expect to see two trends followed by 3G operators [4]:

- In countries with parts of the IMT-2000 spectrum partially or fully in use, a migration path will probably be followed by gradually offering 3G services over the spectrum allocated to 2G.
- In countries where the IMT-2000 spectrum is unused, operators will be allocated new spectrum bands, either paired or unpaired, to deploy their 3G systems.

If the predictions of analysts become true, the evolution and market penetration of 3G mobile networks will lead to a huge number of subscribers and a big traffic increase. The spectrum initially allocated to 3G networks will not be able to support the increased traffic [5], thus new bands will have to be made available for use with 3G networks. The exact bands where this spectrum will be allocated are not yet known, however, the following alternatives are under consideration [6]:

- *470–806 MHz*. Better known as UHF, these frequencies are currently used almost world-wide for analog television broadcasting. Replacement of the analog broadcasting by digital television, which will offer better spectrum efficiency and frequency reuse, may offer the possibility of reusing parts of this band for IMT-2000 services. A benefit of this band is its potential for almost worldwide allocation for IMT-2000. Furthermore, its relatively low frequency provides support for long-range coverage, which is beneficial in cases of sparsely populated areas. However, the transition to digital television is unlikely to be completed before 2010.
- *806–960 MHz*. The lower part of this band is already used for television broadcasting. Above 862 MHz this band is used for 2G systems such as GSM. Benefits of this band are the same as those of the previous one (potential for global availability, long-range coverage). On the other hand, apart from television broadcasting, counties using GSM will face additional problems. The GSM part of the whole band will be allocated for IMT-2000 only in the longer term so the spectrum issue will not be solved in the near future. Furthermore, using the GSM spectrum part for IMT-2000 in Europe will not alleviate the problem of obtaining new spectrum for IMT-2000.
- *1429–1501 MHz*. This band is used for several different services over the world. In particular, satellites and terrestrial digital audio broadcasting use the part from 1452 to 1492 MHz. In Japan, a large part of this band is currently used for 2G systems with the prospect of allocating it for IMT-2000 in the future. This band is considered as an extension band outside of Europe.
- *1710–1885 MHz*. Some parts of this band are already in use by existing mobile systems. Such is the case with GSM1800 in Europe. Other parts are used worldwide for air traffic control. A benefit of this band is that it is already a nearly global mobile allocation near the IMT-2000 spectrum. However, the band will be turned over for sole IMT-2000 use only after 2G system operation is discontinued. Furthermore, since already in use by cellular systems, this band does not solve the problem of additional spectrum for 3G cellular use.
- *2290–2300 MHz*. This is a very small band used by about ten stations worldwide for deep space research. In order to use it for 3G systems, coordination with those stations will have to be achieved with separation distances between them and 3G stations of up 400 km. If used for 3G, this band will probably be combined with the adjacent 2300–2400 MHz band.
- *2300–2400 MHz*. This band is currently used for fixed services and telemetry applications. It benefits by being close to spectrum already allocated to IMT-2000 and being wide enough to offer sufficient additional spectrum for 3G. However, interference problems with current services populating the band need to be solved.
- *2520–2670 MHz*. This is the most probable candidate for additional band globally. It is currently used by several countries for broadcasting applications and fixed services, nevertheless the majority of such applications are deployed in the United States. Benefits of this band are its sufficient width and thus support for increased additional capacity.
- *2700–2900 MHz*. This band is used for radar systems, satellite communications and aeronautical telemetry applications. Although the width of this band is sufficient, deployment of radio navigation and meteorological radar systems is expected to increase in the future, making global use of this band for IMT-2000 difficult.

From the above discussion, it can be easily concluded that differences in IMT-2000 bands among different countries cannot be avoided. In order to enable roaming between (i) 3G service providers that use different standards and (ii) countries with providers using the same 3G standard but different spectrum bands, a 3G handset will have to support a number of different standards and operating frequencies. This fact results in a significant difficulty and thus cost increase in the manufacturing process of 3G handsets. A possible solution to this problem is the concept of software-defined radio. This, along with a number of other enabling technologies that can alleviate problems originating from spectrum shortage, are briefly presented in the next section. Although most of these technologies are still in their early stages, they are believed to be of significant potential for the performance improvement of mobile wireless networks [5].

5.2.2 Enabling Technologies

5.2.2.1 The Need for 3G-handset Flexibility: Software-defined Radios

The Software-Defined Radio (SDR) concept [5,7,8] can provide an efficient and relatively inexpensive means to manufacturing flexible handsets. Current 2G products implement digital technologies for the air interfaces in hardware. As a result, most of them can operate using only a single standard or frequency. The diverse range of cellular standards and operating frequencies, however, often frustrate users who lack the ability to roam between different network types without significant adjustment, or even replacement, of their handsets. SDR offers a potential solution to this problem. SDR is based on a common platform that can be fully re-programmed or modified by downloading software over the air. This is different from the seemingly similar functionality of some present handsets. In these cases, several standards are hardwired into the device and standard activation is made over the air.

The adoption of the SDR idea is enabled by the technology evolution and market acceptance of general purpose Digital Signal Processors (DSP). The performance and manufacturing costs of devices based on software or firmware driven re-programmable DSPs reach that of conventional devices implementing functionality in hardware using Application Specific Integrated Circuits (ASICs). The Software-Defined Radio Forum [9] is closely working with 3GPP to enable the use of SDR technology in 3G products.

However, the acceptance of SDR faces significant problems too. The most important are outlined below [5]:

- Implementation using ASICs is a mature technology. When facing the SDR idea, hardware-based solutions may prove to be more cost efficient. This is especially true in cases of products like base stations and infrastructure systems in general, which will probably be used only inside a single network. Most of the time such products will not need to possess the flexibility to support different standards and bands. Without the use of SDR technology, such systems can be manufactured at a lower cost using ASIC technology. The same may hold for mobile terminals. A significant number of cellular users will remain most of the time under the coverage of the same provider and will thus infrequently, or never, need the flexibility of easy roaming between providers using different standards and bands. As a result, such users can choose a cheaper terminal based on ASIC technology.
- As far as energy consumption is concerned, programmable DSPs tend to consume more energy than ASICs. This is a problem for SDR technology considering the fact that advances in battery life are not made at significant rates.
- SDR-based implementations tend to produce terminals with larger sizes.

The conclusion of the above discussion is that SDR will play a complementary role in future wireless product implementation, possibly increasing its market penetration as time passes. The interested reader can seek information on SDR technology in the scientific journals [10–12].

5.2.2.2 The Need for Increased System Capacity: Intelligent Antennas and Multiuser Detection

The aim of intelligent antennas is to provide increased capacity to terminal-base station links. Research in this field has been going on for years yielding a number of techniques, which either explicitly or implicitly try to increase the Signal to Interference Noise Ratio (SINR) at the receiver. Apart from the classic antenna diversity techniques, more advanced techniques have appeared. Examples are the steered-beam and the switched-beam approaches [5]. Both utilize a set of antenna elements organized in columns. The steered approach uses the antenna elements in order to construct a narrow transmission beam directed to the intended mobile and following it as it moves. The switched-beam approach on the other hand, tries to increase the SIR at the mobile receiver by switching transmission to the appropriate antenna element as the mobile moves. Beyond these, even more intelligent techniques have appeared. For example, Bell Labs Layered Space Time (BLAST) [5,7] addresses the problem of multipath propagation by establishing multiple parallel channels between the transmitter and the receiver in the same frequency band. This results in increased capacity by an order of magnitude over other techniques [5].

Multiuser detection addresses CDMA-based systems. It is a promising technique, which aims to reduce co-channel interference between users in the same cell. This idea of the procedure is based on the observation that the signal of a user is just co-channel interference during the detection process of the signal of other users. Considering the case of two co-channel users, the idea of the technique is as follows: after detection of the strongest signal, subtract it from the aggregate received signal before trying to detect the second (weaker) signal. Once the second signal has been detected, subtracting it from the aggregate received signal can lead to a better estimate of the first signal. It is obvious that iteration of this technique can improve user detection. Many variants of this technique exist, aiming either to detect users one by one, or all of them together. A thorough description both of intelligent antennas and multiuser detection is out of the scope of this chapter. The interested user can seek further information in technical articles [7,13–17].

5.3 Third Generation Service Classes and Applications

When 3G standardization activities were initiated by ITU in 1992, only vague ideas existed regarding the type of services and applications that would be supported. Ten years later thoughts on these subjects have matured, despite the fact that we cannot rule out the possibility of future, yet unforeseen, demands.

The difference between services and applications needs to be defined [18]. Apart from the concept of services and applications, this definition entails the concepts of content and device. Services are combination of elements that service providers may choose to charge for separately or as a package. Applications allow services to be offered users. Applications are invisible to the user and do not appear on the bill. What the user sees and pays for is the

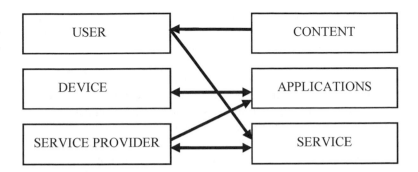

Figure 5.3   Definition of services and applications

content, which is offered through applications running on devices. The definition of services and applications is illustrated in Figure 5.3:

- A user subscribes and pays for services. Services are through applications, which in turn deliver the service content to the user.
- Devices execute the applications needed to deliver the service content.
- The service provider offers services using applications running on devices.

In the remainder of this section we make a brief presentation of the 3G service classes from the point of view of offered capacity. This is followed by a nonexhaustive list of representative 3G applications [18].

5.3.1 Third Generation Service Classes

The deployment of 3G networks does not imply instantaneous change of users demands for certain services. We expect that voice traffic will continue to possess the lion's share in the first years of 3G network operation, with the demand for multimedia services increasing as time passes. In the following, we summarize, in order of increased capacity demand, the main service classes that will be offered by 3G networks [9]. Although none of them are set in hardware yet, they are useful for providers planning coverage and capacity. Furthermore, 3G terminals will probably be rated according to the level of service they offer, providing increased performance/cost ratios to users.

- *Voice and audio.* Demand for voice services was the reason for the big success of 1G and 2G systems. The need for voice communication will continue to dominate the market, accompanied by demands for better quality. Different quality levels for voice communication will be offered, with higher qualities having higher costs. The capacity required by this service class is the lowest, and 28.8 kbps provides substantial support for good quality voice calls.
- *Wireless messaging.* Current 2G systems support rather primitive means of messaging (e.g. the SMS message comprises a maximum of 160 characters). 3G wireless messaging will allow cellular subscribers to use their terminals to read and respond to incoming e-mails, open and process e-mail attachments, and handle terminal-to terminal messages. Depending on the desired speed of message transfer, the capacity demanded by this service class can vary, however, speeds around 28.8 kbps should be more than sufficient.

- *Switched data.* This service class includes support for faxing and dial-up access to corporate LANs or the Internet. As far as file transfer is concerned, speeds like those of today's fast modems (56 kbps) are required in order to shorten the time a user spends on-line and thus the associated cost of file transfers.
- *Medium multimedia.* This should be the most popular service class introduced by 3G. It will enable web browsing through 3G terminals, an application already proving very popular [5]. This service class will offer asymmetric traffic support. This is because in web sessions, the traffic from the network to the terminal (downlink or forward-link traffic) is always much higher than the traffic in the reverse link (uplink or reverse-link traffic). This service class will also support asymmetric multimedia applications such as high-quality audio and video on demand. Speeds up to the maximum (2 Mbps) will be offered at the downlink. Speeds around 20 kbps for the uplink will be enough.
- *High multimedia.* This service class will be used for high-speed Internet access and high quality video and audio on-demand services. It will support asymmetric traffic offering the highest possible bit rates in the downlink. In the uplink, speeds in the order of 20 kbps will suffice.
- *Interactive high multimedia.* This service class will support bandwidth-hungry, high-quality interactive applications offering the maximum speeds possible.

5.3.2 Third Generation Applications

The advanced service classes introduced by 3G networks will enable a wide range of end-user applications that will be either completely new or just mobile versions of applications already running on wired systems. In this subsection, we briefly present some of the 3G applications that will probably be popular among the user community [18].

- *Multimedia applications.* Video telephony and videoconferencing will be typical mobile multimedia applications. The increased capacity offered by 3G systems will enable use of such applications in a cost-efficient manner. Users will be able to participate in virtual meetings and conferences through their 3G terminals. Furthermore, they will have the ability to use audio/visual transport applications that will deliver multimedia content, such as CD-quality music and TV-quality video feeds, from service platforms and the Internet.
- *Mobile commerce applications.* Mobile commerce (m-commerce) is a subset of electronic commerce (e-commerce). m-Commerce will introduce flexibility to e-commerce. As most people keep their handsets with them at all times, they will have the ability to make on-line purchases and reservations upon demand without having to be in front of an Internet-connected PC. Market analysts predict that e-commerce will be a multitrillion dollar industry by 2003. Introducing e-commerce to the mobile platform will be an important source of operator revenue. The increased capacity of 3G systems will offer efficient support for massive use of m-commerce applications.
- *Multimedia messaging applications.* These applications will handle transport and processing of multimedia-enhanced messages. Users will be able to use their 3G terminals to send and receive voice mails and notifications, video feed software applications and multimedia data files. Having a single mailbox on the same terminal for these messages will greatly increase time efficiency for the end user.
- *Broadcasting applications.* Such applications will typically use asymmetric distribution

infrastructures combining high capacities in the downlink with low capacities in the uplink. Multimedia news broadcasting, interactive games and location-based information services, such as flight information in airports are examples of such applications.

- *Geolocation-based applications.* Geolocation technology determines the geographical location of a mobile user. There are two types of geolocation techniques, one based on the handset and the other on the network. The first uses the GPS system to determine user location while in the second, the replicas of the signals from the same handset at different base stations are combined in order to determine user location. Some obvious applications employing geolocation technology include mobile map service and identification of user location for emergency calls.

5.4 Third Generation Standards

5.4.1 Standardization Activities: IMT-2000

ITU is the global organization for development of standards in telecommunications. In the early 1990s, realizing the increased potential of mobile communications, ITU launched a project named Future Public Land Mobile Telecommunications System (FPLMTS), which aimed to unite the world of wireless networks under a single standard. Later, FPLMTS was renamed IMT-2000, with the number 2000 having a three-fold meaning: the year 2000, which was the year that IMT-2000 would become operational according to ITU, data rates of 2000 kbps and global availability of operating frequencies in the 2000 MHz part of the spectrum. None of these goals were entirely fulfilled, nevertheless, the project name remained.

In the beginning, the interest of FPLMTS was solely with advanced next generation cellular networks offering high-speed data services. IMT-2000 was created with the thought of including all wireless technologies, such as Wireless LANs (WLANs), satellite communications and fixed wireless links into a single standard. Despite its elegance and its obvious advantages, this idea was abandoned due to a number of technical issues. For example, fixed wireless links operate with more efficiency in frequencies much higher than those used by 3G mobiles. WLANs, on the other hand, follow their own path to data rates much higher than those offered by 3G standards. As a result, the world of high-speed cellular networks became once again the target of IMT-2000.

In its present version, IMT-2000 aims to be an umbrella for a number of different systems. This concept, known as the 'family of systems' concept was developed in order to ease convergence from existing 2G networks to 3G networks. As different parts of the world are dominated by different 2G standards, the existence of a number of systems under IMT-2000 will enable gradual and cost-efficient transition.

Figure 5.4 shows the various components of the IMT-2000 specification. The Radio Access Network (RAN) comprises a set of interconnected base station controllers each one coordinating a set of base stations. ITU decided not to define the protocol that will be used inside the RAN and the core network in order to allow for reuse of existing infrastructure and evolution of 2G networks according to market needs. Thus, the core networks in Figure 5.4 can be that of GSM, ANSI-41 or an evolved version of either one. The ITU will specify the Network-to-Network Interface (NNI), which is used to connect dissimilar core networks in order to provide roaming capabilities to users moving between cells belonging to different network families.

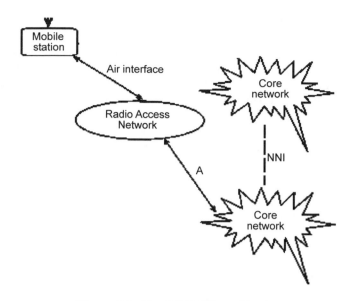

Figure 5.4   The IMT-2000 specification

5.4.2 Radio Access Standards

As far as radio access is concerned, the ITU-Radiocommunication Standardization Sector (ITU-R) issued a call for proposals in 1998 which resulted in ten terrestrial and six satellite proposal submissions by Standards Development Organizations (SDOs) from counties around the world [1]. Although derived from different organizations, several of the proposals where characterized by high commonality. Below, we briefly describe the terrestrial proposals received by the ITU.

The European Telecommunications Standards Institute (ETSI) proposal, also known as the Universal Mobile Telecommunications System (UMTS), calls for use of Wideband CDMA (WCDMA) as the radio access method. The proposal consisted of two WCDMA modes. This dual mode was motivated by the fact that certain frequency bands in Europe are licensed to be used either for uplink or downlink traffic (paired bands) or for both types of traffic using time-sharing (unpaired bands). Thus, ETSI's proposal consisted of Frequency Division Duplex (FDD) WCDMA for paired bands and Time Division Duplex (TDD) WCDMA for unpaired bands.

The Association of Radio Industry Board (ARIB), the SDO in Japan, also proposed WCDMA. The proposal made by ARIB is compatible with that of ETSI. Furthermore, ARIB has halted changes in its WCDMA specification in 1999 so as to enable commercial deployment as soon as possible. As a result, experimental 3G networks are currently starting to deploy in Japan [9].

The United States Telecommunications Industry Association (TIA) made a three-fold proposal: UWC-136, a TDMA-based system which is an evolution of IS-136, cdma2000 as the evolution of IS-95 and a WCDMA system called WIMS. The US T1P1 proposed WCDMA-NA, a FDD WCDMA system.

The Telecommunication Technology Association (TTA) SDO from Korea proposed two systems, one close to the ARIB proposal and the other following closely the cdma200

SDO	Air-interface proposal
ETSI (Europe)	• FDD WCDMA • TDD WCDMA • DECT
ARIB (Japan)	FDD WCDMA
TIA (USA)	• UWC-136 • Cdma2000 • WIMS (WCDMA)
T1P1 (USA)	WCDMA-NA
TTA (Korea)	• WCDMA • Cdma2000
China	TD-SCDMA

Figure 5.5 SDOs and respective radio access proposals to the ITU-R

proposal made by TIA. Finally, China submitted a proposal named TD-SCDMA, which is based on a synchronous TD-CDMA scheme. SDOs and their respective radio access proposals to the ITU-R are highlighted in Figure 5.5.

One can see that the ITU received CDMA-based proposals (WCDMA, TD WCDMA, cdma2000, TD-SCDMA) and TDMA-based proposals (UWC-136 and DECT). In order to facilitate the development of 3G CDMA-based standards, two projects were created. They are the Third Generation Partnership Proposal (3GPP), which deals with WCDMA and 3GPP2, which works on cdma2000. 3GPP and 3GPP2 are working under the coordination of the Operators Harmonization Group (OHG), a group of operators from all the parts of the world who operate different 2G systems (GSM, IS-136, IS-95). The aim of this coordination is to harmonize the IMT-2000 family members and arrive at a point characterized by a smaller number of 3G standards able to operate over different core networks [1,4]. Especially for CDMA-based family members, this harmonization also entails aligning radio parameters as far as possible and developing a combined protocol stack in an effort to enable cost-effective production of dual-mode terminals. Figure 5.6 shows the outcome of the harmonization process. In summary, this effort has resulted in:

- A third-generation TDMA standard being developed for GSM/IS-136 evolution. This is called EDGE/UWC-136.
- A single third-generation CDMA standard with three options: (i) a direct-sequence option based on WCDMA; (ii) a multicarrier option based on cdma2000; and (iii) a TDD direct-sequence mode based on TD-WCDMA.

In general, we can expect to see the following two trends in the coming years:

- Operators following a migration path from 2G to 3G systems. IS-95 and GSM/IS-136 system operators will upgrade their services through introduction of the backwards compatible cdma2000 and EDGE, respectively.

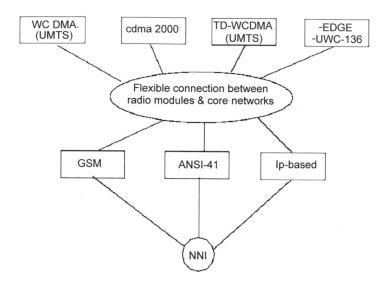

Figure 5.6   Interconnection options

- Operators commencing the deployment of a cellular system from scratch. The dominant interface in this case will be WCDMA.

The EDGE, cdma2000 and WCDMA air access standards are presented in detail in the following subsections. The section ends with a brief discussion on the use of ATM and IP technology in the fixed network part of a 3G system.

5.4.2.1 EDGE

EDGE stands for Enhanced Data Rates for GSM Evolution. It is the only IMT-2000 air interface standard based on TDMA technology. In its initial stages, EDGE was not supposed to be a contender for the CDMA-based 3G standards. Using the same bandwidth and channel structure as GSM, its original purpose was to be the next upgrade for networks supporting GPRS and HSCSD. However, this changed with the adoption of EDGE by the Universal Wireless Communications Consortium (UWCC), an American industry group promoting TDMA technology. Seeing in EDGE the opportunity to upgrade IS-136 to a system of higher speeds, UWCC worked together with ETSI to develop a common TDMA-based 3G air interface. This effort succeeded with the adoption of EDGE as a member of the IMT-2000 air interfaces and its standardization under the name EDGE/UWC-136. The standardization process for EDGE consists of two phases:

- Phase 1 emphasizes increased capacity and spectral efficiency by adopting an enhanced packet-switched mode and an enhanced circuit-switched mode that offer data rates up to 473 and 64 kbps, respectively. In GSM systems, these modes are referred to as Enhanced GPRS (EGPRS) and Enhanced Circuit Switched Data (ECSD). In a IS-136 system, high speed data services are referred to as EGPRS-136HS. The increased capabilities of EDGE modes will be realized through spectrum efficient modulation techniques and radio link control protocol enhancements. Simulation studies [20] have shown that EDGE will enable significantly higher peak rates than GSM and a spectral efficiency three times better than that of GSM.

- Phase 2 will aim to provide support for QoS, real-time and packet-switched voice services as well as interfacing to an all-IP core network.

5.4.2.1.1 EDGE Enhancements over 2G TDMA-based Systems

Physical Layer Enhancements   EDGE uses the GSM bandwidth and 200 kHz channel structure, however, it offers significantly higher spectral efficiency and data rates up to 473 kbps in an effort to enable operators to offer 3G services over the spectrum allocated for 2G TDMA-based systems. If operators of IS-136 and GSM networks lack the ability to use IMT-2000 spectrum, coexistence of 2G and WCDMA-based 3G systems will be difficult, due to the fact that their incompatible channel structures would both have to be supported over the already crowded spectrum. As a result, EDGE is the only candidate to offer support for 3G services in such situations. Of course, EDGE can also be seen as a potential air interface for a new deployment of 3G systems.

The increased performance of EDGE is attributed to modulation techniques of higher level than those of GSM. Apart from reusing the Gaussian Minimum Shift Keying (GMSK) modulation of GSM, EDGE also uses the modulation scheme shown in Figure 5.7, which is known as eight-phase shift keying (8-PSK). In 8-PSK, every transmitted symbol can have eight possible values. It can thus encode three bits per-symbol instead of the one bit per-symbol encoding achieved by GMSK. EDGE maintains the burst format of GSM.

Radio Protocol Enhancements: EGPRS   EGPRS, the packet switched transmission mode of EDGE, will allow for data rates up to 473 kbps. To support higher speeds than GPRS, the EGPRS radio link control mechanism incorporates a number of additional techniques. These techniques are Link Adaptation (LA) and Incremental Redundancy (IR). The aim of LA is to

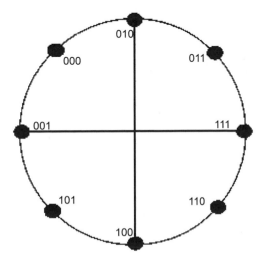

Figure 5.7   8-PSK Modulation

MCS	Slot capacity (Kbps)	FEC overhead	Modulation	Channel capacity (Kbps)
MCS-1	8.8	143%	GMSK	70.4
MCS-2	11.2	91%	GMSK	89.6
MCS-3	14.8	45%	GMSK	118.4
MCS-4	17.6	22%	GMSK	140.8
MCS-5	22.4	187%	8-PSK	179.2
MCS-6	29.6	117%	8-PSK	236.8
MCS-7	44.8	43%	8-PSK	358.4
MCS-8	54.4	18%	8-PSK	435.2
MCS-9	59.2	8%	8-PSK	473.6

Figure 5.8   EGPRS modulation and coding schemes

use estimates of link quality in order to adapt the coding and modulation of the transmitted packets. When a poor link quality is experienced, the radio link control protocol will use GMSK for modulation, which is less susceptible to errors than 8-PSK. However, in cases of links of good quality, the more efficient 8-PSK will be selected. IR is an enhanced ARQ technique. IR initially transmits packets with little coding overhead in an effort to provide higher rates to the user. If the decoding fails, however, packets are retransmitted with additional coding bits and thus higher overhead in an effort to achieve successful reception.

The radio link control protocol of EGPRS supports a combination of LA and IR where the initial Modulation and Coding Scheme (MCS) is based on measurements of the link quality. When the decoding of a packet fails, successive retransmissions for that packet will use an MCS that offers increased protection. This process is repeated until successful decoding of the packet is achieved. As a result, high throughput and robust transmissions will be possible across a diverse range of link conditions. The nine currently defined MCSs are shown in Figure 5.8 along with the respective Forward Error Coding (FEC) overhead, slot and channel capacities. As in GPRS, more than one slot can be allocated to a user in order to meet increased capacity demands.

Radio Protocol Enhancements: ECSD   The ECSD mode of EDGE keeps the existing GSM circuit-switched data protocols intact. The introduction of 8-PSK does not change the data rates offered, however, it enables a more efficient use of the spectrum. For example, while four time slots in GSM serve a 57.6 kbps circuit-switched connection, the same service will use only two slots with ECSD.

5.4.2.1.2 EDGE Classic and EDGE Compact   EDGE development for IS-136 based systems comprises two modes: Compact and Classic. EDGE Compact uses a new 200 kHz control channel structure. By means of base station synchronization and use of a 1/3 frequency reuse pattern, EDGE Compact can be deployed even in only 600 kHz of available bandwidth. EDGE Classic, on the other hand, maintains the traditional GSM control-channel structure used by the ETSI standard with a 4/12 reuse pattern.

EDGE Classic uses the same channel structure as GSM, which typically uses a 4/12 frequency reuse pattern for carriers containing broadcast control channels and a 3/9 pattern for traffic channels. Minor modifications over the ETSI EDGE standard are related to IS-136 information exchange. These modifications enable EDGE Classic to re-use the existing IS-136 30-kHz wide control signaling system for operations such as voice call establishment and termination. As the channel width of EDGE is 200 kHz one can see that operators will need at least 2.4 MHz of available bandwidth to offer EDGE support in their networks. While this fact is not a problem for most countries, operators in North America suffer from spectrum limitations mainly due to the FCC's decision to allocate 3G spectrum to 2G GSM 1900 operators. In those cases, operators have to offer EDGE support using only three 200 kHz channels in a 1/3 reuse pattern. Although efficient in terms of spectrum usage, a 1/3-reuse pattern is too low for control channels to operate reliably. A 4/12 or 3/9 reuse pattern is required for reliable control channel operation.

EDGE Compact solves the problem described above. The changes introduced above EDGE Classic concern only the way bandwidth for control channels is reused. By using synchronization between base stations, EDGE Compact constructs time groups in an effort to turn the 1/3 frequency reuse pattern for the packet common control channels and the broadcast control channels into a 4/12 pattern. Synchronization can be achieved by using Global Positioning System (GPS) receivers.

Figure 5.9 shows an example with four timing groups in addition to 1/3 frequency reuse to obtain a 4/12-reuse scheme for control signaling. In this example, sectors operating on a specific frequency share this frequency in the time domain. Inside the 12-sector cluster outlined in the figure, sectors use frequencies F1, F2 and F3 for control signaling in turn. As far as control information channels are concerned, each frequency–time group combination inside the cluster is unique, resulting in a 4/12 reuse. The adoption of the time group concept by Compact results in modifications for all the packet common control channels of GPRS. These modified channels are known within Compact as:

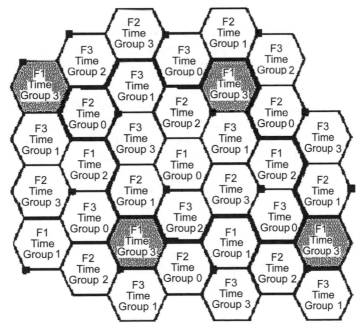

Figure 5.9   Obtaining a 4/12 reuse pattern with three frequency carriers and the use of time groups

- Compact Packet Paging Channel (CPPCH)
- Compact Packet Access-Grant Channel (CPAGCH)
- Compact Packet Random-Access Channel (CPRACH)
- Compact Packet Broadcast Channel (CPBCCH)
- Packet Timing-advance Control Channel (PTCCH).

Finally, it should be noted that the time group structure does not affect the data traffic channels in Compact, which continue to employ a 1/3-frequency reuse pattern. However, when a data transmission will collide with ongoing control information transmission in neighboring sectors, the data transmission will not be performed.

5.4.2.2 cdma2000

Among 2G systems only the IS-95 family, also known as cdmaOne, is based on CDMA technology. This is a significant advantage for IS-95 providers since upgrades to 3G CDMA-based systems will only require software and minor hardware changes to the existing CDMA-based networks. Cdma2000 comprises a family of backwards-compatible standards, a fact that enables smooth transition of 2G CDMA-based networks to 3G networks. Although cdma2000 can be used as the air interface of pure 3G network installations that use the IMT-2000 spectrum, its main advantage is the ability of overlaying cdma2000 and IS-95 2G systems in the same spectrum. This is a very important aspect for IS-95 providers in North America due to fact that the spectrum specified by ITU for IMT-2000 is already in use in these areas. Thus, one can see the reason for cdma2000 backward compatibility: North American providers will offer 3G services by deploying an overlay of cdma2000 and IS-95 in the same bands.

Figure 5.10 shows the two lower layers of the radio interface protocol architecture of cdma-2000. The cdma2000 specification provides protocols and services that correspond to the two lower layers of the OSI model. In the next sections, we cover issues related to the physical (layer 1) and data link layer (layer 2) operation and briefly present the main channels of each layer [21].

5.4.2.2.1 Cdma2000 Physical Layer Issues

The original cdma2000 specification contained two spreading modes, multicarrier and Direct Spread (DS). However, the ongoing harmonization work stated that WCDMA should be used as the DS mode, thus putting an end to work on cdma2000 DS. There are two non-DS cdma2000 modes, 1X and 3X. The 1X mode uses a single cdmaOne carrier, while 3X is a multicarrier system. This means that cdma2000 terminals and base stations based on 3X will use three of the 1.25-MHz wide IS-95 carriers. 1X and 3X are the two modes currently standardized, although modes such as 6X, 9X and 12X may be standardized in the future. As far as carrier chip-rates are concerned, a multicarrier transmission using N cdmaOne carriers de-multiplexes the message signal into N information signals and spreads each of these on a different carrier, at a chip rate of 1.2288 Mcps per-carrier. In this approach, each carrier has an IS-95 signal format. In the DS approach, a chip rate of $N \times 1.228$ Mcps is used ($N = 1, 3, 6, 9, 12$) and the spread signal is modulated onto a single carrier [22].

1X is the simplest version of cdma2000. Despite the use of only one IS-95 carrier, 1X approximately doubles the voice capacity of cdmaOne systems and provides average rates for

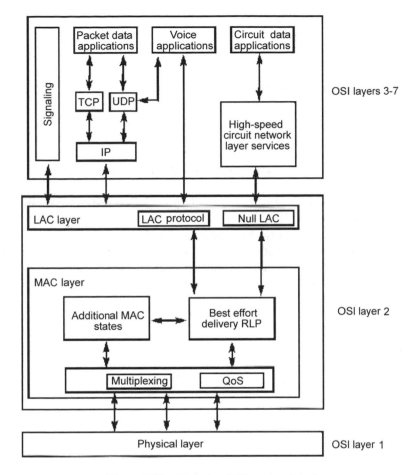

Figure 5.10   The cdma2000 protocol stack

data services up to 144 kbps. This performance gain is attributed to the enhancements of cdma2000 layers 1 and 2 over the corresponding layers of cdmaOne. High Data Rate (HDR) is an enhancement of 1X for data services. Instead of Quadrature Phase Shift Keying (QPSK) used by 1X and 3X, HDR uses the more efficient 16-Quadrature Amplitude Modulation (16-QAM) which codes four bits per transmitted symbol, thus offering speeds up to 621 kbps. However, in cases of heavy interference, HDR modulation drops down to the more robust 8-PSK or QPSK, a fact that decreases the data rates offered.

3X, also known as IS-2000-A, is an enhancement of 1X that uses three cdmaOne carriers for a total bandwidth of approximately 3.75 MHz. It offers greater capacities than 1X and can support data rates up to 2 Mbps. This performance increase is accomplished by multicasting the downlink traffic over the three 1.25 MHz carriers. Due to the terminal complexity induced by multicarrier transmission, 3X uses direct spreading for uplink transmission, which produces a wideband signal that matches the rate of the downlink signaling. ($3 \times 1.2288 = 3.6864$ Mcps). This rate is slightly lower than the 3.84 Mcps rate of the WCDMA-compatible

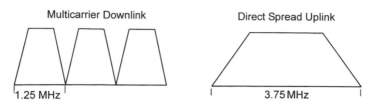

Figure 5.11   Downlink and uplink channels of 3X

cdma2000 DS mode. Figure 5.11 displays the bandwidth of the downlink and uplink channels of a 3X system.

Cdma2000 supports both Frequency Division Duplex (FDD) and Time Division Duplex (TDD) configurations. In the FDD mode, different frequency channels carry the uplink and downlink traffic, whereas in TDD systems, the uplink and downlink are transmitted over the same frequency channels and separated my means of time sharing. TDD systems are valuable in environments where only unpaired frequency bands are available. TDD and FDD systems share the same coding schemes, modulation methods and channel structures described in this section. TDD systems introduce only minor modifications for guard timing. However, as unpaired bands are available mostly in Europe, the vast majority of cdma2000 systems will be of FDD nature.

The cdma2000 physical layer is an enhancement over the corresponding layer of cdmaOne. It supports a number of physical channels for the uplink and downlink. Physical channels can either belong to a specific mobile (dedicated channels), or be of shared access among many mobile stations (common channels). In the remainder of this section, after presenting some of the main characteristics of cdma2000 downlink and uplink [22], we briefly summarize the main cdma2000 physical channels [21].

Downlink Characteristics

- *Transmit diversity.* One enhancement to the downlink design of cdma2000 systems is the use of transmit diversity at the base station. Of course this approach will be effective if diversity is employed on mobile receivers too. It applies to both the multicarrier and the DS approaches. In the multicarrier approach, the base station will use different, spatially separated antennas, to transmit the multiple subcarriers. Signals originating from the different antennas will fade independently, thus increasing frequency diversity. In the DS approach, the base station spreads the data stream into two substreams, which in turn are transmitted via separate antennas. Orthogonality between the two output streams is maintained, since each antenna uses a different code for spreading the transmitted data stream.
- *Fast power control.* Cdma2000 calls for use of closed-loop fast power control in the downlink. Mobile stations measure the power of downlink traffic and issue 'power-up' or 'power-down' commands to the base station according to the measurements. The closed loop power control compensates for medium to fast fading and achieves significant performance improvements for high-speed transfers in low-mobility environments.
- *Common pilot and auxiliary pilots.* Cdma2000 uses a common code multiplexed pilot for all users on the downlink. Mobile nodes within a cell share this channel in order to obtain

information about multipath fading and channel condition. Optional use of auxiliary pilots supports smart antenna systems.

- *Synchronized base station operation.*Synchronization among base stations, as in cdmaOne, enables fast handovers[1] between cdmaOne and cdma2000 networks.

Uplink characteristics

- *Pilot-based coherent detection.* While a pilot channel for coherent demodulation and power control measurements by the mobiles also exists in the downlink of cdmaOne, such a structure is not available in its uplink. The incorporation of the Reverse Pilot Channel (R-PICH) in cdma2000 enhances its uplink performance. The R-PICH provides a means for the base station to perform coherent demodulation of received traffic, a clear advantage over the noncoherent modulation of the reverse link of cdmaOne.
- *Use of open or closed power control.* The uplink can use both open loop and fast closed loop power control, which are features inherited from cdmaOne.

Common Characteristics

- *Double number of Walsh codes.* Channel multiplexing for cdmaOne uses 64 available Walsh codes. As a result, up to 64 transmissions can be carried out simultaneously over the same frequency carrier in cdmaOne. In contrast, cdma2000 employs variable spreading, featuring a maximum number of Walsh codes of 128. This can rise the per carrier capacity of cdma2000 to twice that of cdmaOne.
- *Turbo codes.* Cdma2000 utilizes turbo coding for the SCHs [23], which provides additional robustness for high-speed data services.
- *Independent data channels.*Two types of physical data channels exist, fundamental channels (FCHs) and supplemental (SCH) channels (described below). The two physical channel types are separately coded and interleaved and can be set to have different frame error rates and QoS requirements in order to optimize cdma2000 use with multiple simultaneous services. The addition of SCHs enables support for high-rate data services [22,23].
- *5-ms frame option.* Common frames have a duration of 20 ms. However, a 5 ms option is also defined allowing for low latency transmission of signaling information.
- *Backward compatible chip rates and frame structure.* Cdma2000 chip rates are multiples of cdmaOne in order to enable simplified design of dual mode cdma2000 and cdmaOne terminals. The same holds for the frame structure of cdma2000.

Data Traffic Physical Channels

- In order to meet different QoS requirements, two kinds of data traffic channels are defined, fundamental channels (FCHs) and supplemental (SCH) channels. The FCHs and SCHs are code-multiplexed in both the uplink and downlink. The encoding and modulation parameters of traffic channels are specified by radio configurations (RCs) [21]. Nine RCs exist, with the first two being specified to provide compatibility with cdmaOne. Some of the RCs

---

[1] Both the terms handover and handoff appear in the literature. In this book, these terms are used synonymously.

include some additional traffic channels, such as the Forward/Reverse Dedicated Control Channels (F-DCCH, R-DCCH).

The FCHs are of similar structure to that of cdmaOne and are used for variable rate transmission. Each F-FCH is spread with a different orthogonal code and supports frame sizes of 20 ms and 5 ms. The 20 ms frame structure supports data rates of 9.6 kbps, 4.8 kbps, 2.7 kbps, and 1.5 kbps for RC1, and 14.4 kbps, 7.2 kbps, 3.6 kbps, and 1.8 kbps for RC2.

The SCHs are used for data traffic in circuit or packet mode and can support a wide range of applications with different QoS requirements. SCHs can be operated in two modes. In the first, variable rates up to 14.4 kbps are provided, according to the cdmaOne-compatible RC1 and RC2. In the second, the higher data rates of cdma2000 are supported. SCHs support 20 ms frames.

As far as handover procedures are concerned, it is expected that for FCHs, soft handover will operate similarly to the soft handover in IS-95. We recall that in soft handover the mobile is connected to more than one base station while examining which one to use. In hard handover, the mobile station terminates the connection with a base station before connecting to a new one. For SCH handover in cdma2000, the list of the base stations the mobile communicates with (Active Set) can be a subset of the Active Set for the fundamental channel. This approach has the advantages of increased capacity and simplification of control processes [18].

Downlink (Forward Link) Physical Channels

- *The forward pilot channel (F-PICH).* This channel is produced by spreading with Walsh code 0 of an all-zero sequence and is continuously broadcast throughout the cell providing timing and phase information. Mobile nodes within a cell share this channel in order to obtain information about multipath fading and channel condition.
- *The forward auxiliary pilot channels (F-APICHs).* A number of optional APICHs exist. They are used in combination with smart antennas in beam-forming applications. Using this approach, the coverage to a specific geographical point can be increased, thus creating hot-spots that can be shared by several mobiles. APICHs are code-multiplexed on the downlink by assignment of a different Walsh code to each APICH.
- *The transmit diversity pilot channel (F-TDPICH) and auxiliary transmit diversity pilot channel (F-ATD-PICHs).* These channels are used for synchronization by mobiles inside a specific cell.
- *The forward common control channel (F-CCCH).* This channel is used by the base station to transmit MAC sublayer or higher-layer messages to the mobiles.
- *The forward sync channel (F-SYNCH).* Mobile stations acquire initial synchronization information through use of this channel. There are two types of forward sync channels, the shared sync channel and the wideband sync channel. The shared sync channel is provided in overlay configurations of cdma2000 over IS-95 systems and is provided to both IS-95 channels and cdma2000 wideband channels. The wideband synch channel can exist in both overlay and nonoverlay configurations and is modulated across the entire cdma2000 wideband channel.
- *The forward paging channel (F-PCH).* Base stations use this channel to transmit overhead information and mobile station specific messages. There are two types of forward paging channels, the shared paging channel and the wideband paging channel. The shared paging

channel is provided in overlay configurations of cdma2000 over IS-95 systems and is provided to both IS-95 channels and cdma2000 wideband channels. The wideband paging channel can exist in both overlay and nonoverlay configurations and is modulated across the entire cdma2000 wideband channel.

- *The forward broadcast channel (F-BCH).* This channel is used to send control information to mobile stations that have not been assigned a traffic channel.
- *The forward quick paging channel (F-QPCH).* The base station conveys control information to mobile stations by using this channel.
- *The forward common power control channel (F-CPCCH).* The base station conveys information for the power control of uplink common control channels by using this channel.
- *The forward common assignment channel (F-CACH).* The base station provides quick assignment of the reverse (uplink) common control channel by using this channel.
- *The forward data traffic channels.*

Uplink (Reverse Link) Physical Channels

- *The reverse pilot channel (R-PICH).* This is an unmodulated spread spectrum signal channel that helps the base station to detect a mobile's transmission. The mobile station inserts a reverse power control channel in R-PICH to transmit power control commands to the base station.
- *The reverse access channel (R-ACH).* Mobile stations use this channel to initiate communication with the base station and to respond to paging messages.
- *The reverse enhanced access channel (R-EACH).* Mobile stations use this channel to initiate communication with the base station and to respond to MS-directed messages.
- *The reverse common control channel (R-CCCH).* This channel conveys user and signaling information to the base station when reverse traffic channels are not in use.
- *The reverse data traffic channels.*

5.4.2.2.2 Cdma2000 Data Link Control Layer Issues The Cdma2000 Data Link Control (DLC) layer uses a logical channel structure to enable information exchange. In cdmaOne, mobiles make channel access requests over reverse physical access channels. However, in cdma2000 the request is made using the efficient 5 ms frame option. After submitting the request, the mobile expects an assignment reply from the base station in the downlink. If such an assignment is not received, the mobile executes an exponential backoff procedure and retries.

The cdma2000 DLC layer evolves further over the corresponding layer of cdmaOne in order to support a wide range of high-rate services running in the upper layers. The DLC layer comprises two sublayers, the MAC and Link Access Control (LAC). The LAC provides in-sequence reliable frame delivery utilizing, if necessary, Automatic Repeat Request (ARQ) protocols. In this section, after presenting the main enhancements of the MAC sublayer of cdma2000 over that of cdmaOne [22], we briefly summarize the main cdma2000 logical channels [21].

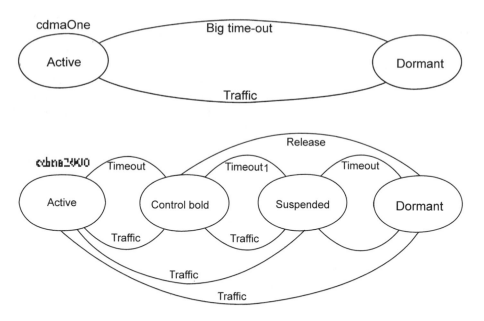

Figure 5.12   The cdmaOne and cdma2000 MAC sublayer state machines

MAC Sublayer Enhancements

- *QoS support.* Apart from best-effort delivery through a Radio Link Protocol (RLP), the cdma2000 MAC sublayer supports a QoS negotiation mechanism through the Multiplex and QoS mechanism. QoS negotiation is accomplished by appropriate prioritization of conflicting requests from contending services. The multiplexing mechanism combines information from various sources according to QoS demands and hands the resulting frames to the physical layer for transmission through the appropriate physical channel. Multiplexing can be performed on both common and dedicated channels. In the first case, control and data traffic concerning applications running on different mobiles can be multiplexed, whereas in the second case information regarding different applications of the same mobile are multiplexed.

- *Additional MAC states.* The finite-state machine of the cdma2000 MAC sublayer comprises four stages, two more than the corresponding two-state machine of cdmaOne (Figure 5.12). This machine reflects the status of packet or circuit data transmissions and a different machine is maintained for each ongoing transmission. While the MAC sublayer approach of cdmaOne works well for low-rate data services, it provides inefficient support for high-speed data services. This is due to the excessive interference incurred by traffic channels of idle mobile users in the Active state and the high overhead associated with dormant-to-active stage transition. The addition of the two extra states alleviates these problems. Particularly, in the Control Hold state the traffic channel is released, however a dedicated MAC logical channel (described below) is provided to idle mobile users. Over this channel, MAC commands, such as the request for a traffic channel establishment to serve a high-speed data burst, can be transferred almost immediately. In the suspended state, idle users do not possess dedicated channel. Nevertheless, state information is stored

1ˢᵗ letter	2ⁿᵈ letter	3ʳᵈ letter
f=forward	d=dedicated to a specific mobile	t=traffic
r=reverse	c=common	m=MAC
		s=signaling

Figure 5.13   cdma2000 logical channel naming rules

both in the mobile and the base station in order to enable fast assignment of a dedicated channel when packet events for the mobile occur. Finally, the dormant state is updated with the addition of a short data burst mode that enables delivery of short messages without the costly transition to the active state. This mode uses the Radio Burst Protocol (RBP) and the Signaling Radio Burst Protocol (SRBP) to provide a mechanism for delivering relatively short data and control messages over logical common traffic channels (ctch, described below), respectively.

Cdma2000 Logical Channels   A logical channel name comprises three or four lowercase letters followed by 'ch' (which stands for 'channel'). The fourth letter is applied only in cases of common channels used in the dormant or suspended states. Logical channels can either belong to a specific mobile (dedicated channels), or shared access among many mobile stations (common channels). Figure 5.13 shows the naming rules for the cdma2000 logical channels. The main logical channels are summarized below:

- *The forward/reverse dedicated MAC logical channel (f/r-dmch).* This channel is allocated in the active and control-hold states and is used to carry MAC-related messages. Forward channels are mapped to F-FCH or F-DCCH. Reverse channels are mapped to R-FCH or R-DCCH.
- *The forward/reverse dedicated traffic logical channel (f/r-dtch).* This channel is allocated in the active state and is used to carry user data. Forward channels are mapped to F-FCH, F-SCH or R-DCCH. Reverse channels are mapped to R-FCH, R-SCH or R-DCCII.
- *The forward/reverse common traffic logical channel (f/r-ctch).* This channel is used to carry short data bursts in the short data burst mode of the dormant state. It is mapped to R-CCCH or R-ACH.
- *The forward/reverse common signaling channel (f/r-csch) and the forward/reverse dedicated signaling channel (f/r-dsch).* These channels are used to carry signaling information. For csch, forward channels are mapped to F-CCCH or F-PCH and reverse ones to R-CCCH or R-ACH. For dsch, forward channels are mapped to F-FCH or F-DCCH and reverse ones to R-FCH or R-DCCH.

5.4.2.3 WCDMA

Wideband CDMA (WCDMA) is the second 3G air interface standard based on CDMA technology. In contrast to the requirement for synchronous operation of the base stations in cdma2000, inherited from its cdmaOne ancestor, WCDMA is an asynchronous scheme. This enables easier installation/integration of indoor WCDMA components with outdoor infrastructure. As mentioned before, during the 3G standardization process, several SDOs

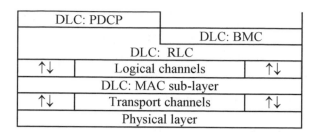

Figure 5.14   WCDMA radio interface protocol architecture

submitted WCDMA proposals. The 3GPP WCDMA standard is based on the ETSI and ARIB WCDMA proposals, with the main parameters in the uplink and downlink from the ETSI and ARIB proposals, respectively. The ETSI proposal for the 3G WCDMA standard is also known as the Universal Mobile Telecommunications Subsystem (UMTS). Despite the fact that the WCDMA proposal to ITU was developed first by ETSI, Japan developed its WCDMA standard more quickly. As a result, trial WCDMA system deployments began in Japan in 2000.

In the WCDMA specification, the term 'wideband' denotes use of a wide carrier. WCDMA uses a 5 MHz carrier; four times that of cdmaOne and 25 times that of GSM. The use of a wider carrier aims to provide support for high data rates. However, using wider carriers requires more available spectrum. This poses a significant difficulty in cases of spectrum shortage, as is the case with North American operators. As a result, WCDMA is likely to be favored for greenfield cellular deployments where sufficient IMT-2000 spectrum is available. However, WCDMA-based systems can also coexist with older generation systems if the corresponding spectrum can be spared.

Figure 5.14 shows the two lower layers of the radio interface protocol architecture of WCDMA. It consists of the physical layer and the DLC layer. The DLC layer is split into the following sublayers: Medium Access Control (MAC), Radio Link Control (RLC), Packet Data Convergence Protocol (PDCP) and Broadcast/Multicast Control (BMC). The physical layer offers different transport channels to the MAC sublayer. MAC offers different logical channels to the Radio Link Control (RLC) sublayer of Layer 2.

In the next sections, we cover issues relating to physical (layer 1) and data link layer (layer 2) operation and briefly present the main channels of each layer [21].

5.4.2.3.1 WCDMA Physical layer issues   The WCDMA physical layer offers information transfer services to MAC and higher layers. It introduces an air interface based on direct spread CDMA over a 5 MHz channel bandwidth. The original WCDMA proposals called for a chip rate of 4.096 Mcps, however, in order to enable easy manufacturing of terminals supporting both WCDMA and cdma2000, this rate was later reduced by harmonization activities to 3.84 Mcps. This is the chip rate for the DS mode of cdma2000 and is also very close to the 3.68 Mcps rate of multicarrier cdma2000.

WCDMA supports a number of physical channels for the uplink and downlink. These channels serve as a means of transmitting the data carried over logical channels. WCDMA uses 10 ms frames and has two operating modes, FDD and TDD, for use with paired and unpaired bands, respectively. The TDD mode is especially useful for European providers, due to the existence of unpaired frequency bands in Europe. The basic structure of TDD and FDD

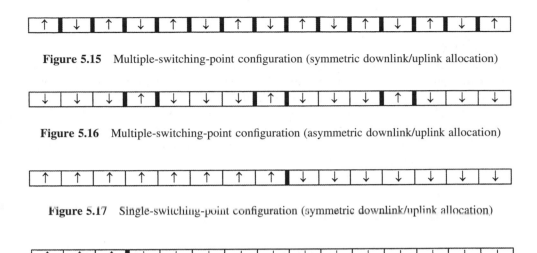

Figure 5.15   Multiple-switching-point configuration (symmetric downlink/uplink allocation)

Figure 5.16   Multiple-switching-point configuration (asymmetric downlink/uplink allocation)

Figure 5.17   Single-switching-point configuration (symmetric downlink/uplink allocation)

Figure 5.18   Single-switching-point configuration (asymmetric downlink/uplink allocation)

frames is the same, however, TDD frames contain switching points for uplink/downlink traffic separation. The ratio of uplink/downlink slots within a frame can vary in order to support asymmetric traffic requirements with downlink/uplink ratios ranging from 15/1 up to 1/7 [1]. Possible structures of TDD WCDMA frames are shown in Figures 5.15–5.18.

FDD mode requires the allocation of two frequency bands, one for the uplink and another for the downlink. FDD advantages are the ability to transmit and receive at the same time. However, FDD is not very efficient in allocating the available bandwidth for all types of services. Consider the example of Internet access. Such a service requires more throughput on the downlink than on the uplink. Of course by adjusting the spreading factor, FDD makes it possible to use only the required data rate, however, trading uplink capacity for downlink is not possible.

TDD, on the other hand, uses the same frequency band both for uplink and downlink by allocating time slots to each direction. Therefore, FDD can efficiently allocate capacity between the uplink and downlink and offer support to asymmetric traffic demands. However, it requires better time synchronization than FDD in order to guarantee that mobile and base station transmissions never overlap in the time domain.

The asynchronous nature of base station operation must be taken into consideration when designing soft handover algorithms for WCDMA. In an effort to support increased capacities through Hierarchical Cell Structures (HCS), WCDMA also employs a new handover method, called interfrequency handover. In HCS several different frequency carriers are simultaneously used inside the same cell in an effort to serve increased demands in hot spots. To perform handover in HCS situations, the mobile station needs to possess the ability to measure the signal strength of an alternative carrier frequency while still having the connection running on the current frequency. Two methods for interfrequency measurements exist for WCDMA [18]: The first, called dual receiver mode, is used when antenna diversity is employed. It uses different antenna branches for estimating different frequency carriers. The second, called slotted mode, uses compression of transmitted data (possibly using a lower spreading ratio during a shorter period) to save time for measurements on alternative frequency carriers.

The WCDMA physical layer provides two types of packet access using random access and dedicated (user) channels. Random access is based on a slotted ALOHA approach and is used only on the uplink for short infrequent bursts. The random access method is more efficient in terms of overhead, as the channel is not maintained between bursts. Dedicated access serves more frequent bursts both on the uplink and downlink. Furthermore, the WCDMA physical layer provides broadcasting and paging capabilities to the upper layers. In the remainder of this section, we outline the major characteristics of the WCDMA physical layer [2,19,23–26] and we briefly summarize the main WCDMA physical channels [21].

WCDMA Physical Layer Characteristics

- *Wideband.* The use of 5 MHz channels provides support for increased capacity. WCDMA has double the capacity of narrowband CDMA in urban and suburban environments [2].
- *Spreading.* Orthogonal Variable Spreading Factors (OVSFs) are used for channel separation. These factors range from 4 to 256 in the FDD uplink, from 4 to 512 in the FDD downlink, and from 1 to 16 in the TDD uplink and downlink. Depending on the spreading factor (SF), it is possible to achieve different data rates. For cell separation, the FDD uses 10 ms period gold codes of length $2^{18} - 1$, and the TDD scrambling codes of length 16. For user separation, the FDD uses 10 ms period gold codes of length 2^{41} and the TDD codes with period of 16 chips. The modulation method used is QPSK.
- *Adaptive antenna support.* Support for adaptive antenna arrays improves spectrum efficiency and capacity by optimizing antenna performance for each mobile terminal.
- *Channel coding and interleaving.* Depending on the BER and delay requirements, different coding schemes may be applied. Convolutional coding, turbo coding or no coding at all is supported. In order to randomize transmission errors, bit interleaving is also performed.
- *Downlink/Uplink coherent demodulation and fast power control.*
- *Support for downlink transmit diversity and multiuser detection techniques.*

Downlink (Forward Link) Physical Channels

- *Physical Synchronization Channel (PSCH).* The PSCH provides timing information and is used for handover measurements by the mobile station.
- *Downlink Dedicated Physical Channel (Downlink DPCH).* Within one downlink DPCH, data and control information generated at layer 2 and layer 1, respectively, are transmitted in a time-multiplexed manner. The downlink DPCH can thus be seen as a time multiplex of a *Downlink Dedicated Physical Data Channel (DPDCH)* and a *Downlink Dedicated Physical Control Channel (DPCCH)*.
- *Common Pilot Channel (CPICH).* CPICH is used as a reference channel for downlink coherent detection and fast power control support.
- *Primary and Secondary Common Control Physical Channel (P-CCPCH, S-CCPCH) and Physical Downlink Shared Channel (PDSCH).* These are used to carry data and control traffic.

Uplink (Reverse Link) Physical Channels

- *Uplink Dedicated Physical Data Channel (Uplink DPDCH).* This channel is used to carry the data generated at layer 2 and above.
- *Uplink Dedicated Physical Control Channel (Uplink DPCCH).* This channel is used to carry control information, such as power control commands, generated at layer 1. Layer 1 control information consists of known pilot bits to coherent detection, transmit power-control commands, etc.
- *Physical Random Access Channel (PRACH) and Physical Common Packet Channel (PCPCH).* These channels are used to carry user data traffic. The WCDMA random access scheme is based on a slotted ALOHA technique. More than one random access channel can be used if demand exceeds capacity.
- *Physical Uplink Shared Channel (PUSCH) (TDD mode).* This channel is used to carry user data traffic.

5.4.2.3.2 WCDMA Data Link Control Layer Issues The DLC layer of WCDMA offers services to upper layers. It comprises the MAC, RLC BMC and PDCP sublayers. The MAC sublayer provides services to upper layers through the use of logical channels. The MAC sublayer accesses services offered by the physical layer through the use of transport channels. The services offered by the MAC, RLC BMC and PDCP sublayers to upper layers and the functions performed by these sublayers in order to offer these services are briefly presented below [27]. A brief presentation of the cdma2000 transport and logical channels follows.

MAC sublayer services to upper layers

- *Data transfer:*It offers unacknowledged transfer of MAC frames between peer MAC entities. This service does not provide any data segmentation. Segmentation/reassembly procedures are the responsibility of the upper RLC sublayer.
- *Resource and MAC parameter reallocation.* This service serves upper-layer requests for reallocation of resources and changing of MAC parameters. Such requests concern the identity of a terminal, the change of transport channels for the traffic, etc.
- *Measurement reports.*The MAC sublayer also offers measurements such as traffic volume and channel quality to upper layers.

MAC Functions

- *Mapping between logical channels and transport channels.* This MAC sublayer function is responsible for mapping logical channels to the appropriate transport channels.
- *Transport channel format selection.*Given the transport format requirements from upper layers, this MAC sublayer function manages the selection of the most appropriate transport format in order to ensure efficient use of transport channels.
- *Priority handling.*This MAC sublayer function handles the mapping of data flows to transport channels by taking into account data flow priorities.
- *Identification of terminal identities on common transport channels.* When a terminal is addressed on a common downlink channel or is using the Random Access Channel

(RACH, described later), the identification of the terminal identity is a responsibility of this MAC sublayer function.

- *Multiplexing/demultiplexing support.* This MAC sublayer function performs multiplexing/ demultiplexing of both common and dedicated transport channels. In the second case, this function enables efficient merging of several upper layer data flows onto the same transport channel.
- *Monitoring traffic volume.* This MAC sublayer function measures traffic volume on logical channels and reports results to upper layers in order to enable transport channel switching decisions.
- *Ciphering.* This MAC sublayer function prevents unauthorized acquisition of data. Ciphering is performed in the MAC layer for transparent RLC mode.
- *Access service class selection for transport Random Access Channel (RACH) transmission.* The resources of the RACH (e.g. access slots), a transport channel described below, can be divided between different access service classes in order to provide different priorities of RACH usage. Each access service class can have a set of back-off parameters associated with it, some or all of which may be broadcast by the network. This MAC sublayer function applies the appropriate back-off parameters to packet transmission procedures.

RLC Services to Upper Layers

- *Connection establishment and release.* This service provides establishment and release of connections between RLC peer entities.
- *Transparent data transfer.* The RLC sublayer provides for transmission of higher layer PDUs possibly employing segmentation/reassembly functionality, without the overhead of adding any RLC protocol information.
- *Unacknowledged data transfer.* The RLC sublayer provides for transmission of higher layer PDUs without guaranteed delivery. During this unacknowledged data transfer mode, the RLC sublayer uses the sequence check function, to deliver to upper layers only unique copies of error-free frames. The receiving RLC sublayer delivers frames to higher layer receiving entities as soon as they arrive at the receiver.
- *QoS setting.* The RLC sublayer offers different levels of QoS to higher layers.
- *Notification of unrecoverable errors.* The RLC sublayer notifies upper layers about errors that cannot be resolved by the layer itself.

RLC Functions

- *Segmentation and reassembly.* This RLC sublayer function performs segmentation and reassembly between the variable-length higher layer PDUs and the smaller RLC PDUs. When the remaining data to be sent does not fill an entire RLC PDU of a given size, the RLC sublayer fills the remaining data field with padding bits.
- *User data transfer options.* This RLC sublayer function performs acknowledged, unacknowledged and transparent data transfer with or without QoS requirements.
- *Error correction.* This RLC sublayer function supports error correction by retransmission mechanisms (e.g. Go Back N, Selective Repeat) in the acknowledged transfer mode. Error

correction includes detection of duplicate received PDUs. In this case, the RLC sublayer guarantees that only one copy of the PDU will be handed to the upper layer.

- *In/out of sequence delivery of higher layer PDUs.* This RLC sublayer function manages both in-sequence and out-of-sequence Protocol Data Unit (PDU) delivery between peer RLC sublayers. In the second case, it is up to higher layers to restore the order of the received PDUs.
- *Flow control.* This RLC sublayer function at the receiver can control the transmission rate of the peer RLC entity.
- *Protocol error detection and recovery.* This RLC sublayer function can detect and recover from errors occurring during its operation.
- *Ciphering.* This RLC sublayer function prevents unauthorized acquisition of data. Ciphering is performed in the RLC sublayer when the nontransparent RLC data transfer service is offered to higher layers.

PDCP Services to Upper Layers

- *Network layer PDU transmission/reception.* The PDCP sublayer is responsible for the transmission and reception of higher layer PDUs in the acknowledged, unacknowledged and transparent RLC modes.

PDCP Functions

- *PDU mapping.* This PDCP sublayer function maps the incoming network PDUs to PDUs of the RLC sublayer.
- *Compression-decompression.* This PDCP sublayer function performs efficient transmission and reception of layer 3 PDUs using compression and decompression of redundant network-layer PDU control information (e.g. header) at the transmitting and receiving entities, respectively.

BMC Services to Upper Layers

- *Broadcasting-multicasting.* The BMC sublayer provides broadcast and multicast transmission capabilities to upper layers for common user data in transparent or unacknowledged transfer mode.

BMC Functions

- *Storage of Cell Broadcast Messages.* This BMC sublayer function stores messages to be broadcast to all mobiles within a cell (cell broadcast messages).
- *Scheduling of BMC messages.* This BMC sublayer function based upon the scheduling information of each cell broadcast message schedules them accordingly.
- *Transmission of BMC messages to mobiles.* This BMC sublayer function transmits the BMC messages according to schedule.
- *Delivery of broadcast messages to upper layers.* This BMC sublayer function in the terminal side is responsible for delivery of received broadcasts to the upper layer. Corrupted broadcasts are not delivered to the upper layer.

The term transparent transmission characterizes the case where a protocol, does not require any protocol control information. However, the existence of the peer protocol at the receiving entity is required since some protocol functions may still be executed. In the case of RLC, for example, segmentation and reassembly operations can be performed without segmentation headers when a higher layer PDU fits into a fixed number of RLC PDUs to be transferred in a given transmission time interval. In this case, segmentation and reassembly operations follows predefined rules known to both peer RLC entities.

The data flows through layer 2 are characterized by data transfer modes employed by the RLC sublayer in combination with the data transfer types of the MAC sublayer. The RLC sublayer provides three transfer modes: acknowledged, unacknowledged and transparent. Acknowledged and unacknowledged RLC transmissions both require a RLC header. In unacknowledged mode, only data PDU is exchanged between peer RLC entities, while in the acknowledged mode, both data PDUs and control PDUs are exchanged between peer RLC entities. The MAC sublayer offers the ability of transparent MAC transmission in which the addition of a MAC header is not required.

The MAC sublayer of WCDMA operates on the channels defined below. The transport and logical channels convey information between the MAC-physical layer and MAC-RLC sublayer interfaces, respectively. The remainder of this section provides a brief overview of those channels [21].

Transport Channels

- *Random Access Channel (RACH) (uplink).* A contention-based channel used for transmission of relatively small amounts of data, such as nonreal-time control information. This channel is mapped to PRACH.
- *Forward Access Channel(s) (FACH) (downlink).* Used for transmission of relatively small amounts of downlink data. This channel is mapped to S-CCPCH.
- *Broadcast Channel (BCH) (downlink).* Used for broadcast of system information within a cell. This channel is mapped to P-CCPCH.
- *Paging channel (PCH) (downlink).* Used for broadcast of control information to mobiles in power-saving mode. This channel is mapped to P-CCPCH.
- *Synchronization channel (SCH) (TDD downlink).* Used for broadcast of synchronization information into an entire cell in TDD mode. This channel is mapped to PSCH.
- *Downlink Shared Channel (DSCH).* Shared by mobiles for carrying control or traffic data. This channel is mapped to PDSCH.
- *Common Packet Channel(s) (CPCH) (FDD uplink).* A contention channel used for transmission of bursty data traffic in the uplink of the FDD mode. This channel is mapped to PCPCH.
- *Uplink Shared Channel(s) (USCH) (TDD).* Shared by several mobiles for carrying dedicated control or traffic data, used in TDD mode only. This channel is mapped to PUSCH.
- *Dedicated Channel (DCH) (uplink/downlink).* A channel dedicated to a specific mobile. This channel is mapped to DPDCH.
- *Fast Uplink Signaling Channel (FAUSCH).* This channel is used to allocate dedicated channels (in conjunction with FACH).

Logical Channels

- *Synchronization Control Channel (SCCH) (downlink TDD).* Used for broadcasting synchronization information. This channel is mapped to SCH.
- *Broadcast Control Channel (BCCH) (downlink).* Used for broadcasting system control information. This channel is mapped to BCH and may also be mapped to FACH.
- *Paging Control Channel (PCCH) (downlink).* Used for transfer of paging information when the network does not know the location cell of the mobile, or the mobile is in sleep mode. This channel is mapped to PCH.
- *Common Control Channel (CCCH).* A bi-directional channel used for transmitting control information between the network and the mobiles. This channel is mapped to RACH and FACH.
- *Dedicated Control Channel (DCCH).* A point-to-point bi-directional channel that transmits dedicated control information between the network and the mobiles. This channel is mapped to either RACH and FACH, to RACH and DSCH, to DCH and DSCH, to a DCH, a CPCH (FDD only) to FAUSCH, CPCH (FDD only), or to USCH (TDD only).
- *Shared Channel Control Channel (SHCCH).* A bi-directional channel used to transmit control information for uplink and downlink shared channels between the network and the mobiles. This channel is mapped to RACH and USCH/FACH and DSCH.
- *Dedicated Traffic Channel (DTCH) (uplink/downlink).* Used for transfer of user information. DTCH channels are dedicated to specific mobiles. This channel is mapped to either RACH and FACH, to RACH and DSCH, to DCH and DSCH, to a DCH, a CPCH (FDD only) or to USCH (TDD only).

5.4.3 Fixed Network Evolution

The many 2G systems deployed in different regions of the world will form the basis for the evolution and migration towards 3G systems [1]. While this migration entails a revolutionary path for the air interface standards, the fixed network evolution will be more conservative. The goal is to reuse as much of the fixed network infrastructure as possible, in an effort to provide seamless migration from 2G to 3G systems and lower the accompanying costs.

As reference architecture for our discussion, we use a simplified version of the UMTS Release '99 [4]. This architecture is shown in Figure 5.19 and besides the air interface between the base stations and the mobiles, it also comprises the following parts:

- 3G-capable base stations.
- Radio Network Controllers (RNC), which in GSM terminology correspond to the Base Station Controllers (BSC). RNCs and base stations are connected through the Iub interface which corresponds to GSM's Abis interface [28]. RNCs control the operation of several 3G-capable base stations each and are interconnected through a new interface, the Iur interface which supports handover functionality.
- The RNCs and 3G-capable base stations form the Radio Access Network (RAN), also known as the UMTS terrestrial RAN (UTRAN), which corresponds to the Base Station Subsystem (BSS) of GSM. RNCs are connected to the Core Network (CN) through the Iu interface, which corresponds to GSM's A interface.

The convergence of wireline and wireless networks and the increasing demand for wireless

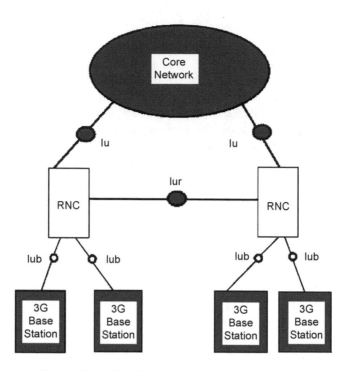

Figure 5.19   Simplified UMTS network architecture.

services of performance equal to that of wireline services lead to studies regarding the applicability of ATM and IP in the UTRAN and CN parts of the cellular architecture [28–30]. ATM is a promising solution for integrated support of voice, data and multimedia services with stringent QoS and delay requirements. In fact, 3GPP decided to use ATM in the RAN interfaces specified in the UMTS Release '99 specification. Specifically, UMTS Release '99 supports use of ATM Adaptation Layer 2 (AAL2) in the UTRAN. AAL2 was designed to meet the requirements of low bit-rate and delay-sensitive applications. It can efficiently handle bandwidth issues and QoS requirements with reports on simulation results [30] mentioning that a balance needs to be maintained between bandwidth utilization efficiency and stringency in delay requirement precision. Overall, with careful network and resource management, ATM/AAL2 is capable of meeting the delay requirements of 3G traffic in the UTRAN.

IP-based solutions for use in the UTRAN are also being studied by the 3GPP. However, a number of challenges regarding IP need to be overcome. For example, there is a need for QoS support including delay, jitter and loss requirements. Furthermore, as the IP header is much larger than that of an ATM cell, there is a significant increase in overhead for voice support. Nevertheless, work is being done to solve those problems [28] with IETF trying to support QoS in IP and also to increase its bandwidth efficiency by multiplexing low bit-rate connections over the same IP connection.

The selection of a transport architecture for the fixed parts of future 3G cellular networks will be affected by many factors, such as the 3G services offered, backward compatibility, market penetration and provider policies. A couple of years ago, ATM was thought to be the only choice for transport technology. However, during the last few years IP has gained more importance. Keeping in mind that the evolution of the fixed part of the cellular network will not be made quickly and the fact that work on the subject is still under way, one can realize that several options for this evolution may exist. Studies [28] have indicated that four different options are possible:

- *Use of ATM in the UTRAN and TDM/frame relay in the CN.* In this option, ATM technology is used in the UTRAN in order to meet requirements for QoS, high-speed soft-handoff and scalability. The well-established GSM technology will continue to dominate the CN. The obvious advantage of this option is smooth evolution towards 3G networks while retaining existing investments.
- *Use of ATM both in the UTRAN and the CN.* This option will probably be favored by new operators entering the market and for operators that already own a public or private ATM network. This choice, offers seamless integration of wireless and wireline networks.
- *Use of ATM in the UTRAN and IP in the CN.* This option will exploit the ATM QoS capabilities in order to provide support for time-critical services in the UTRAN. The use of IP in the CN will support the growth of packet-data services in wireless networks.
- *Use of IP both in the UTRAN and the CN.* This option leads to an all-IP-based infrastructure. However, efficient solutions on IP QoS issues need to be found. Since the UTRAN sets even more stringent delay requirements than the CN, IP QoS issues must be first solved for the CN before IP is introduced in the UTRAN.

5.5 Summary

The goal of third generation (3G) wireless networks is to provide efficient support for both voice and high bit-rate data services. In this chapter we covered third generation wireless networks by focusing on a number of issues:

- *3G spectrum requirements.* The enhanced capabilities of 3G networks call for use of additional spectrum. However, spectrum assignment for 3G systems has proven to be a difficult task due to the fact that spectrum is not identically regulated in every country. Spectrum shortage is especially evident in North America where the entire frequency region that ITU regulated for 3G systems is already in use. As the market penetration of 3G systems increases, the need for more spectrum will arise. Furthermore, the development and commercial use of efficient technologies that can alleviate problems attributed to nonuniform worldwide spectrum regulation and spectrum shortage will be highly beneficial. Such techniques are software radio, intelligent antennas and multiuser detection.
- *Service classes.* Apart from supporting traditional voice calls, 3G systems will offer support for file transfer, web browsing, multimedia and videoconferencing applications. The requirements of those applications in terms of capacity span the entire range of the data rates offered by 3G systems, from several kbps, up to 2 Mbps. Several 3G service classes have been identified based on capacity demands. Furthermore, the enhanced abilities of these service classes will enable widespread use of advanced multimedia, m-commerce and geolocation-based applications.
- *Standardization procedures.* 3G standardization activities originated in 1992 by ITU. The outcome of the standardization effort, IMT-2000, comprises a number of different 3G

standards for the air interface. Work has been done to harmonize those standards. ITU decided not to define the protocol that will be used inside the fixed part of a 3G network in order to allow for flexible evolution of 3G systems.

- *Air interface standards.* The standardization activities resulted in three main 3G air interface standards. EDGE is a TDMA-based system that evolves from GSM and IS-136 and offers data rates up to 473 kbps. Being a descendant of 2G TDMA-based standards, EDGE can be easily integrated with those systems in order to provide support for data applications with high-rate demands. Cdma2000 is a fully backwards-compatible descendant of IS-95 enabling smooth transition of a 2G IS-95 system to a 3G cdma2000 system. Cdma2000 supports data rates up to 2 Mbps. Finally, WCDMA is a CDMA-based system that introduces a new 5-MHz wide channel structure. WCDMA is also capable of offering speeds up to 2 Mbps.
- *Fixed part of the network.* The selection for a transport architecture for the fixed parts of future 3G cellular networks comprises several alternatives. ATM and IP impact these alternatives resulting in a number of possible transport architectures.

WWW Resources

1. *www.itu.int/imt2000*: the IMT-2000 official web page. It contains both introductory and technical information relating to 3G standardization.
2. *www.umts-forum.org*: the European forum that supports WCDMA development contains useful information on 3G deployment worldwide.
3. *www.3gpp.org*: the page of the Third Generation Partnership Proposal (3GPP), which deals with the WCDMA standard.
4. *www.3gpp2.org*: the page of the Third Generation Partnership Proposal no. 2 (3GPP2), which deals with the cdma2000 standard.
5. *www.etsi.org*: the page of the European Telecommunications Standards Institute (ETSI), a nonprofit organization that produces European standards in the telecommunication industry.
6. *www.uwcc.org*: the page of the Universal Wireless Communications Consortium, a group which represents the TDMA industry.
7. *www.sdrforum.org*: the page of the Software Defined Radio Forum is a useful source of information on this enabling technology.
8. *http://www.ericsson.com/review/*: on-line reviews of networking topics, including several interesting articles on 3G systems.
9. *http://www.lucent.com/minds/techjournal/findex.html*: the *Bell Labs Technical Journal* publishes several articles on 3G systems.

References

[1] Chaudhury P., Mohr W. and Onoe S. The 3GPP Proposal for IMT–2000, *IEEE Communications Magazine*, December, 1999, 72–81.
[2] Nilsson T. Toward Third-Generation Wireless Communication, *Ericsson Review*, 2, 1998.
[3] Bos L. and Leroy S. Toward an All-IP-Based UMTS System Architecture, *IEEE Network*, January/February, 2001, 36–45.

[4] Nilsson M. Third-Generation Radio Access Standards, *Ericsson Review*, 3, 1999.

[5] Bi Q., Zysman G. I. and Menkes H. Wireless Mobile Communications at the Start of the 21st Century, *IEEE Communications Magazine*, January, 2001.

[6] The UMTS Forum. *Report on Candidate Extension Bands for UMTS/IMT–2000 Terrestrial Component*, Second Edition, March, 1999.

[7] Zysman G. I., Tarallo J. A., Howard R. E., Freidenfelds J., Valenzuela R. A. and Mankiewich P. M. Technology Evolution for Mobile and Personal Communications, *Bell Labs Technical Journal*, January–March, 2000, 107–127.

[8] The Software-Defined Radio Forum. http://www.sdrforum.org

[9] Dornan A. *The Essential Guide to Wireless Communications Applications, Prentice Hall*, 2001.

[10] Special Issue on Software Radios. *IEEE Communication Magazine*, May, 1995.

[11] Special Issue on Software Radios. *IEEE Journal on Selected Areas in Communications,* April, 1999.

[12] Special Issue on Software Radios. *IEEE Personal Communications*, August, 1999.

[13] Buehrer R M., Kogiantis A. G., Liu S. C., Tsai J. A. and Uptegrove D. Intelligent Antennas for Wireless Communications-Uplink, *Bell Labs Technical Journal*, July–September, 1999, 73–103.

[14] Special Issue on Active and Adaptive Antennas, *IEEE Transactions on Antennas and Propagation*, March, 1964.

[15] Verdu S. *Multiuser Detection*, Cambridge University Press, 1998.

[16] Hallen D. A., Holtzman J. and Zvonar Z. Multiuser Detection for CDMA Systems, *IEEE Personal Communications*, April, 1995, 46–58.

[17] Moshavi S. Multiuser Detection for DS-CDMA Communications, *IEEE Communication Magazine*, October, 1996, 124–136.

[18] The UMTS Forum. Enabling UMTS/Third Generation Services and Applications, October, 2000.

[19] Eldstahl J. and Nasman A. WCDMA evaluation system - Evaluating the Radio Access Technology of Third-Generation Systems, *Ericsson Review*, 2, 1999.

[20] Furuskar A., Mazur S., Muller F. and Olofsson H. EDGE, Enhanced Data Rates for GSM and TDMA/136 Evolution, *IEEE Personal Communications,* June, 1999, 56–66.

[21] Sarikaya B. Packet Mode in Wireless Networks: Overview of Transition to Third Generation, *IEEE Communications Magazine*, September, 2000, 164–172.

[22] Knisely D., Li Q. and Ramesh N. S. Cdma2000: a Third Generation Radio Transmission Technology, *Bell Labs Technical Journal*, July–September, 1998, 79–97.

[23] Prasad R. and Ojanpera T. An Overview of CDMA Evolution Toward Wideband CDMA, *IEEE Communications Surveys* , Fourth Quarter, 1998.

[24] Steinbugl J. J. *Evolution Toward Third Generation Wireless Networks.*

[25] Ojanpera T. and Prasad R. An Overview of Third Generation Wireless Personal Communications: a European Perspective, *IEEE Personal Communications*, December, 1998, 59–65.

[26] ETSI TS 125, 201 V3.0.1. Universal Mobile Telecommunications System (UMTS); Physical Layer - General Description, 2000–01.

[27] ETSI TS 125 301 V3.3.0. Universal Mobile Telecommunications System (UMTS); Radio Interface Protocol Architecture.

[28] Subbiah B. and Raivio Y. Transport Architecture Evolution in UMTS/IMT–2000 Cellular Networks, *International Journal of Communication Systems*, Wiley, August, 2000, 371–385.

[29] Huber J. F., Weiler D. and Brand H. UMTS, the Mobile Multimedia Vision for IMT-2000: a Focus on Standardization, *IEEE Communications Magazine*, September, 2000, 129–136.

[30] Dixit S., Guo Y. and Antoniou Z. Resource Management and Quality of Service in Third Generation Wireless Networks, *IEEE Communications Magazine*, February, 2001.

Further Reading

[1] Chan M. C. and Woo T. Y. C. Next Generation Wireless Data Services: Architecture and Experience, *IEEE Personal Communications*, February, 1999, 20–33.

[2] Hu J. Applying IP over wmATM Technology to Third Generation Wireless Communications, *IEEE Communications Magazine*, November, 1999, 64–67.

[3] Chuang J. and Sollenberger N. Beyond 3G: Wideband Wireless Data Access Based on OFDM and Dynamic Packet Assignment, *IEEE Communications Magazine*, July, 2000.

[4] Huber J. F., Weiler D. and Brand H. UMTS, the Mobile Multimedia Vision for IMT-2000. A Focus on Standardization, *IEEE Communications Magazine*, September, 2000, 129–136.

[5] Schmidts H. and Visser J. Framework for IMT-2000 networks, *Computer Networks*, 34, 2000, 705–715.

[6] Blanchard C. Security for the Third Generation (3G) Mobile System, *Information Security Technical Report*, 5(3), 2000, 55–65.

[7] Nilsson T. Toward Third-Generation Mobile Multimedia Communication, *Ericsson Review*, 3, 1999, 122–131.

[8] Lindheimer C., Mazur S., Molny J. and Waleij M. Third-Generation TDMA, *Ericsson Review*, 2, 2000.

[9] Larsson G. Evolving from cdmaOne to Third-Generation Systems, *Ericsson Review*, 2, 2000.

[10] Lindheimer C., Mazur S., Molny J. and Waleij M. Third-Generation TDMA, *Ericsson Review*, 2, 2000.

[11] Almers P., Birkedal A., Kim S., Lundqvist A. and Milen A. Experiences of the Live WCDMA Network in Stockholm, Sweden, *Ericsson Review*, 4, 2000.

[12] Pittampalli E. Third Generation CDMA Wireless Standards and Harmonization, *Bell Labs Technical Journal*, July–September, 1999.

[13] Dell White Paper, Wireless Technologies, August, 1999.

[14] TIA, The cdma2000 ITU-R RTT Candidate Submission, June, 1998.

[15] ARIB, Japan's Proposal for Candidate Radio Transmission Technology on IMT-2000:W-CDMA, June, 1998.

[16] ETSI TS 125 211 V3.1.1, Universal Mobile Telecommunications System (UMTS); Physical Channels and Mapping of Transport Channels onto Physical Channels (FDD).

[17] ETSI TS 125 221 V3.1.1. Universal Mobile Telecommunications System (UMTS); Physical Channels and Mapping of Transport Channels onto Physical Channels (TDD).

[18] ETSI TS 125 321 V3.2.0. Universal Mobile Telecommunications System (UMTS); MAC Protocol Specification.

[19] Pirhonen R., Rautava T. and Penttinen, S. TDMA Convergence for Packet Data Services, *IEEE Personal Communications*, June, 1999, 68–73.

6

Future Trends: Fourth Generation (4G) Systems and Beyond

6.1 Introduction

By looking back to the history of wireless systems, one can reach the conclusion that the industry follows a ten-year cycle. First generation systems were introduced in 1981 followed by the deployment of second generation systems in 1991, ten years later. Moreover, third generation systems are due for deployment in 2001–2002. From the point of view of services, 1G systems offered only voice services, 2G systems also offered support for a primitive type of low-speed data services and 3G systems will enable a vast number of advanced voice and high-speed data services. The trend is towards support for even advanced data services.

3G networks, although having the advantage of support for IP and enhanced mobility, will suffer from a divergence between several standards. This divergence will limit easy roaming between 3G networks based on different standards, thus limiting user mobility. Furthermore, 3G networks will have, in the best case, an upper capacity limit of 2 Mbps. Although more than enough for the application demands of the years to come, 3G networks will most likely need to evolve in order to meet the mobile application demands of the next decades. As in all areas of technology, the quest for better and more efficient systems never ends and as soon as the time for deployment of a system comes, research on the next generation is usually under way. Consequently, the imminent deployment of 3G systems is accompanied by initiation of research on the next generation of systems. If the ten-year cycle continues, it is logical to expect that the next generation of wireless systems, known as Fourth Generation (4G), will reach deployment stage somewhere around 2010.

As seen later in the chapter, the vision for 4G and future systems is towards unification of various mobile and wireless networks. However, there is a fundamental difference between wireless cellular and wireless data networks, such as WLANs. The difference is that cellular systems are commonly circuit-switched, meaning that for a certain call, a connection establishment has to take place prior to the call. On the contrary, wireless data networks are packet-switched. It is expected that the evolution of wireless networks towards an integrated system will produce a common packet-switched (possibly IP-based) platform for wireless systems, thus enabling the 'wireless Internet'. However, in order for such an integration to take place research is needed in order to provide interoperability between wireless cellular networks and wireless data networks. The envisioned unified platform for the next generations of wireless networks will provide transparent integration with the wired networks and enable users to seamlessly access multimedia contents such as voice, data and video, irrespective of the access methods of the various wireless networks involved.

The next generations of wireless networks target the market of 2010 and beyond, aiming to offer increased data rates with reports mentioning from 50 Mbps to 155 Mbps. In the course of their development many different types of issues (technical, economical, etc.) must be studied and resolved. Some of them, such as the development of even more efficient modulation techniques, identification of new spectrum, and developments in battery technology/power consumption, are quite straightforward and have been identified during 2G and 3G research and development stages. Other issues are not so clear and are heavily dependent on the evolution of the telecommunications market and society in general. These issues need to be identified and resolved at the earliest possible stage in order to unsure market success for 4G and beyond wireless systems.

6.1.2 Scope of the Chapter

This chapter provides a vision of some of the characteristics of 4G and future systems. Section 6.2 describes the design goals and corresponding research issues for 4G systems. Section 6.3 presents a preliminary set of possible 4G service classes. Section 6.4 identifies the challenge of predicting the future of wireless communications and provides three possible scenarios for the future. Finally, the chapter ends with a brief summary in Section 6.5.

6.2 Design Goals for 4G and Beyond and Related Research Issues

Since 4G systems target the market of 2010 and beyond, there is time for 4G research and standards development. So far, no 4G standard has been defined and only speculations have been made regarding the structure and operation of 4G systems. The question to ask here is what will be the desired advantages and new features of 4G systems over their predecessors. Due to the fact that related research is under way, 4G is still an acronym without a generally accepted meaning. However, research efforts [1–3] agree more or less on the following targets:

- *System interoperability.* 4G and future systems should bring something that is missing from their predecessors: flexible interoperability of the various kinds of existing wireless networks, such as satellite, cellular wireless, WLAN, PAN and systems for wireless access to the fixed network. Alternatively, this can be thought of as an ability to roam between multiple wireless and mobile standards (e.g. moving from a cellular network to a WLAN while maintaining connections). If the target of system interoperability is met, the whole worldwide communications infrastructure will be turned into a transparent network allowing users to use it independent of a specific access method. Due to the requirement for interoperability of different mobile and wireless networks, a big challenge will be how to access several different mobile and wireless networks through the same terminal. We can identify the three possible configurations described below [3]:

– *Multimode terminals.* This option provides for further development of older generation systems and has also been applied in the past (e.g. dual AMPS-CDMA cellular phones). It calls for a single terminal which is capable of accessing several different wireless networks. This is obviously achieved by incorporating multiple interfaces to the terminal, one for the access method of every different kind of wireless network. The ability to use many access methods will enable users to use a single device to access the 4G network irrespective of the particular access method used. The option of multimode terminals will offer increased coverage and reliable wireless access in the case of failure of one or more networks in an area. Furthermore, the multimode terminal option lowers the complexity of the fixed part of the network due to the fact that the additional complexity is incorporated into the device [3].

– *Overlay network.* In this architecture users will access the 4G network through the Access Points (APs) of an overlay network. Upon connection with a terminal, an AP will select the wireless network to which the terminal will be connected. This choice will be made based on user-defined choices, resource availability, QoS requirements, etc. The AP will perform protocol translation and QoS negotiation for the connections. Since APs can monitor the resources used by a user, this architecture supports single billing and subscription.

– *Common access protocol.* This choice calls for use of one or two standard access protocols by the wireless networks. A possible option is for the wireless networks to use either ATM cells with additional headers or WATM cells.

- *Terminal bandwidth and battery life.* Terminals of next generation networks will be characterized by a wide range of supported bandwidths, ranging from several kbps to about 100 Mbps or beyond. The battery life of these devices is expected to be around one week. This advance will be accompanied by reduction in the weight and volume of batteries.

- *Packet-switched fixed network.* According to studies, the 4G architecture will use a connectionless packet switching (possibly IP-based) fixed network to interconnect the several different mobile and wireless networks.

- *Varying quality of bandwidth for wireless access.* The mixing and internetworking of different networks on a common platform will provide a set of, possibly overlapping, layers with different access technologies complementing each other. Depending on their geographical location, users will be served by different layers and enjoy different qualities of wireless access in terms of bandwidth. Possible layers will be [1]:

– *Distribution layer.* This will support digital video and broadcasting services at moderate speeds over relatively large cells. This layer will support full coverage and mobility and will cover sparsely populated rural areas.

– *Cellular layer.* This layer will comprise 2G and 3G systems. It will provide high capacities in terms of users and data rates inside densely populated areas such as cities. This layer will offer support for rates up to 2 Mbps. The cell size will obviously be smaller than that used in the distribution layer. This layer will also support full coverage and mobility.

– *Hot-spot layer.* This layer will support high-rate services over short ranges, like offices or buildings. It will comprise WLAN systems, such as IEEE 802.11 and HIPERLAN.

This layer is not expected to provide full coverage, due to its short range, however, roaming should be provided.

- *Personal network layer.* This layer will comprise very short-range wireless connections, such as Bluetooth. Due to the very short range, mobility will be limited, however, roaming should also be provided in this layer.

- *Fixed layer.* This will comprise the fixed access systems, which will also be part of the 4G network of the future.

- *Advanced base stations.* Base stations of future generation networks will utilize smart antennas to increase system capacity. Furthermore, base stations will employ self-configuring functionality in an effort to reduce operating costs. Finally, these devices will obviously support a multitude of air interfaces in order to accommodate a wide range of terminals.

- *Higher data rates.* 3G systems will have, in the best case, an upper capacity limit of 2 Mbps. Although more than enough for the application demands of the years to come, 3G systems will most likely need to evolve in order to meet the mobile application demands of the next decades. 4G systems aim to provide support for such applications. Although there exists some vagueness regarding the maximum number for data rates of 4G systems, with reports mentioning from 50 Mbps [3,4] to 155 Mbps [2], 4G systems will surely offer significantly higher speeds than 3G systems.

In order to support the higher data rates new air interfaces will obviously be introduced. An ideal air interface should be spectrum efficient and provide the flexibility to offer different bit rates. Furthermore, such an interface should be resistant to frequency-selective fading and require little equalization; Orthogonal Frequency Division Multiplexing (OFDM) is an air interface that can meet such requirements and is expected to be greatly used in the wireless systems of tomorrow. It is described in the next subsection.

6.2.1 Orthogonal Frequency Division Multiplexing (OFDM)

Orthogonal Frequency Division Multiplexing (OFDM) is a form of multicarrier modulation, which splits the message to be transmitted into a number of parts. The available spectrum is also split into a large number of low-rate carriers and the parts of the message are simultaneously transmitted over a large number of low-rate frequency channels. By recalling that (a) the phenomenon that dominates the error behavior of wireless channels is fading; (b) fading is frequency-selective; and (c) delay spread must be very long to cause significant interference to a carrier, one can realize the inherent robustness of OFDM to fading. Thus, by splitting a message into parts and slowly sending (due to low-carrier bandwidths) these parts in parallel over a number of low-rate carriers, signal reflections due to multipath propagation will probably be late at the receiver only by a small amount of a bit time. This, together with the fact that overall message transmission is made over a large number of low-rate carriers in the same time, results in a high-capacity, multipath-resistant link.

OFDM resembles FDMA in that they both split the available bandwidth into a number of carriers. The obvious difference of course is that FDMA is a multiple access technique whereas OFDM is a form of multicarrier transmission. Another difference concerns efficiency: FDMA is inefficient in terms of spectrum utilization, since it wastes a significant amount of bandwidth as guard interval between neighboring channels in order to ensure that they do not interfere with one another. This bandwidth overhead allows signals from neighboring channels to be filtered out correctly at the receiver. TDMA systems which allow a single user to utilize the entire channel capacity for a specific time period are also subject to a bandwidth overhead since TDMA systems need to be synchronized. As a result, guard time periods occur at the beginning of each user's slot in order to compensate for synchronization problems between stations. Thus, TDMA systems also waste some bandwidth to ensure their proper operation.

Such bandwidth overheads are not desirable in future generations of wireless systems. This is because spectrum is expected to be a scarce resource, and given a certain amount of spectrum this will need to be utilized to the highest extent possible in order to accommodate as many users as possible. OFDM tries to solve this problem by significantly reducing the amount of wasted spectrum by dividing the message to be transmitted into a number of frequency carriers and spacing these carriers very close to each other. In order to ensure that OFDM carriers do not interfere, they are made orthogonal to one another. Orthogonality ensures that although carriers are very close in frequency and their spectra overlap, messages in different carriers do not interfere with one another since detection for one carrier is made at the point where all other carriers are null.

In an OFDM system, detection is performed in the frequency domain. The actual signal transmission, however, occurs in the time domain. To better understand this, Figure 6.1 illustrates the operation of a simple OFDM system. As can be seen, OFDM transmission/reception comprises the following states:

- *Transmitter: serial to parallel conversion.* The data stream to be transmitted takes the form of the word size required for transmission. For instance, if QPSK is used, the stream is split into data words of two bits each. Then each data word is assigned to a different carrier.
- *Transmitter: modulation of each carrier.* The data word that forms the input of each carrier is modulated.
- *Transmitter: Inverse Fourier Transform (IFT).* After the actual contents of the various frequency carriers have been defined, the contents of these carriers form the input to an IFT in order to obtain a representation of the OFDM signal in the time domain. The IFT can be performed using the Fast Fourier Transform (FFT), which nowadays can be implemented at low cost.
- *Transmitter: Digital to Analog Conversion (DAC).* The output of the IFT is converted into an analog form suitable for radio transmission.
- *Receiver.* In order to receive the message, the receiver performs the reverse operation to

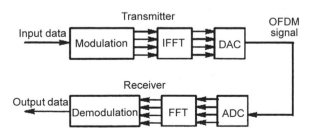

Figure 6.1   A simple OFDM system

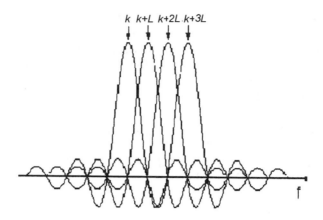

Figure 6.2   Detection of OFDM symbols

the transmitter. It digitizes the received signal (the ADC box in Figure 6.1) and performs an FFT on the received signal in order to obtain its representation in the frequency domain. The output of this is the actual content of the carriers, which are then demodulated in order to obtain the data words transmitted in each carrier. The data words are then combined to produce the original message.

One can realize from the above discussion that before OFDM modulation the data on each carrier is considered to be in the frequency domain. Figure 6.2 shows the various carriers of an OFDM transmission. The spectrum of each OFDM carrier has a $\sin(x)/x$ form and is modulated at a certain symbol rate. For the purposes of this discussion we assume $l = 1$ kHz symbol rate. Assuming that the main lobe of the signal on the first carrier is at k kHz, this signal will have the first null at $k + l$,with subsequent nulls occurring every l kHz. If we modulate the second carrier at a frequency exactly l kHz (the symbol rate) higher than the first using the same symbol rate, the mail lobe of the second carrier occurs at a null of the first one. Using this approach, the main lobe of each carrier occurs at nulls of the other carriers. Thus, at the point of detection there is no interference from any other carriers.

In an effort to increase the robustness of each carrier to inter-symbol interference (ISI) caused by multipath propagation, the transmitted symbols can be prolonged by adding a guard interval between successive symbol transmissions. The existence of a guard interval allows for delayed components of a symbol's transmission to reach the receiver before the energy of the next symbol is received. The actual content of the guard interval is produced by repeating the 'tail' of the symbol and placing that tail before the actual symbol transmission. Provided that delayed echoes of a signal carrying a symbol k are within the guard interval, multipath propagation does not affect detection of the next symbol, $k + 1$. However, by preceding the useful part of the symbol's transmission time by the guard interval, we lose some bandwidth that cannot be used for transmitting information. Figure 6.3 illustrates the transmission of OFDM symbols in the time domain with use of guard intervals. The arrows in cases 'a' to 'c' represent the energy of symbol 2 at the receiver, in the time domain. In case 'a', there is obviously no intersymbol interference, thus decoding of symbol 3 produces the correct symbol. Decoding is also successful in case 'b', where delayed echoes of symbol 2

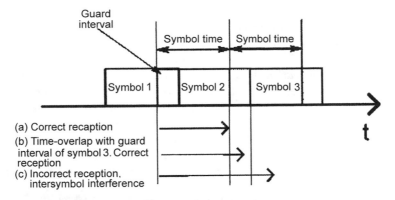

Figure 6.3   Adding a guard interval to transmitted symbols

overlap with the guard interval of symbol 3. However, in case 'c', decoding of symbol 3 will be affected by intersymbol interference since echoes of symbol 2 overlap in time with symbol 3.

Variants of OFDM also exist. COFDM stands for Coded OFDM. COFDM enables further resistance to errors due to fading. This is due to the fact that a carrier suffering one or more bit errors can be corrected by the error-correcting code which is transmitted on a different carrier, which may be error-free since fading is frequency selective. However, since coding for error correction is used in most of today's OFDM systems, the 'C' is redundant. Wideband OFDM (WOFDM) is a variant of OFDM where the spacing between carriers is wider in an effort to alleviate the problem of frequency errors between a transmitter and a receiver. The larger spacing ensures that such an error falls in the spacing and thus have a negligible effect on the performance of the system. Thus, an offset occurring at a transmitter will be perceived by the receiver only as a sampling error, which can be tolerated.

6.3 4G Services and Applications

The applications and service classes that will dominate the 4G-market are not yet known, however, some trends are emerging from ongoing research [5–8]. A nonexhaustive but indicative list of service classes is as follows:

- *Tele-presence.* This class will support applications that use full stimulation of all senses to provide users with the illusion of actually being in a specific place. These will be real-time virtual reality services and will offer virtual meetings, an evolution of today's teleconferencing applications. The conference attendants, although in different places, will have the illusion of participating in a conference in the very same room. Such applications, coupled with efficient compression techniques, will require capacities in the order of 100 Mbps. Furthermore, extremely strict delays and QoS levels will be demanded due to the real-time nature of these applications. The concept of a virtual meeting will be one of the major applications foreseen in 4G and future systems.
- *Information access.* This class will call for the ability of instantaneous access to large volumes of data such as large video and audio files. Compared to tele-presence, such

applications will be less delay sensitive, since real-time delivery of data is not needed here. As far as data rates are concerned, this class will demand the highest rates possible. However, the traffic pattern will probably be asymmetrical, with 50/1 ratios or more characterizing the downlink/uplink data rate ratio.

- *Inter-machine communication.* This service class will offer devices the ability to communicate with one another either for maintenance or for intelligence purposes. An example application of this type is car engine equipment that contains wireless interfaces enabling parts to contact the respective vendors when malfunctions occur.
- *Intelligent shopping.* This will offer users access to information regarding prices and products offered by shops they visit. Upon entering a shop, the user terminals will automatically tune to the shop's service providers and display information regarding the products sold by the shop.
- *Security.* Secure services will be an indispensable feature of the future generations of networks. Integrity of data is bound to be a crucial factor that will enable the proliferation of banking and electronic payment applications. Furthermore, security services will protect the privacy of users' personal information.
- *Location-based services.* It is envisioned that 4G and future systems will have the ability to determine the location of users with a high level of accuracy. This cannot be made true with today's systems which can only report the cell servicing the user, thus being accurate to within a few city blocks at best. Emergency applications will greatly benefit from location-based services. For example, if a person with a health problem calls an ambulance from his handset but is unable to report his location to the operator, his position can be determined with high accuracy by querying the user's handset for its location.

6.4 Challenges: Predicting the Future of Wireless Systems

In the course of research on 4G and future systems many issues of different types (technical, economical, etc.) must be studied and resolved. Some, such as the development of even more efficient modulation techniques, identification of new spectrum, and developments in battery technology/power consumption, are quite straightforward and have been identified during 2G and 3G research and development stages. Other issues are not so clear and are dependent on the evolution of the telecommunications market and society in general. Although the aim of 4G research will obviously be towards better performance, certain aspects of the telecommunications market and society's perception of communications may significantly influence the market penetration of products for the next generation of mobile and wireless networks.

As already mentioned, 4G and future systems target the market of 2010 and beyond. Since we cannot reliably foresee the state of telecommunications and society after such a time period, it is practical to study possible evolution scenarios in order to identify issues that may impact the future market for such systems and thus affect the related research. Three such scenarios have been identified [5–8]. In the remainder of this section, we provide a short overview of the concepts of these scenarios, how these three scenarios were created and finally present the three scenarios.

6.4.1 Scenarios: Visions of the Future

The concept of scenarios as tools for prediction future situations was first used after World War II to evaluate the significance of development in various technological areas. In order to keep up with the increasing pace of development, the two superpowers needed to set certain priorities. The problem was which priorities to set. A possible solution was to spy on the other side, understand its priorities and act accordingly. The other option was to act independently by predicting the developments and set priorities according to the predictions. Since a single prediction is not accurate, more than one possible prediction for the future was preferable in order to prepare for more than one different alternative situation. Each of these different predictions is called a scenario.

Scenarios are basically stories that express assumptions about the future. These assumptions are the result of different individuals' and groups' beliefs about the future. Scenarios are usually produced by posing specific questionnaires to, possibly, specialized groups of people. The individual opinions combine to produce a set of trends for the future. By identifying the trends that are sure to play an important role in the future and varying the relative impact of other trends, several scenarios are produced.

Scenarios are useful in cases where limited knowledge on a future situation exists, however, a decision regarding the situation has to be made. There are, of course, inherent vulnerabilities of the scenario-based approach: one cannot predict what will really happen, but only speculate based on present situations. Furthermore, in the process of identifying the trends that make up the scenarios, several factors that influence the situation might be overlooked or misinterpreted. Furthermore, as we approach the time of the situation under study, visions on the situation may change and thus some trends may vanish and new ones may appear.

6.4.2 Trends for Next-generation Wireless Networks

In the process of the research mentioned in Ref. [8], several trends regarding next generation wireless networks (2010 and beyond) were identified. These are briefly summarized below:

- *Globalization of products, services and companies.* Globalization has affected peoples' lives ever since the time ancient civilizations started to come in contact with each other. However, globalization show a surge with the invention of television, Internet and telecommunications in general. According to the survey, the impact of globalization will continue to exist and will surely affect the telecommunications scene of the future.
- *Communicating appliances.* This trend states that future consumer devices, such as TV sets, videos and stereos, will employ 'intelligence'. Although this is also true for the present, future consumer devices are expected to make certain kinds of decisions on their own and have the necessary equipment to communicate with other devices.
- *Services become more independent of the underlying infrastructure.* This trend states that future services are expected to be more separated from the infrastructure they use. This will enable many different devices to use the same network infrastructure.
- *Information trading/overflow.* Communications in the society of the future will be an integral part of peoples' lives. Computers will be the primary means for accessing information, thus diminishing the importance of printed versions of mass communications like newspapers. This trend also identifies the possibility of individuals receiving large

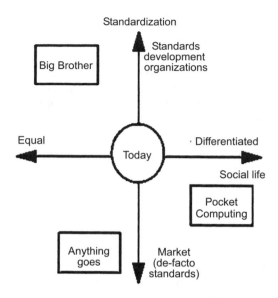

Figure 6.4   The three scenarios' dependence on standardization and social issues.

amounts of information, much more than they can handle. This trend identifies the need for refining and controlling information exchanges.

- *Standardization diversification.* This trend identifies the possibility of companies taking over control of the market and forcing their own de facto standards. This could be either due to political issues inside standards development organizations or market success giving power to some companies.

The following sections provide three scenarios for the future of telecommunications that were identified by research. Figure 6.4 shows the way social issues and standardization affect the generation of those scenarios.

6.4.3 Scenario 1: Anything Goes

This scenario has the following characteristics:

- High development rate for telecommunications.
- Transparent access to the network.
- Manufacturing companies have a strong market power.
- Large number of de facto standards.
- Generic hardware equipment will run software enabling specialized services.
- Self-configuring systems.

In this scenario, telecommunications technology is envisioned to achieve a deep market penetration and become an essential part of peoples' everyday life. This will lead to fierce industrial competition and decreased cost of product manufacturing and service offering. The reduced cost of products and services will enable almost everyone to have the ability to seamlessly access the services of the next generations of networks regardless of what access system is used. The increased acceptance of 4G and future systems will raise research to extreme levels with crucial aims being the identification of techniques offering efficient management of the scarce spectrum (possibly altering regulation processes) and efficient handling of the high number of subscribers. Furthermore, since telecommunications will become an integral part of peoples' lives, a high degree of mobility is expected to appear. This will require research for flexible and fully automated dynamic resource allocation and flexible roaming schemes.

The increased popularity of telecommunication systems will make companies manufacturing such products a dominant player in the telecommunications world. This will possibly change the way standardization work is conducted in the future. Companies that enjoy a big market share will probably establish their own de facto standards bypassing standards developments organizations. This means that the significance of such organizations will diminish.

The deep penetration of telecommunications in peoples' lives will serve a very diverse range of needs. Users will demand availability of ready-to-use systems, tailored for their needs. Thus, it would be desirable to research towards intelligent ad hoc systems, able to either automatically deploy and configure themselves or demand little such knowledge and intervention by users. Furthermore, personal adaptation of services based on user preferences would also be desirable. This will lead to individual applications adapted to specific users. Such intelligent systems will use a generic set of hardware and employ all the necessary functionality to support different networks and services in software.

6.4.4 Scenario 2: Big Brother

This scenario has the following characteristics:

- Privacy is the first priority.
- Governmental organizations ensure privacy.
- Limited telecommunications market.
- Low development rate of telecommunications.
- Very few operators.

This scenario foresees a limited telecommunications development speed. This is due to the fact that the rapid development of telecommunications in the earlier decade has led to a point where it will be easy to find almost any information about a person or a company, by directly eavesdropping on data exchanges, through the WWW, or by buying it from information thieves and traders. This, of course, is illegal, but the inherent freedom of the WWW provides the means to post and trade such information. Society will be very reluctant to use telecommunications services unless a high level of security is guaranteed.

To solve security problems, governments will form agencies responsible for certifying operators to be trusted and secure. These agencies will eventually come up with a mandatory security standard and act as Orwell's Big Brother, by making sure that all companies either follow this standard or are shut down. Every company that either manufactures telecommunication products or offers services will be tested to ensure compliance with the security standard. This will possibly lower the number of legal operators and product manufacturers since a number of companies may not pass certification and go out of business. The decreased number of companies and the reluctance of users to embrace telecommunications due to fears related to security problems will obviously limit the telecommunications market. The smaller market will make operating companies offer less money for research, thus lowering the speed of telecommunications development.

In such a scenario, the most important research issues will concern security and privacy. Since a lot of bandwidth will be consumed for security purposes, a 'security overhead' will characterize the performance of all telecommunication systems. Thus, development of efficient techniques offering high channel capacities over the same amount of spectrum will need to be addressed.

6.4.5 Scenario 3: Pocket Computing

This scenario has the following characteristics:

- Social and political differences.
- Existence of highly differentiated service and pricing categories.
- Service providers offering specialized services also provide equipment for specialized purposes.

This scenario envisions a world in which technological development is fast, however, the customer base is divided into two parts due to economical and sociological factors. The first part will comprise those customers who possess the financial ability to keep up with technology developments while the second part will comprise those who do not. The customers in the latter category will be ordinary people who prefer to pay for reduced services at minimum price. These people will use evolved versions of legacy 2G/3G systems. Evolved variants of GSM will still possess a significant market share due its low pricing, however, it will remain inappropriate for supporting multimedia needs due to lack of bandwidth. DECT, IS-95 and other legacy systems will also continue to exist. The second part of the customer base will comprise those users who will be able to afford the increased cost of advanced services. Such services will use the different wireless networks in combination and will be relatively expensive. Consequently apart from other research issues, the issue of smooth integration and interoperability of 4G and future systems and 2G/3G legacy systems will have to be efficiently solved. Furthermore, despite the fact that the customer base will be divided, telecommunications is bound to become an integral part of peoples' lives. Thus, as in the case of the 'anything goes' scenario, a high degree of mobility is expected to appear. This will require research for flexible and fully automated dynamic resource allocation and flexible roaming schemes.

To support such a divided customer base, the service providers are likely to offer a wide range of different services, addressing the needs of various user groups. Furthermore, equipment developers will need to provide specialized terminals for each user group. Finally, spectrum regulation issues will need to be resolved, as new spectrum will be needed for the advanced services.

6.5 Summary

This chapter provides a vision of 4G and future mobile and wireless systems. Such systems target the market of 2010 and beyond, aiming to offer support to mobile applications demanding data rates of 50 Mbps and beyond. Due to the large time window to their deployment, both the telecommunications scene and the services offered by 4G and future systems are not known yet and as a result aims for these systems may change over time. However, as 3G systems move from the research to the implementation stage, 4G and future systems will take their place as an extremely interesting field of research on future generation wireless systems. This chapter has covered a number of issues:

- *4G design goals and related research issues.* 4G and future systems aim to provide a common IP-based platform for the multiple mobile and wireless systems and possibly offer higher data rates. The desired properties of 4G systems are identified. OFDM, a promising technology for providing high data rates, is presented.
- *4G services and applications.* Although the applications and service classes that will dominate the 4G market are not yet known, research has identified some possibilities. Tele-presence, information access services, inter-machine communication and intelligent shopping will be enabled by 4G and future systems.
- *The challenge of predicting the future of wireless systems.* The exact state of 4G and future systems cannot be reliably foreseen, due to the large time window until their deployment. Many issues of these systems are not so clear and are dependent on the evolution of the telecommunications market and society in general. Scenarios are tools for predicting future situations and setting research priorities. Three different scenarios for the future generations of wireless networks are presented, along with possible research issues for each scenario.

WWW Resources

- *http://www.s3.kth.se/radio/4GW:* this is the home page of the Personal Computing and Communication research group of the Swedish Royal Institute of Technology. The group's effort is towards development of a 4G system.
- *http://www.ofdm-forum.com:* this is the home page of the industry-initiated OFDM Forum. The Forum is open to anyone interested in OFDM and its aim is to achieve market acceptance of OFDM through the establishment of a single high-speed OFDM standard.

References

[1] Mohr W. Development of Mobile Communications Systems Beyond Third Generation, Wireless Personal Communications, Kluwer, June 2001, pp. 191-207.
[2] Lilleberg J and Prasad R. Research Challenges for 3G and Paving the Way for Emerging New Generations, Wireless Personal Communications, Kluwer, June 2001, pp. 355-362.
[3] Varshney U and Jain R. Issues in Emerging 4G Wireless Networks, IEEE Computer, June 2001, pp. 94-96.
[4] Wideband Orthogonal Frequency division Multiplexing (W-OFDM), Wi-LAN Inc., version 1.0, 2000.
[5] Flament M, Gessler F, Lagergren F, Queseth O, Stridh R, Unbehaun M, Wu J and Zander J. Key Research Issues in 4th Generation Wireless Infrastructures, In Proc. of the PCC Workshop, Stockholm, Sweden, 1998.
[6] Flament M, Gessler F, Lagergren F, Queseth O, Stridh R, Unbehaun M, Wu J and Zander J. Telecom scenarios for the 4th Generation Wireless Infrastructures, In Proc. of the PCC Workshop, Stockholm, Sweden, 1998.
[7] Flament M, Gessler F, Lagergren F, Queseth O, Stridh R, Unbehaun M, Wu J and Zander J. An Approach to 4th Generation Wireless Infrastructures-Scenarios and Key Research Issues, In Proc. of IEEE VTC 1999.
[8] M. Flament, F. Lagergren, R. Stridh, O. Queseth, M. Unbehaun, J. Wu, J. Zander, Telecom Scenarios : a wireless infrastructure perspective, PCC Group report (2010) 1998.

7

Satellite Networks

7.1 Introduction

7.1.1 Historical Overview

The first reference to satellite communication systems was made in the mid-1940s by Arthur Clarke [1]. In this paper, Clarke described a number of fundamental issues relating to the building of a satellite network that entirely covers the earth including issues related to spectrum use, the power needed to run the network and the way of bringing the satellites to orbit. Clarke also introduced the concept of geostationary satellites, which – as explained later – orbit the earth in a radius that allows them to appear stationary from the earth's surface. These ideas seemed to be too ambitious at the time of the publication due to the fact that technology was not advanced enough to allow for reliable transceivers and easy deployment of satellites. Nevertheless, after no more than 40 years, satellites have emerged to be a significant industry. This is mainly due to (a) the introduction of transistors, which enabled the construction of small and reliable devices, and (b) the advancements made in rocket technology, which now allows for easier and less costly deployment of satellites as well as easier access by the astronauts for maintenance purposes.

The evolution of satellite technology did not occur over a small time period but rather followed an evolutionary path. Satellite technology was enabled by the advances in radio, telemetry and rocketry technology during World War II and the cold war era. The first attempts to establish communications via objects orbiting the earth commenced in 1956 by the US Navy. The orbiting object that was used was the natural satellite of Earth, the moon. This project used 26-m antennae in two base stations in Washington and Hawaii, which exchanged messages by bouncing signals off the moon's surface. Two years later the ECHO project offered single hop radio coverage of the entire US area through a passive reflector that was carried by a balloon at an altitude of 1500 km.

However, the era of true satellites began in 1957 with the launch of Sputnik by the Soviet Union. Nevertheless, the communication capabilities of Sputnik were very limited. The first real communication satellite was the AT&T Telstar 1, which was launched by NASA in 1962. This satellite enabled real-time two-way communications and had the ability to relay either 600 voice channels or a single television channel. Telstar 1 was enhanced in 1963 by its successor, Telstar 2.

From the Telstar era to today, the satellite industry has enjoyed an enormous growth offering services such as data, paging, voice, TV broadcasting and a number of mobile services. However, the position of satellites in the communications scene turned out to be quite different from that envisioned a couple of decades ago. At that time, the high bandwidth and wide coverage offered by satellite systems led to the conclusion that the future of communications lay with satellites. Nevertheless, the introduction of high-bandwidth fiber-based links changed this and the biggest application of satellites turns out to be as a wireless local loop technology with great coverage. There are a number of issues that favor the use of satellites in certain applications [2]. These issues are briefly summarized below:

- *Mobility.* Satellites favor applications that demand mobility, whereas fiber networks are limited in this sense.
- *Broadcasting.* Satellites offer the capability of easy broadcasting of messages to a very large number of ground stations. This is easier than implementing broadcasting on a wired network.
- *Hostile environments.* Satellites can easily provide coverage to areas where installation of wires is either very difficult or costs a lot. Such is the case of providing telephony services in Indonesia, where wiring the large number of islands was impractical and thus a dedicated satellite serves domestic telephone communications.
- *Rapid deployment.* By using satellites, a network can be deployed far more quickly than a wired-based one. This is very important in disaster situations or military applications.

7.1.2 Satellite Communications Characteristics

Satellite communications typically comprise two main units, the satellite itself and the Earth Station (ES). The satellite, which is also known as the space segment of the system, essentially acts as a wireless repeater that picks up uplink signals (signals from the ES to the satellite) from an ES and, after amplification, transmits them on the downlink (from the satellite to the ES) to, possibly more than one, other ESs. Due to this functionality, satellites are also known as bent-pipes. The uplink occupies a different frequency band than that of the downlink. Furthermore, there may exist more than one uplink channel. Thus, satellites typically contain many transponders, each of which contains receiver antennae and circuitry in order to listen to more than one uplink channel at the same time. Using the above scheme, communications between two or more ESs that are substantially far away from one another is established over ES-satellite links. The uplink is a highly directional, point-to-point link using a dish antenna at the ground station. The downlink can cover a wide area or alternatively focus its transmission on a small region which will reduce the size and cost of ESs. Some satellites can also dynamically redirect their focused transmissions and thus alter their coverage area. Moreover, as seen in later sections, satellites exist that employ the functionality that enables them to communicate directly with one another either for control or data message exchanges.

Satellite communication systems have a number of characteristics that differentiate them both from wired and other kinds of wireless links. These characteristics are briefly summarized below:

- *Wide coverage.* Due to the high altitudes used by satellites, their transmissions can be picked up from a wide area of the Earth's surface. The area of coverage of a satellite is known as its footprint.
- *Noise.* It is known that the strength of a radio signal reduces in proportion to the square of the distance between the transmitter and the receiver. Thus, the large distances between the ESs and the satellite makes the received signal very weak, typically in the order of a few hundred of picowatts). This problem is typically combated by employing FEC and ARQ techniques.
- *Broadcast capability.* As mentioned above, satellites are inherently broadcast devices. This means that a transmission can be picked up by an arbitrary large number of ESs within the satellite's footprint without an increase in either the cost or complexity of the system.
- *Long transmission delays.* Due to the high altitude of satellite orbits, the time required for a transmission to reach its destination is substantially more than that in other communication systems. Such propagation delays, which can be between 250 and 300 ms can cause problems in the design of satellite communication systems. An example of this situation is the inefficiency of using the CSMA/CD MAC protocol in satellite systems: It is known that in order for the carrier sensing mechanism of a CSMA protocol to perform satisfactorily, the propagation delay τ must be comparable to the frame transmission time t (in IEEE 802.3 LAN $t = 2\tau$). Since it is not possible for satellite frame transmissions to last at least 500–600 ms each, it is obvious than in satellite systems $t \ll \tau$, thus CSMA application will be inefficient. In most satellite systems, the access method used is FDMA- or TDMA-based.
- *Security.* As in all kinds of wireless communication systems, security is also a major concern in satellite systems.
- *Transmission costs independent of distance.* In satellite systems, the cost of a message transmission is fixed and does not depend on the distance traveled.

7.1.3 Spectrum Issues

As in other wireless communication systems, satellite systems are also subject to international agreements that regulate frequency use. Such agreements also regulate the use of the various orbits, which is described in the next section. Figure 7.1 shows the three bands that are commonly used. It can also be seen from this figure that different frequencies are used for the uplink and downlink channels. The 'C' bands were the first to be used for satellite traffic. The frequency range of this band leads to dish diameters of 2–3 m. However, the 'C' band is overcrowded nowadays due to the fact that it is also being used by terrestrial microwave links. As a result, the trend is towards use of the higher-

Band	Downlink Frequency (GHz)	Uplink Frequency (GHz)
C	3.7 – 4.2	5.925 – 6.425
Ku	11.7 – 12.2	14.0 – 14.5
Ka	17.7 – 21.7	27.5 – 30.5

Figure 7.1 The main frequency bands for satellite systems

frequency Ku and Ka bands. The Ku band is typically used for broadcasting and Internet connections and enables antenna diameters as low as 0.5 m. This band typically suffers less interference than the 'C' band, however, its higher frequency makes it susceptible to interference. Specifically, this band is subject to interference from rain, however, this can be combated by using a large number of widely separated interconnected ESs. As storms appear over relatively small geographical areas, they are likely to cause interference only to a small number of ESs and the system will be able to adapt by switching between ESs. The above problem also concerns the Ka band, which also has a disadvantage in terms of cost, since the equipment needed to operate at this band is more expensive than that for the other bands. Plans to use frequency bands higher than Ka, such as the V band (40–75 GHz) also exist. These offer the advantages of higher bandwidths and smaller antenna size, however, the technologies needed to use these bands are still under development.

From Figure 7.1, it can be seen that in all bands the lower part is the one that serves downlink traffic while the upper part serves uplink traffic. This is because higher frequencies suffer greater attenuation than lower ones and consequently demand increased transmission power to compensate for the loss. By using low frequency channels for the downlink, satellites can operate at lower power levels and thus preserve energy. On the other hand, ground stations are not subject to power limitations and thus use the higher parts of the bands.

7.1.4 Applications of Satellite Communications

There are a number of applications where satellite communication systems are involved. An indicative list is briefly outlined below.

- *Voice telephony.* Satellites are a candidate system for interconnecting the telephone networks of different countries and continents. Although the alternative of cables also exists, satellite use for interconnecting transoceanic points has sometimes been preferred rather than installing submarine cables.
- *Cellular systems.* Satellite coverage can be overlaid over cellular networks to provide support in cases of overload. When cells in the cellular network experience overload, the satellite can use a number of its channels to serve the increased traffic in the cell.
- *Wordwide coverage systems.* Satellite systems can provide connectivity even to places where no infrastructure exists, such as deserts, oceans, unpopulated areas, etc.
- *Connectivity for aircraft passengers.* This is a service that is provided by geostationary satellites. Aircraft can be equipped with transceivers that can use such satellites to provide connectivity to passengers while airborne.
- *Global Positioning Systems (GPS).* The well-known GPS system offers the ability to determine the exact coordinates of the GPS receiver. This is achieved with the help of multiple satellites through triangulation.
- *Internet access.* Satellite communication systems possess a number of characteristics that enable them to effectively provide efficient Internet access to globally scattered users. Such characteristics are the broadcast capability of satellite systems, their potentially worldwide coverage independent of terrestrial infrastructure and support for mobility. This issue is described in a later section.

7.1.5 Scope of the Chapter

The remainder of this chapter is organized as follows. Section 7.2 presents the various possible orbits of satellite systems and describes their characteristics. Section 7.3 presents the VSAT approach and describes its topology and operation. Section 7.4 presents Iridium and Globalstar, which are primarily voice-oriented satellite systems. Satellite-based Internet access is discussed in Section 7.5. Various architectures are identified along with a discussion on routing and transport techniques. Finally, the chapter ends with a brief summary in Section 7.6.

7.2 Satellite Systems

Satellite communication systems comprise two main parts: the ground segment and the space segment. The ground segment consists of gateway stations, a network control center (NCC) and operation control centers (OCCs). Gateways interface the satellite system to terrestrial networks, perform protocol translation, etc. NCCs and OCCs deal with network management and control of satellite orbits. The space segment comprises the satellites themselves, which are often classified by the orbit they use. Thus, satellite orbits are an essential characteristic of a satellite communication system. They are characterized by the following properties:

- *Apogee*: the orbit's farthest point from the Earth.
- *Perigee*: the orbit's closest point to the Earth. This has to be significantly outside the Earth's atmosphere in order to avoid severe friction.
- *Orbital period*: This is the time it takes to go around the Earth once when in this orbit and is determined by the apogee and perigee.
- *Inclination*: This stands for the angle between the orbital plane and the equatorial plane of Earth.

Many characteristics of artificial satellites can be studied with the help of the laws of Kepler. Originally developed to describe planetary motion, these laws also apply to satellites. According to Kepler's First Law, orbits are generally elliptical, however, satellites usually target orbits that are almost circular in an effort to minimize the variance of their height. Thus, in the following discussion, assume circular orbits unless stated otherwise.

An important characteristic of a satellite is the time it is visible to a given position on the surface of the Earth. This characteristic is defined by the orbital radius of the satellite and its inclination to the equator. For a circular orbit of distance D from the center of Earth, T can be calculated with the help of the Third Law of Kepler:

$$T^2 \propto D^3 \tag{7.1}$$

Circular orbits can be categorized in ascending order into low, middle and geosynchronous. These are shown schematically in Figure 7.2. A discussion on the characteristics of the various orbit categories is given below followed by a discussion on the characteristics of systems that employ elliptical orbits.

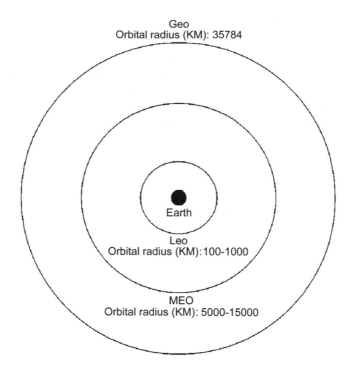

Figure 7.2   Low, middle and geosynchronous circular earth orbits

7.2.1 Low Earth Orbit (LEO)

LEO orbits are those that lie in the area between 100 and 1000 km above the Earth's surface. The small radius of a LEO orbit gives it a small period of rotation T (typically between 90 and 120 min), which of course translates into a high orbiting speed (high angular velocity). The main characteristics of LEO orbits are the following:

- *Low deployment costs.* Lower orbits are easier to reach by rocket systems. This translates into reduced cost for satellite deployment.
- *Very short propagation delays.* Due to their low distance from the Earth's surface, LEO systems exhibit very short propagation delays. This is a very useful property that simplifies the development of satellite communication systems, especially voice-related ones. Typical propagation delays for LEO are between 20 and 25 ms, which are comparable to that of a terrestrial link.
- *Very small path loss.* As we have seen, the received signal strength at distance r follows a kr^{-n} characteristic. This of course means that lower orbits are characterized by a smaller path loss and thus a smaller BER. Thus, LEO-based systems have low power requirements. Furthermore, for a given transmission power, LEO systems can receive the signal more easily than higher-orbit systems, a fact that lowers the complexity of terminals. This lower complexity allows for portable terminals.
- *Short lifetime.* The Earth's atmosphere extends to several thousands of kilometers above its surface and becomes thinner with increasing height. At the altitudes of LEO systems,

friction with atmospheric molecules is more intense than in higher orbits. This fact causes LEO satellites to quickly lose height and eventually fall back to Earth. Some satellites contain small boosters that regularly re-adjust their height in order to compensate for the loss. However, these boosters require fuel and cannot operate using solar power. Thus, when the satellite runs out of fuel the problem still exists. Of course, LEO satellites could be brought back to proper orbit by a space shuttle, as happens in the case of the Hubble telescope. However, this approach is more costly than deploying a new LEO satellite and is thus not followed. Consequently, LEO systems have a small lifetime and must be replaced every few years.

- *Small coverage.* The low height of a LEO satellite means that it has a decreased footprint. This fact is a disadvantage of LEO systems due to the fact that many satellites are required for worldwide coverage (e.g. the Iridium project that is covered later called for a constellation of 66 LEO satellites). As a consequence, both the complexity and cost of a LEO system to cover the entire Earth is increased.
- *Small Line of Site (LOS) times.* LEO systems are characterized by angular orbiting speeds. This is problematic from the point of view of the time the satellite remains visible from a given location on the Earth's surface. For LEO systems this time is very small. This means that terminals will need to possess steerable antennae in order to track the satellites as they move. Furthermore, the high angular speed raises the need for efficiently combating large Doppler shifts. These facts of course raise terminal complexity.

7.2.2 Medium Earth Orbit (MEO)

MEO orbits are those that lie in the area between 5000 and 15,000 km above the Earth's surface. These orbits are higher than those of LEO systems, thus the orbital period T also increases (typical values of T are several hours). At such distances, the characteristics considered as advantages of LEO systems, fade to become disadvantages for MEO systems. Similarly, the characteristics considered as disadvantages of LEO systems, become advantages for MEO systems. Some of them are briefly summarized below:

- *Moderate propagation delay.* Although not much higher than that of LEO systems, the propagation delay in MEO systems is higher.
- *Greater lifetime.* The atmosphere is thinner at higher orbits. Thus, MEO systems experience lower friction with atmospheric molecules, a fact that translates into higher lifetimes.
- *Increased coverage.* The relatively high orbits of MEO systems give them an increased footprint. Compared with lower orbits, fewer satellites are needed to achieve worldwide coverage. A typical number is around ten. However, the exact number depends on the orbit radius.

Theoretically, MEO satellites can be deployed as high as 35,000 km or more. However, few MEO satellites use orbits above 10,000 km. This is due to the fact that at distances greater than this, deployment costs and propagation delay become significant without additional advantages. The most well-known system that uses MEO orbits is the Global Positioning System (GPS).

7.2.3 Geosynchronous Earth Orbit (GEO)

The Geosynchronous Earth Orbit (GEO) was discovered by Arthur Clark in his work [1]. If a satellite is placed at approximately 36,000 km above the Earth's surface, then its angular velocity will be the same as that of the Earth.

A special case of GEO is the Geostationary Earth Orbit. In this, the satellite rotates at an inclination of 90°, which means that it remains in the same spot above the Equator. In such a case the satellite will appear to remain fixed at the same position in the sky. This is very useful for communications systems since ESs antennae do not have to track the satellite as it moves but rather remain focused on a specific point.

Contrary to common belief, the Geosynchronous Earth Orbit has a period of 23 h and 56 min, not 24 h. This is because Earth makes a complete rotation around its axis in 23 h and 56 min. On the other hand, 24 h is the duration of the so-called solar day, which stands for the duration of a complete rotation of the Earth relative to the Sun. This difference of about 4 min stems from the Earth's motion around the Sun. Due to this motion, Earth has to rotate slightly more than 360° so that a given place on its surface points exactly towards the Sun. Consequently GEO satellites have an orbital period of 23 h and 56 min to match the angular speed of the Earth.

The main characteristics of GEO are the following:

- *No atmospheric friction.* At such a high altitude, atmospheric friction is nearly nonexistent. As a result GEO satellites remain in orbit for a very long time.
- *Wide coverage.* Due to their high altitude, GEO systems exhibit a wide coverage. By using three GEO satellites spaced 120° from one another, almost worldwide coverage can be achieved with obvious advantages for multicasting applications.
- *High deployment costs.* Due to the high altitude of GEO systems, the construction of rockets in order to deploy or reach the satellite for repair is high.
- *High propagation delay.* The high altitude of the geostationary orbit incurs a significant propagation delay. This causes problems for applications that require low delays, such as voice-related and interactive applications. Typical values of this delay for GEO systems are between 250 and 280 ms.
- *High path loss.* The high altitude of the geostationary orbit also translates into increased path loss. This translates into a need for increased transmission power and antennae sizes, which of course makes the construction of portable, low-cost mobile devices that communicate with GEO satellites difficult. The same problem applies to satellites, which also need to employ large antennae and powerful transmitters.

Geostationary satellites also have the following properties:

- *Static position.* Geostationary satellites appear to remain fixed at the same position in the sky, thus ESs only need to point their antennas at the satellite position once and leave them there.
- *Reduced coverage at high latitudes.* Geostationary satellites rotate above the Equator. This means that coverage at regions in the north and south is problematic due to the fact that a clear LOS must exist between the satellite and the ES. In regions of the Earth in the north and south the satellite will appear low in the horizon and LOS may be obstructed by buildings, hills, etc. This is shown schematically in Figure 7.3. Furthermore, the received signal power at these areas will be less, as for such latitudes it will have to travel through a

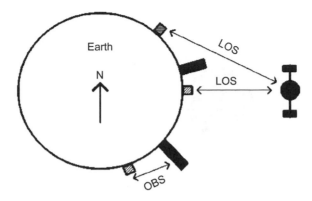

Figure 7.3   Line of Site (LOS) and Obstructed Line of Sight (OBS) situations at different latitudes for a gcostationary satellite

longer path in the atmosphere. This is shown in Figure 7.4. Thus, the dish size of ESs at such latitudes has to increase in order to compensate for the weakening of the signal.

- The geosynchronous orbit above the equator seems to be a valuable resource. As in the general case of GEOS, satellites at this orbit must be placed apart by at least 2°, meaning that there is room only for 180 geostationary satellites. As with frequencies, orbits are also handled by the ITU, which originally used a first-come first-served approach to assign geostationary orbit 'slots' to interested countries. As a result, such slots were mostly awarded to technologically advanced countries, a fact that irritated equatorial countries. Thus, ITU decided to allocate to these countries slots of their own. However, since few could actually use them, these slots remained unused until ITU stated that slot owners must either launch a satellite or give up their rights on the slot.

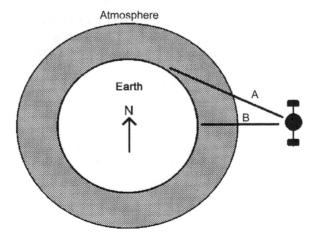

Figure 7.4   In situation A the path traveled through the atmosphere is longer than for B

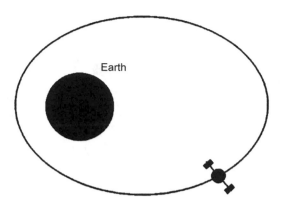

Figure 7.5   An elliptical-orbit satellite

7.2.4 Elliptical Orbits

Apart from the LEO, MEO and geostationary orbits, which are all very close to circular, there are satellites that employ elliptical orbits. Such an orbit is shown in Figure 7.5. The elliptical nature of the orbit results in a variation of both the altitude and the speed of the satellite. Near the perigee, the satellite altitude is much lower than that near the apogee. The opposite applies for the orbital speed. Near the perigee the speed is much higher than that near the apogee. As a result, from the point of view of an observer on the surface of the Earth, an elliptical-orbit satellite remails visible for only a small period of time near the perigee but for a long period of time near the apogee.

Elliptical-orbit satellites combine the low propagation delay property of LEO systems and the stability of geostationary systems. Thus, such a satellite has the properties of a LEO system near the perigee of its orbit (low delay, low LOS times) and the properties of a geostiationary system near the apogee (high LOS times, high propagation delays). Elliptical-orbit satellites are obviously easier to access near their apogee because their high LOS times and low speeds permits ESs to track them without having to perform very frequent antenna readjstments. Thus, systems that employ such orbits have found use in systems that provide high LOS times for regions of the Earth far in the north or south. Since such areas cannot be effectively serviced by geostationary satellites as they orbit above the equator, elliptical orbits can provide high LOS times for such areas. This approach was followed by the former USSR in the Molniya satellites; since most of USSR is located far too north for geostationary satellite coverage, three elliptical-orbit satellites at an inclination of 63.4° have been used. The orbits were chosen in such a way so that at least one satellite covered the entire region of the country at any time instant. The parameters[1] of the Molniya system are depicted in Figure 7.6, along with those of other elliptical-orbit systems.

---

[1] In this figure, eccentricity describes the form of the elliptical orbit. The higher the eccentricity, the more elliptical is the orbit. The circle has an eccentricity of zero.

	Molniya	Tundra	Loopus
Orbital period (hours)	12	24	24
Eccentricity	0.65	0.2	0.6
Apogee (Km)	39400	44220	41700
Perigee (Km)	2900	27350	5642

Figure 7.6 Parameters of elliptical-orbit satellite systems

7.3 VSAT Systems

As mentioned above, the design of ESs in satellite-based systems is quite complicated. This increases both construction and maintenance costs. An innovation in data communication satellites was brought about by the development of highly directional antennae which can focus transmission on a certain area of the Earth's surface. If such a directional antenna is integrated into the satellite, then ESs can afford to employ a smaller antenna in order to reduce their size and cost. This approach is known as Very Small Aperture Terminals (VSAT).

A VSAT system is typically organized into a star architecture, as shown in Figure 7.7. The system comprises the following elements:

- *A number of relatively small-sized terminals.* The small size of VSAT terminals allows easy installation at user premises and even mobility. However, as the system uses a geostationary satellite, the VSAT antenna size depends on the latitude of the terminal. Furthermore, it depends on the frequency used, since higher frequencies typically demand a smaller antenna.
- *An ES acting as a hub.* This ES has a very powerful antenna, employs routing capabilities and has a high-speed connection to a wired backbone in order to serve as a gateway to/from the VSAT network.
- *A geostationary satellite equipped with a directional antenna.* This satellite is used to connect the VSAT terminals to the hub.

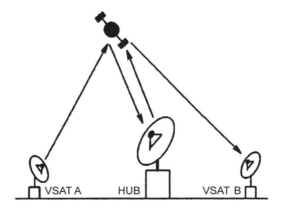

Figure 7.7 VSAT architecture

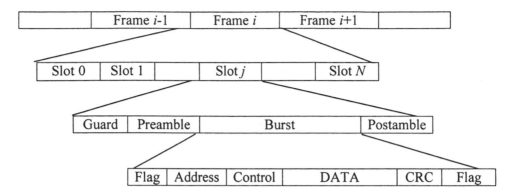

Figure 7.8   VSAT-to-hub channel structure

Using the architecture of Figure 7.7, the VSAT terminals transmit data to the satellite by using a random access technique. Most of the time this is ALOHA-based with typical examples being pure ALOHA, slotted ALOHA or an ALOHA/TDMA combination, like the dynamic TDMA schemes that were covered in Section 6.2. The organization of the VSAT to hub channel is shown in Figure 7.8. After receiving VSAT traffic, the satellite transmits it back to the hub. The hub performs collision checks and upon successful reception of a packet, uses the satellite to route the packet to the intended destination.

Contrary to traffic from the VSAT terminals to the hub, traffic in the opposite direction is delivered via a TDM scheme. This scheme is shown in Figure 7.9. It comprises a number of frames which in turn comprise slots that are used to transmit packets. As can be seen from the figure, every frame comprises a synchronization pattern which is used to keep the VSATs reliably synchronized. Every VSAT uses the address field to extract from the TDM scheme the packets which are destined for it and filter out all other packets. Of course, special addresses can be used in order to enable a message in the uplink to be broadcast to all VSATs or multicast to a specific group of VSATs. As far as network protocols are concerned, the most commonly used in VSAT systems is X.25.

VSAT systems are especially useful for interconnecting large numbers of users residing in remote areas. Furthermore, in some cases a VSAT system is likely to be more econom-

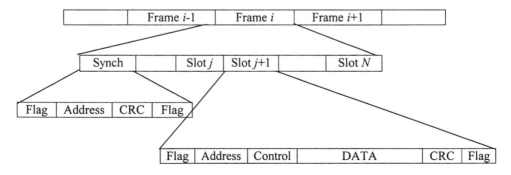

Figure 7.9   Hub-to-VSAT channel structure

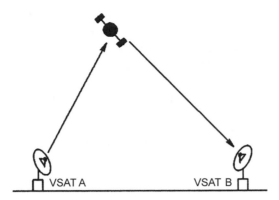

Figure 7.10 A VSAT communication system via an intelligent satellite

ical than a wired-based system. However, the main disadvantage of a VSAT-based system is that terminal traffic has to be relayed through the ES, a fact that results in a delay at least twice that of the propagation delay from a VSAT to the satellite. However, in recent years, technology has enabled incorporation of the functionality of the ES hub into the satellite. Thus, VSATs can now be connected directly via the intelligent satellite, as shown in Figure 7.10, with an obvious decrease in the propagation delay.

7.4 Examples of Satellite-based Mobile Telephony Systems

In the late 1980s, satellite systems appeared to be a promising approach for constructing telephony systems with worldwide coverage. At that time, conventional cellular telephony was not very widespread and its cost was relatively high. These facts made room for satellite-based systems. However, by the time satellite-based systems were ready for deployment, the market penetration of cellular telephony was so big that little space was left for satellite phones. However, satellite telephony is not completely without future or benefit: It is still an efficient way to link mobile users existing in regions of the world without communications infrastructure. Furthermore, it may be the only available mobile telephony system in many regions of the world, as there are countries in which conventional cellular systems have a limited coverage.

In this section, we study two examples of satellite-based mobile telephony systems: Iridium and Globalstar. Iridium was an ambitious project aiming for worldwide coverage using a dense constellation of LEO satellites. However, the project was finally abandoned in 2000. Globalstar, which on the other hand had a better fate than Iridium, is a simpler system and its coverage also depends on the existence of ES.

7.4.1 Iridium

The Iridium project [3–5] was initiated by Motorola in the early 1990s. The project aimed to offer coverage to every place on the planet through a dense constellation of LEO satellites. The Iridium satellites employ significantly richer functionality than simple 'bent-pipe' satellites by enabling intra-satellite communication for relaying of control signaling and phone calls. The project initially called for use of 77 LEO satellites. This was the fact that gave it the name Iridium, since Iridium is the chemical element with an atomic number of 77. Despite the fact that the number of satellites was later reduced to 66, the name Iridium stayed probably due to the fact that the marketing people preferred it to Dysprosium, which is the chemical element with an atomic number of 66. Nevertheless, this decision did not seem to favor the project's fate, as Iridium was finally abandoned for economic reasons in 2000.

The Iridium system comprised four main components: the satellite constellation, the system control facilities, the gateways and the subscriber units. These are described below, along with a number of issues relating to the operation of Iridium.

7.4.1.1 The Iridium Satellite Constellation

Iridium employs 66 LEO satellites, about 700 kg each, that orbit the Earth at an altitude of 780 km above its surface. This altitude was chosen in an effort to minimize delay (which is discussed later), enable portable ground units and provide for an acceptable satellite lifetime (8 years for an Iridium satellite). This is the time it takes for a satellite to consume the fuel, which powers the engines that combat friction with the atmosphere and keep the satellite in proper orbit. Iridium satellites have an orbital period of 100 min.

Iridium satellites are divided into six polar orbital planes with each plane having 11 satellites. Each orbit has an inclination of 86.5° with respect to the equator. In each plane satellites rotate in the same direction with the exception of planes 1 and 6 which are counter-rotating. Co-rotating and counter-rotating planes are spaced 31.6° and 22° apart. Each satellite is able to maintain up to four Inter-Satellite Links (ISLs), two of which are permanent and involve the two adjacent satellites in its plane. The other two ISLs are dynamically established with the two satellites in the adjacent orbital planes. Exceptions to this fact are the satellites in planes 1 and 6. These maintain only three ISLs, due to the fact that the rapid relative angular speed of a pair of counter-rotating satellites from these planes does not allow them to establish ISLs between each other. Finally, ISLs operate at frequencies between 22.5 and 23.5 GHz at a link speed of 25 Mbps.

Each Iridium satellite is equipped with an antenna comprising three panels: the first is perpendicular to the direction of the satellite's travel, and the next two are 120° and 240° displaced relative to the first one. As the satellite moves in its orbit, the footprint of each panel obviously moves on the Earth's surface. Each panel transmits 16 beams resulting in a total of 48 beams per satellite. Combining this with the total of 66 satellites used by the system, one can see that Iridium provides 3168 beams overall. However, only 2150 beams are used to provide global coverage, due to the fact that that there is a significant overlap among the beams of satellites from adjacent orbital planes when these satellites are above areas near the poles. Since global coverage can be achieved without such an overlap, a satellite's beams are reduced near those areas in order to conserve power.

7.4.1.2 Frequency Reuse

Iridium employs frequency reuse like conventional mobile telephony systems. It divides bands into groups, called clusters. Each cluster contains beams that can use the same frequency. The principle of operation is the same as that of frequency reuse schemes of cellular systems. Adjacent beams are not allowed to use the same frequency. Iridium uses a frequency reuse factor of 12. Beams that use the same frequency channels can be found as follows: (a) starting from the center of a beam, move two beam centers; (b) make a turn of 60°; and (c) move two cells.

7.4.1.3 MAC

Iridium employs a combination of TDMA and FDMA as its multiple access technique both for uplink and downlink. These use QPSK for modulation. The FDMA component is attributed to the above-mentioned frequency reuse scheme. The system uses the spectrum from 1616 MHz to 1625.5 MHz. Of this bandwidth, 10 MHz are used to constitute a total of 240 41.67 kHz channels. The bandwidth of these channels totals 10 MHz as the additional 500 kHz are used for establishing guard bands between adjacent channels. Each guard band has a width of 2 kHz.

The TDMA scheme comprises 90 ms frames each of which contains four pairs of slots supporting four full-duplex channels at a rate of 4800 bps. Additionally, half-duplex data channels of 2400 bps are supported. The specific details of the TDMA frame structure were not published in open literature [5]. The same holds for the nature of the voice codec used.

7.4.1.4 System Control Facilities

The operation of Iridium is assisted by two System Control Facilities (SCFs) which are responsible for maintaining control of the constellation of the 66 satellites. Each satellite is monitored via the SCFs which manages their operation in order to ensure correct performance within orbit. Furthermore, the network formed by these satellites is also monitored by the SCFs, which informs the constellation in the event of a malfunctioning node.

7.4.1.5 Gateways

Gateways are ESs that interface Iridium to external communication networks, such as the PSTN for voice calls. Such an interface extends the coverage of Iridium since it enables Iridium subscribers to place/receive calls from PSTN users. Gateways perform a number of operations, such as subscriber location, billing and call setup.

7.4.1.6 Numbering

As in cellular systems, Iridium subscribers are assigned a home gateway which contains a permanent record regarding the subscriber's identity. The numbers that can identify an Iridium subscriber are the following [4]:

- *Mobile Subscriber Integrated Services Digital Network Number (MSISDN)*. This number is the permanent number assigned to the Iridium subscriber. In order to dial a number to establish a voice call with the subscriber, the MSISDN is preceded by two more fields: (a)

The Iridium country code (ICC), which is a four-digit number that identifies the Iridium network; and (b) a three-digit geographical code that is used to identify a user's home country in cases of gateways that serve more than one country.

- *Temporary Mobile Subscriber Identification (TMSI).* This number is used to achieve confidentiality of the user's MSISDN. In order not to send the MSISDN over the airwaves, this is mapped to the TMSI which is sent instead. The TMSI changes periodically to increase security.
- *Iridium Mobile Subscriber Unit (MSI).* This number is permanently stored in the user's Subscriber Identity Module (SIM) that resides within his phone and uniquely identifies the subscriber.

7.4.1.7 Call Management: Subscriber Location

In order for Iridium to be able to serve roaming users, a method for determining the location of subscribers is needed. This is made possible through the concept of home and visitor gateways. The process of subscriber location is depicted schematically in Figure 7.11. For the purpose of illustration, assume a user registered in Europe travels to North America. The user's home gateway will be the European one, while his visiting gateway will be the North American one. When the user arrives in North America and switches on his phone, communications will be established with the closest Iridium satellite (point A in Figure 7.11). The satellite will connect the user to the local North American gateway (point B) which will of course recognize the user as a visitor and create a relative entry in its database. Furthermore, the visiting gateway will determine the subscriber's home gateway via his TMSI. Next, the visiting gateway will instruct the satellite above it to contact the home gateway (point C) and (a) inform it of the new location of the subscriber, (b) ask for permission to allow call access to the subscriber. The latter is necessary in cases of subscribers with pending bills or stolen phones. If the home gateway grants call access to the subscriber (point D), the latter is ready to make/accept calls via the North American gateway.

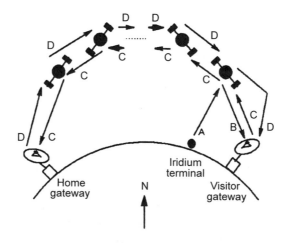

Figure 7.11   Subscriber location in Iridium

7.4.1.8 Call Management: Call Setup

When a call arrives for a user outside the area of its home gateway, then a joint operation of the home and visitor gateways ensures call setup. Returning to the case of the previous example, assume that a European PSTN user makes a call to the Iridium subscriber while the latter is still in North America. The PSTN user will obviously dial the Iridium ICC, which will lead the call to the European Iridium gateway. This is the home gateway of the Iridium subscriber. Thus, it will check its database and determine that the subscriber's current location is in North America. Consequently, it will use the satellite above it to contact the North American gateway. This operation will probably pass through one or more ISLs. The North American gateway is the subscriber's visitor gateway. It will check in its database, determine that the user is in its area and use the satellite above it to relay the call to the subscriber's terminal. When the latter goes off hook, a corresponding message is relayed to the visitor gateway via the satellite above. The visitor gateway then sends this message via the satellite constellation to the home gateway. Upon reception of this message, the home gateway starts sending voice packets to the satellite above the subscriber's location, which in turn relays them directly to the subscriber. Thus, the call is established. It can be seen that the call setup process in Iridium is very similar to that of the AMPS system.

7.4.1.9 Handoffs

There are three types of handoffs in Iridium: intra-beam, inter-beam and inter-satellite. Intra-beam handoff occurs when the satellite beam that serves a subscriber has to change its operating frequency as it approaches another geographical region. This could be due either to (a) regulatory issues that do not allow use of this frequency in the specific geographical region or (b) interference reasons. The latter is true in situations when the beam is too close to that of another satellite that uses a beam of the same frequency. In any of the above cases, the satellite will inform the user to change to the new beam frequency. In this handoff scenario, the intelligent unit is obviously the Iridium satellite.

Inter-beam handoff involves two different beams of the same satellite. Recall that in conventional cellular systems, terminals operating within a specific cell constantly monitor link quality to adjacent cells. Upon finding a link with better quality, a handoff to this cell is made. The same principle describes inter-beam handoff in Iridium: an Iridium terminal in operation constantly monitors the link quality of two adjacent beams. When the terminal detects an alternative beam with a better signal quality than that of the current beam, the terminal initiates a handoff to the new beam. In this handoff scenario, the intelligent unit is obviously the Iridium terminal.

Inter-satellite handoff involves two satellites. Due to the rapid movement of Iridium satellites relative to ground units (typical LOS times are 10 min) handoffs between satellites are very common. When an Iridium terminal goes out of coverage of the satellite above it (e.g. satellite A), it is approached by a new satellite (e.g. B) to which it will be handed. The handoff procedure is a responsibility of the local gateway since it knows both the satellite and terminal movements. The gateway will thus send a message to satellites A and B asking for release and acceptance of the terminal, respectively. After the handoff is

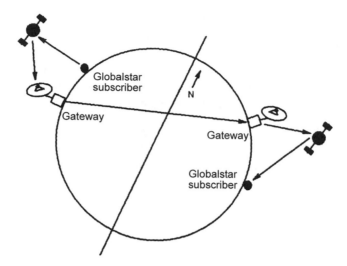

Figure 7.12   Operation of Globalstar

made, satellite B contacts the Iridium terminal in order to notify it of the frequency to use. In this handoff scenario the intelligent unit is obviously the gateway.

7.4.2 Globalstar

Globalstar [6,7] is a satellite-based telephony system that aims to enable users to talk from virtually anyplace in the world. However, the word 'virtually' has a definite meaning. This is due to the fact that contrary to Iridium, which enables true worldwide coverage through the use of ISLs for call routing, the operation of Globalstar depends on the presence of a Globalstar gateway in range of the satellite that serves the user. This is because gateways are necessary in order to connect users, since no ISLs are used, as in Iridium. This fact, which is shown in Figure 7.12, limits the coverage of Globalstar. On the other hand, it constitutes an advantage in terms of cost and system simplicity. Furthermore, since typical Globalstar gateways can have a range of many kilometers, few such stations are needed to support the system. When every gateway is operational, Globalstar can cover most of the Earth's surface, except for the regions in the middle of the oceans where ESs deployment is not possible or costs a lot and those near the poles, for reasons that are described later. The frequency bands used by Globalstar are shown in Figure 7.13.

Link type	Frequency (GHz)
Mobile-satellite	1.61-1.63
Satellite-mobile	2.48-2.5
Gateway-satellite	5.09-5.25
Satellite-gateway	6.7-7.08

Figure 7.13   The frequencies used by Globalstar

The Globalstar system comprises three main components: the satellite constellation, the gateways and the subscriber units. These are described below, along with a number of issues relating to the operation of Globalstar.

7.4.2.1 The Globalstar Satellite Constellation

The Globalstar system comprises LEO satellites that operate in eight planes. These satellites use LEO orbits with an altitude of 1400 km. Each plane contains six satellites and has an inclination of 52° with respect to the equator. This fact explains the system's inability to cover regions with latitudes beyond 70° in both hemispheres [7]. However, this is not as bad as it sounds, since most of the population (thus subscribers) are located outside those areas.

Each Globalstar satellite employs 16 beams and the same frequencies are reused within each beam. Satellite orbit is monitored and maintained by using a GPS system, which also supplies accurate time to the satellite [6]. The satellite is powered by rechargeable solar batteries and the system controlling the satellite's altitude is driven by small thrusters. Of course, when a satellite runs out of fuel it will eventually fall back to Earth. The life expectancy of a Globalstar satellite is around 8 years. Finally, both satellites and ground units in Globalstar implement antenna diversity in an effort to increase performance.

7.4.2.2 MAC

Globalstar uses CDMA as a MAC technique. The forward link (uplink from gateway to satellite and downlink from the latter to the user terminal) uses a chip rate of 1.2288 Mcps and a spreading factor of 256, which results in a peak information transmission rate of 4800 bps [6]. The forward link also employs a pilot channel that is defined by the same spreading code for all gateways. The pilot channel is used by Globalstar terminals to synchronize with gateways. The reverse link (uplink from the user terminal to the satellite and downlink from the latter to the gateway) uses a spreading code of length 215. Finally, the CDMA nature of Globalstar demands that the signal of all users reaches a gateway with the same power. This has led to use of closed-loop power control in order to combat the near-far problem.

7.4.2.3 Gateways

A gateway is a special fixed ES. Apart from its main functionality which is to link satellites, it contains HLRs and VLRs for user location management and performs operations relating to security, billing and interfacing to PSTN and GSM.

7.4.2.4 Handoffs

As was mentioned above, Globalstar satellites use the same frequency in adjacent beams or overlapping beams of different satellites. This fact enables soft handoff, which is similar to that of conventional CDMA-based cellular systems. When a Globalstar terminal is covered by another beam or satellite, it reports this to the gateway. Due to satellite movement, the Globalstar terminal is soon likely to experience a better link quality at the new beam. The gateway is thus informed that a soft handoff may take place and starts transmitting the same information at both beams. When the terminal is within coverage of the new beam, the system is informed of this fact and it drops the connection to the beam of the departing satellite.

7.4.2.5 Subscriber Units

Globalstar terminals can be either single or dual mode units that also support conventional cellular systems, such as GSM.

7.5 Satellite-based Internet Access

The amount of penetration of the Internet, both for personal and commercial use, in recent years is well known. A constantly increasing number of users use Internet services such as e-mail, web-browsing, file transfer, as well as QoS demanding services such as videoconferencing. Satellite communication systems possess a number of characteristics that enables them to provide efficient Internet access to globally scattered users. Such characteristics are the broadcast capability of satellite systems, their potentially worldwide coverage independent of terrestrial infrastructure and support for mobility. The true potential of satellite-based Internet systems relies on their availability to interoperate with existing terrestrial infrastructure in order to seamlessly provide service to users. There are a number of issues that makes the design of satellite-based data networks a challenging task. These stem not only from the use of wireless transmission but also from the relatively large distances between ESs and satellites. In the next subsections, we describe the main issues of satellite-based Internet access: possible architectures and routing in satellite constellations. Next, we discuss the inefficiency of conventional TCP for use as a transport protocol in satellite-based systems and present proposed enhancements that combat this inefficiency. Finally, we briefly cover commercial satellite systems that offer the capability of Internet access.

7.5.1 Architectures

Satellite systems can act either as high-speed parts of the Internet backbone, interconnecting a number of other networks, as Internet access networks, or a combination of the above [8,9]. Presently, the first two architectures are commonly used. These are described below.

7.5.1.1. Access Network

An access network has the architecture of Figure 7.14. In this scenario, subscriber terminal transmissions are picked up by satellites which relay these transmissions to the nearest gateway which interfaces the satellite system to the Internet. After reaching the gateway, user traffic is forwarded to its destination, which can be an Internet host either inside the terrestrial core network or a user terminal of another satellite system. It is obvious that in this architecture satellites do not employ significant intelligence and simply act as 'bent-pipes'.

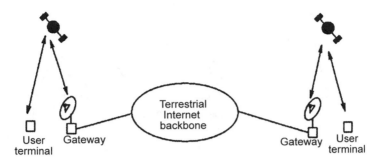

Figure 7.14   Satellite system for Internet access

7.5.1.2 Access/Core Network

A satellite-based access/core network has the architecture of Figure 7.15. In this scenario subscriber terminal data transmissions are again picked up by satellites. However, these are not necessarily sent to the gateway in order to reach their destination. In such an architecture, satellites have the ability to perform onboard processing and switching. This enables them to maintain ISLs which can be used to relay user transmissions to their destinations. Thus, for a satellite mobile receiver, packets can reach their destination either entirely through ISLs or through a combination of ISLs and the terrestrial Internet backbone. For a ground-based destination station, the packet is forwarded to the gateway which relays it to the terrestrial Internet backbone. Thus, in this architecture the network formed among the satellites acts as a part of the Internet backbone. It is obvious that in this architecture satellites employ significant intelligence which is required for operating ISLs.

7.5.1.3 Asymmetric Access Architecture

In the two architectures mentioned above, terminal-satellite links are assumed to operate

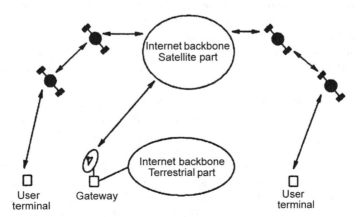

Figure 7.15   Satellite system acting as an Internet access and core network

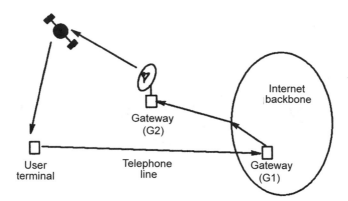

Figure 7.16 Internet access through a hybrid system

in both directions. However, terminal to satellite transmissions raises the complexity and thus the cost of the terminal. Thus, in an effort to make satellite-based Internet access more appealing to the consumer, the terminal to satellite links can be substituted by a low-rate terrestrial link, such a telephone link. The low rate of the terrestrial link is not a problem, since Internet traffic is highly asymmetric, with most of the traffic concerning data coming from an Internet server to the user, while the reverse link needs only to relay mouse clicks and user commands. Such an architecture can be considered a hybrid system.

The idea mentioned above is depicted in Figure 7.16. Users send their commands (such as requests for web pages) through the low-rate telephone line. This line is connected to the ISP's gateway (G1), which examines the request and sends the corresponding data to the user through the satellite network. A system following this architecture is DirecPC.

7.5.2 Routing Issues

Routing is a challenging task in all wireless systems due to their inherent mobility. The same holds for satellite-based systems, especially LEO-based systems employing ISLs. In such systems, the main technical issue is dynamic topology. This is due to the rapid movement of LEO satellites which remain visible only for a small amount of time from a specific point on the Earth's surface. In such a system, careful planning of the satellite constellation is needed, as there must always be at least one satellite within LOS of a specific user.

The rapid movement of the constellation continuously drops inter-plane ISLs and creates new ones in their place. On the other hand, inter-plane ISLs are maintained permanently. Thus, routing schemes should be able to handle such topology changes. Although these changes occur frequently, the good thing is that the strict movement of satellites into certain orbits makes these changes periodic and thus predictable. Two routing schemes that are considered as good candidates for routing in a satellite communication system are Discrete Time Dynamic Virtual Topology Routing (DT-DVTR) and Virtual Node-based (VN) schemes [9]. These are briefly described below along with a discussion regarding routing in the asymmetric architecture described in Section 7.5.

7.5.2.1 DT-DVTR

This scheme works by acknowledging the periodic nature of the satellite constellation's movement. The rotation period of the constellation is divided into a number of segments with each segment being identified by a single topology change, which takes place at its start. Thus, in each segment the routing problem is treated as a static routing problem and can be easily solved. Furthermore, the periodicity in constellation movement makes it possible to store predetermined solutions for the static routing problems and avoid costly computations.

7.5.2.2 VN

The VN concept tries to hide the constellation's movements from upper layers. To this end, a set of Virtual Nodes (VNs) is defined and VNs are mapped to the actual satellites. As the constellation rotates, this mapping of course changes. Each VN keeps routing information regarding users in each coverage area. As the actual satellite that this VN is mapped to is replaced by another one, this information is transferred from the first satellite to the second one. Routing is performed based on the VNs and is thus independent of the satellite constellation's movement.

7.5.2.3 Routing in an Asymmetric System

The asymmetric architecture described in Section 7.5 possesses a significant problem for routing due to the fact that traditional routing schemes assume bidirectional links. An example is distance vector routing, where upon receiving the distance vector tuple {destination, cost} from its neighbor, a router assumes that it can reach the destination through that neighbor. However, due to the absence of the reverse link to the satellite, this is not true in satellite-based systems.

This problem can be solved through tunneling. Tunneling works at the link layer and hides network asymmetry from the upper layers responsible for performing routing. It works by establishing a virtual bidirectional link (tunnel) between the satellite and the user. The tunnel is used to relay packets from the user to the satellite. Tunneling works by encapsulating the packet and passing it to the routing protocol through the terrestrial link. Then the encapsulated packet is directed to the satellite where, after decapsulation, it is forwarded to the routing protocol. Thus, for the routing protocol, the whole procedure appears to operate on bidirectional links.

7.5.3 TCP Enhancements

Satellite-based Internet will continue to serve applications using the conventional TCP/IP protocol stack. However, the performance of both these protocols is greatly affected by the characteristics of satellite links. In this section, we briefly describe these characteristics along with some techniques to enhance the performance of TCP over such links [9,10].

7.5.3.1 Satellite Link Characteristics

The long latency of satellite links increases the propagation delay. This is a problem for TCP since acknowledgements will be delayed and with a corresponding degradation in rate and congestion control. Furthermore, satellite links are characterized by a large fluctuation in round-tri times, a fact that results in false TCP timeouts and TCP performance degradation.

Satellite links are characterized by increased BER compared to wired links. This has a negative effect on TCP since the latter is likely to mistake packet losses due to transmission errors for losses due to congestion. Upon damage to a packet, TCP window size is reduced to half and TCP takes precautions to combat congestion although this does not exist.

Another problem stems from the fact that in some cases satellite-based systems for Internet access are likely to be asymmetric. In such a case, the backlogged acknowledgements over the slow terrestrial link will slow down refreshing of the TCP window. Furthermore, loss of acknowledgements due to the congested terrestrial link will cause unnecessary retransmissions thus degrading TCP performance.

7.5.3.2 Enhancements for TCP Use in Satellite-based Systems

Apart from techniques that operate at different layers (such as FEC), there are a number of TCP-related techniques that increase the performance of TCP over satellite links. These include: (a) TCP selective acknowledgement (SACK), which enables the sender to retransmit only those packets actually lost; (b) TCP for transaction (T/TCP) which aims to reduce the latency of a connection from two round-trip times (RTT) to one; (c) persistent TCP connection, which enables small transfers to be performed over the same persistent TCP connection. Furthermore, in an effort to combat the problems of long end-to-end delay and asymmetry, a solution is to divide TCP connections into smaller ones. This division is performed at the gateways connecting the satellite network to the terrestrial network. There are three such dividing approaches:

- *TCP spoofing.* The split connections are isolated by gateways. Premature acknowledgements are sent to the source stations upon reception of the packets by destination stations in order to prevent unnecessary TCP timeouts and retransmissions.
- *TCP splitting.* In this case, connections are fully split and a proprietary protocol is used at the satellite network. Of course this calls for a protocol converter at the splitting points.
- *Web caching.* This approach uses a web cache. Satellite users that access Internet resources are first directed to the cache. If the requested item is cache-resident it is retrieved from the cache and there is no need to establish a TCP connection to the actual server that contains the requested data.

7.6 Summary

Although facing competition from terrestrial technologies and having faced market problems, satellite-based systems seem to be promising for voice and especially Internet services to globally scattered users around the world. This chapter provides an overview of satellite communication systems by focusing on a number of issues:

- Spectrum issues for satellite-base systems have been discussed. The characteristics of the three bands mainly used, 'C', 'Ka' and 'Ku' are presented. The 'C' bands were the first to be used for satellite traffic, however, it is nowadays overcrowded due to the fact that it is also being used by terrestrial microwave links. The Ku band suffers less interference than the 'C' band, however, its higher frequency makes it susceptible to interference, which can be combated by using a large number of widely separated interconnected ESs. Interference concerns the Ka band as well, which also has a disadvantage in terms of cost.
- The various possible orbits of satellite systems and their characteristics are described. These include LEO, MEO, GEO and elliptical orbits. LEO and MEO orbits are characterized by relatively short propagation delays. They form constellations that orbit the Earth at a speed greater than its rotation speed. Thus, careful planning of satellite orbits is needed to ensure continuous coverage. Furthermore, overall system design needs to take into account satellite movement and frequent handoffs. GEO satellites rotate at geostationary orbit over the equator. GEO systems experience higher propagation delays than LEO/ MEO systems. However, they have the advantage of rotating at a speed equal to that of the Earth's rotation, thus eliminating the need to track the satellite as it moves. Rather, a GEO satellite appears fixed at a certain point in the sky. Using as few as three GEO satellites, almost worldwide coverage can be achieved (except for areas near the poles). Finally, elliptical orbits try to combine the low propagation delay property of LEO systems and the stability of geostationary systems. Thus, such satellites have the properties of a LEO satellite near the perigee of their orbit and the properties of a geostiationary satellite near the apogee.
- The VSAT approach along with its topology and operation is presented. VSAT systems are especially useful for interconnecting large numbers of users residing in remote areas. They can operate either by using an ES as a 'hub', or by using an intelligent satellite that incorporates the hub's functionality.
- The Iridium and Globalstar, voice-oriented satellite systems were highlighted. Iridium, which was abandoned in 2000 for economic reasons targets worldwide coverage through a LEO constellation of 66 satellites, orbiting at 11 different planes with six satellites per plane. Satellites are able to communicate with each other through ISLs. Globalstar is a relatively simpler system, which demands the presence of a Globalstar gateway in range of the satellite that serves the user. ISLs are not supported.
- A number of issues relating to satellite-based Internet access are also discussed. These include possible architectures, routing and transport issues.

WWW Resources

1. *www.ee.surrey.ac.uk/Personal/L.Wood/constellations*: this web site contains a vast amount of information regarding satellite constellations, including information on voice, data, messaging and navigation satellite systems.

References

[1] Clarke A. C. Extra-Terrestrial Relays, *Wireless World*, October, 1945, 305–308; Republished in *Ascent to Orbit*, Wiley, 1984, pp. 60–63.
[2] Tannenbaum A. *Computer Networks*, Third Edition, Prentice Hall.
[3] Leopold R. J. and Miller A. The Iridium Communications System, *IEEE Potentials*, April, 1993, 6–9.
[4] Hubbel Y. C. A Comparison of the Iridium and AMPS Systems, *IEEE Network*, March/April, 1997, 52–59.
[5] Pratt S. R., Raines R. A., Fossa C. E. and Temple M. A. An Operational and Performance Overview of the IRIDIUM Low Earth Orbit Satellite System, *IEEE Communications Surveys*, Second Quarter, 1999, 2–10.
[6] Dietrich F. J., Monte P. and Metzen P. The Globalstar Cellular Satellite System, *IEEE Transactions on Antennas and Propagation*, June, 1998, 935–942.
[7] Black U. *Second Generation Mobile and Wireless Networks*, Prentice Hall, 1999.
[8] Bem D. J., Wieckowski T. W. and Zielinski R. J. Broadband Satellite Systems, *IEEE Communications Surveys*, First Quarter, 2000.
[9] Hu Y. and Li V. O. K. Satellite-Based Internet: a Tutorial, *IEEE Communications Magazine*, March, 2001, 154–162.
[10] Ghani N. and Dixit S. TCP/IP Enhancements for Satellite Networks, *IEEE Communications Magazine*, July, 1999, 64–72.

Further Reading

[1] Satellite Communications - A Continuing Revolution, *IEEE Aerospace & Electronic Systems Magazine, Jubilee Issue*, October, 2000, 95–107.
[2] Chakraborty D. VSAT Communication Networks - an Overview, *IEEE Communications Magazine*, May, 1998, 10–24.
[3] Ghani N. and Dixit D. TCP/IP Enhancements for Satellite Networks, *IEEE Communications Magazine*, July, 1999, 64–72.
[4] Farserotu J. and Prasad R. A Survey of Future Broadband Multimedia Satellite Systems, Issues and Trends, *IEEE Communications Magazine*, June, 2000, 128–133.
[5] Choi K.-K., Qadan O., Sala D., Limb J. O. and Meyers J. Interactive Web Service via Satellite to the Home, *IEEE Communications Magazine*, March, 2001, 182–190.
[6] Peyravi H. Medium Access Control Protocols Performance in Satellite Communications, *IEEE Communications Magazine*, March, 1999, 62–71.

8

Fixed Wireless Access Systems

The goal of this chapter is to review the main techniques used for Wireless Local Loop (WLL) including the Multichannel Multipoint Distribution Service (MMDS), and the Local Multipoint Distribution Service (LMDS). We also present the main aspects, advantages and disadvantages, and applications of these techniques. The chapter also deals with the wireless local loop subscriber terminals, Wireless Local Loop Interfaces to the Public Switched Telephone Network (PSTN), and the IEEE 802.16 standards on Broadband Wireless Access. Then a final section is given to summarize the main points presented in this chapter.

8.1 Wireless Local Loop versus Wired Access

Fixed Wireless Access (FWA) systems, which can also be called Wireless Local Loop (WLL) systems, are intended to provide primary access to the telephone network; that is, wireless services supporting subscribers in fixed and known locations. In general, WLL is a system that connects subscribers to the public switched telephone network (PSTN) using radio signals as a substitute for copper transmission media for all or part of the connection between the subscriber and the switch. This may include cordless access systems, proprietary fixed radio access, and fixed cellular systems. There are two alternatives to WLL: narrowband and broadband schemes. Narrowband WLL offers a replacement for existing telephone system while broadband WLL can provide high speed voice and data service. Some authors call WLL, Radio In The Loop (RITL), Fixed-Radio Access (FRA), or Fixed Wireless Access (FWA) [1,2].

It is expected that the global WLL market will exceed 202 million subscribers by the year 2005. Much of this growth will be in the developing countries where over half the world's population lacks Plain Old Telephone Service (POTS). This approach is cost effective and can save burying tons of copper wire. WLL networks can be deployed very quickly and in a cost-effective manner. This is a key advantage in a market where multiple service providers are competing for the same user base. In developed countries, WLL will help unlock competition in the local loop, enabling new operators to bypass existing wireline networks to deliver traditional and data access [3,4] (Figure 8.1).

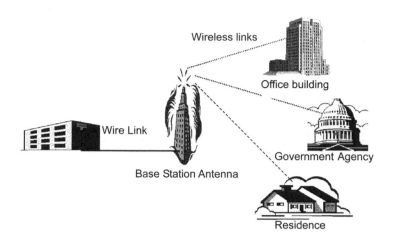

Figure 8.1 A wireless local loop configuration

WLL systems have a number of advantages over wired systems to subscriber local loop support. Among these are [1–10]:

- *Time of installation.* The time required to install a WLL system is much less than that for a wired system. The major issues are having a permission to use a given frequency band and finding an elevated site for the base station antenna. Once these issues are resolved, a WLL system can be installed very quickly.
- *Cost.* Despite the fact that the electronics of a wireless transceiver is more expensive that for a wired system, overall total cost of wireless system components, installation, and maintenance is less than for a wired system.
- *Scale of installation.* In WLL, radio transceivers are installed only for those subscribers who need the service at a given time. In wired systems, a cable is usually laid out in anticipation to serve an entire block or area.

The communications regulatory commissions in most countries have set aside frequency bands for use in commercial fixed wireless service. For example, the Federal Communications Commission (FCC) in the United States has set aside fifteen frequency bands for use in this application.

Many WLL systems are based on Personal Communication Systems (PCS) or cellular technology. These PCS or cellular technologies can be analog such as AMPS, TACS, and ETACS, or digital such as GSM, DECT, PDC, CDMA, W-CDMA, and CDMA2000. WLL systems based on such technologies can benefit from the associated economies of scale as well as the reduced costs. There are specific characteristics of fixed wireless systems that are not fully addressed by mobile wireless technologies without explicit consideration. While mobile technologies can readily be used for WLL systems, the ideal WLL system for a given market will be designed and adapted for fixed rather than mobile services. The major distinctions between mobile and fixed technologies are very clear from the WLL network deployment, the WLL subscriber terminals, and the WLL interface to the PSTN.

Due to the fact that the subscriber locations in a WLL system are fixed and not mobile, the initial deployment of radio base stations need only provide coverage to areas where immediate demand for service is apparent. Service for a town or city, for example, can begin area by area as the WLL base stations are deployed. This distinction between mobile and fixed applications does not in itself imply much difference in the technology, but other dissimilarities associated with network deployment do. It is important to note that the capacity needed for a WLL system is different from that for a fixed WLL system. A mobile system's base stations must provide adequate capacity to support worst-case traffic, while a fixed system's base stations must only provide the capacity needed to support a known number of subscribers. In a fixed system, the Quality of Service (QoS) may need to be better, and the traffic generated per subscriber may be higher due to lower charging rates than those for premium mobile service and the different usage patterns of homes and offices. Therefore, the ideal fixed WLL system should be fully scaleable and modular in order to be able to add any necessary additional capacity to the base stations. This allows the redistribution of additional capacities among existing base stations. Also, this means that base stations can be redeployed as needed in order to meet new demands in traffic. There is also a difference in the nature of coverage of WLL and wired access loop applications. While in the former applications a mobile system must effectively provide communications to all areas within signal range of the base station, a wired access system can assume that the subscriber terminal has been positioned to obtain the best possible signal.

In a fixed subscriber environment, a terminal is oriented for the greatest signal strength upon installation, and, if needed, a directional antenna pointing to the nearest WLL base station can be used to improve the signal quality in terms of the carrier-to-interference ratio or range extension. Furthermore, a fixed subscriber terminal will not experience the same magnitude of fading effects seen by a mobile terminal.

Due to the differences between fixed and mobile propagation environments, the transmit power levels of a fixed WLL system can be reduced compared to that of a mobile system, assuming the same range of coverage and all other variables hold constant. Directional antennas may be used at the base station to further improve the system's link margins if the fixed subscribers are localized [5–7].

8.2 Wireless Local Loop

In this section, we look at the main characteristics of the two well-known types of wireless local loop techniques: the Multichannel Multipoint Distribution Service (MMDS), and the Local Multipoint Distribution Service (LMDS).

8.2.1 Multichannel Multipoint Distribution Service (MMDS)

In the United States, the FCC has allocated five frequency bands in the range of 2.15–2.68 GHz for fixed wireless access using the Multichannel Multipoint Distribution Service (MMDS). Table 8.1 shows the fixed wireless communication bands that have been allocated by FCC.

The first two bands were licensed in the 1970s for TV broadcasting and they were then called Multipoint Distribution Services (MDSs). In 1996, the FCC increased the allocation and allowed for Multichannel Multipoint Distribution Services (MMDS). This new service has become a strong competitor to cable TV providers for offering services in rural and

Table 8.1 Frequency bands allocated by the United States FCC for fixed wireless communications bands

Frequency range (in GHz)	Application
2.150–2.162	Licensed Multichannel Distribution Service (MDS), 2 bands of 6 MHz each
2.4000–2.4835	Unlicensed Industrial, Scientific, and Medical (ISM)
2.596–2.644	Licensed Multichannel Multipoint Distribution Service (MMDS), 8 bands of 6 MHz each
2.650–2.656	Licensed MMDS
2.6620–2.6680	Licensed MMDS
2.6740–2.6800	Licensed MMDS
5.7250–5.8750	Unlicensed National Information Infrastructure (ISM-UNII)
24.000–24.250	Unlicensed ISM
24.250–25.250	Licensed
27.500–28.350	Licensed LMDS/Block A
29.100–29.250	Licensed LMDS/Block A
31.000–31.075	Licensed LMDS/Block B
31.075–31.225	Licensed LMDS/Block A
31.225–31.300	Licensed LMDS/Block B
38.600–40.000	Licensed

remote areas that cannot be reached by broadcast or cable TV. It is due to this specific application that MMDS is also called wireless cable.

The FCC does not allow the transmitted power of the base station of an MMDS to service an area beyond 50 km. The subscriber antennas of the transmitter and receiver must be in the line of sight. The main advantages of MMDS, over the Local Multipoint Distribution Service (LMDS) are [2,3,7]:

- Due to the fact that equipment operating at lower frequencies is less expensive, the cost of the subscriber and base station is lowered.
- Since the wavelengths of MMDS signals are larger than those for LMDS, MMDS signals can travel farther without suffering from power losses. This means that MMDS can operate in considerably larger cells, which lowers the cost of the electronics for base stations.
- Because MMDS signals have relatively longer wavelengths, they are less susceptible to rain absorption. Moreover, MMDS signals do not get blocked easily by objects, which allow them to be sent for longer distances.

The main drawback of MMDS systems compared to LMDS systems is that they offer much less bandwidth. Due to this drawback, it is expected that MMDS services will be used mainly for residential subscribers and small businesses [1–6].

8.2.2 Local Multipoint Distribution Service (LMDS)

The Local Multipoint Distribution Service (LMDS) is the broadband wireless technology used to deliver voice, data, Internet, and video services in the 25 GHz and higher spectrum. It is considered a relatively new service. Due to the propagation characteristics of signals in this

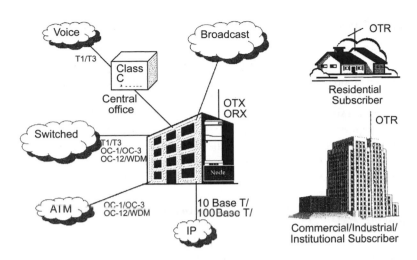

Figure 8.2 An example of Local Multipoint Distribution Service (LMDS)

frequency range, LMDS systems use a cellular-like network configuration. In the United States, 1.3 MHz of bandwidth has been allocated for LMDS to deliver broadband services in point-to-point or point-to-multipoint configurations to residential and commercial customers; near 30 GHz. In Europe and most developing countries, frequencies near 40 GHz will be used for this purpose. Table 8.1 depicts the frequency bands that have been allocated by the FCC for fixed wireless access in the United States including LMDS. Figure 8.2 shows a general configuration on the Local Multipoint Distribution Service (LMDS).

In LMDS systems, the propagation characteristics of the signals limit the potential coverage area to a single cell site. In metropolitan areas, the range of an LMDS transmitter can go up to 8 km. Signals are transmitted in a point-to-multipoint or broadcast method. The wireless return path, from subscriber to the base station, is a point-to-point transmission. It is important to note that the services offered through an LMDS network are entirely dependent on the operator's choice of service.

The main advantages of LMDS systems are [7–10]:

- It is easy and fast to deploy these systems with little disruption to the environment.
- As a result of being able to deploy these systems rapidly, realization of revenue is fast.
- LMDS are relatively less expensive, especially if compared with cable alternatives.
- Easy and cost-effective network maintenance, management, and operation.
- Data rate is relatively high, in the Mbps range.
- Scalable architecture with customer demand, which makes them cost effective.

It is expected that the LMDS services will be a combination of voice, video, and data. Therefore, both Asynchronous Transfer Mode (ATM) and Internet Protocol (IP) transport technologies will be practical when viewed within a country's telecommunications infrastructure.

The major drawback of LMDS is the short range from the base station, which necessitates the use of a relatively large number of base stations in order to service a specific area [10].

8.3 Wireless Local Loop Subscriber Terminals (WLL)

A wireless local loop subscriber terminal can be a handset that allows good mobility. It also can be an integrated desktop phone and a radio set or may be a single or multiple line unit that can connect to a standard telephone. These terminals can be mounted outdoors or indoors with or without battery back-up depending on the need.

In a WLL system, subscribers receive phone service through terminals linked by radio to a network of base stations. As mentioned above, there are different types of terminal types. The difference between them reflects the use of different radio technologies in wireless local loop systems and the varying levels of services that can be supported.

Single and multiple line units that connect to standard wireline telephones are very well suited for fixed wireless services. Multiple line subscriber terminals provide more than one independent channel of service, where each line is routed as needed to support an office building, an apartment complex, or a group of payphones. By using such single and multiple line designs, the WLL subscriber terminal virtually becomes the analog of a wireline phone jack. WLL capabilities should be above and beyond those for many mobile systems. For example, the WLL and its subscriber terminal should support data and fax services as well as voice without requiring any special external digital modem adapters. Moreover, the subscriber terminals should support the signaling needed for payphone service.

8.4 Wireless Local Loop Interfaces to the PSTN

As mentioned earlier, subscribers to a wireless local loop (WLL) system are linked using radio to a network of radio base stations. The latter are tied by a backhaul network to allow connection to the Public Switched Telephone Network (PSTN). The WLL system's interface to the telephone network can be supported through direct connection to the local exchange or by the use of its own switch. It is important to note that the way in which a WLL interconnects to the telephone network represents a key distinction between systems based on mobile wireless techniques or adapted to fixed wireless systems. The requirement to have mobile switch centers as part of wireless local loop systems means additional cost to the network operator. On the other hand, direct connection of a WLL system to existing central office switches effectively makes the wireless local loop system a direct extension of the wireline network. Moreover, it allows the use of switching resources that are underutilized.

It is possible to have the WLL system itself rely on the PSTN to provide all main switching functions. Moreover, it is desirable to have the WLL network adapted to fixed wireless services as a cost-effective extension of the wireline network. It should be possible to connect to existing local exchanges in a cost-effective manner that preserves the advanced features provided by the exchange.

In order to have direct connection to PSTN switches, an analog or digital interface is needed. If the local loop is copper, then the central office switches can provide two- or four-wire interfaces. On the other hand, digital interfaces using 64 kbps Pulse Code Modulation (PCM) voice channels can be more convenient and less expensive. The European Telecommunications Standard Institute (ETSI) has standardized the V5.2 landline digital interconnect as the recommended open digital interface between a WLL system, and a remote switch unit or Private Branch eXchange (PBX). However, the V5.2 standard adaptation by vendors has just begun. WLL equipment manufacturers have developed special proprietary digital interfaces to match specific switches as needed by their markets [3,5].

8.5 IEEE 802.16 Standards

The IEEE 802.16 Working Group on Broadband Wireless Access Standards develops standards and makes recommendations to support the development and deployment of broadband Wireless Metropolitan Area Networks. The IEEE 802.16 is a unit of the IEEE 802 LAN/MAN Standards Committee. This committee is working on developing interoperability standards for fixed broadband wireless access. A similar standard called HIPERACCESS is being developed in Europe by the standardization committee for Broadband Radio Access Networks (BRAN). While the US LMDS bands are 27.5–28.35 GHz, 29.1–29.25 GHz, and 31.075–31.225 GHz, the European standard band is 40.5–43.5 GHz.

The Broadband Wireless Access (BWA) industry is following a similar path to that of IEEE 802.3, IEEE 802.11 through the IEEE Working Group on Broadband Wireless Access, which is developing the IEEE-802.16 wireless MAN standard for wireless metropolitan area networks. This standard, which covers licensed and license-exempt bands from 2 to 66 GHz worldwide, is creating a good foundation for the development of this industry. The Working Group 802.16 began its work in July 1999. This group currently has about 200 members and some observers from over 100 companies. The charter of the group is to develop standards that: (a) use licensed spectrum; (b) use wireless links with microwave or millimeter wave radios; (c) are capable of broadband transmission at a rate greater than 2 Mbps; (d) are metropolitan in scale; (e) provide public network service to fee-paying customers; (f) provide efficient transport of heterogeneous traffic supporting quality of service (QoS); and (g) use point-to-multipoint architecture with stationary rooftop or tower mounted antennas.

The IEEE 802.16 group's work has primarily targeted the point-to-multipoint topology with a cellular deployment of base stations, each tied to core networks and in contact with fixed-wireless subscriber stations. Initial work has focused on businesses applications; small- to medium-size enterprises. However, attention has increasingly turned toward residential applications, especially as the lower frequencies have become available for two-way service.

Three subgroups have been established to produce standards for:

- *IEEE 802.16.1*. Air interface for 10–66 GHz.
- *IEEE 802.16.2*. Coexistence of broadband wireless access systems.
- *IEEE 802.16.3*. Air interface for licensed frequencies in the 2–11 GHz range.

Figure 8.3 illustrates the IEEE 802.16 Protocol Architecture.

The Working Group 802.16 is now completing a draft of the IEEE-802.16 Standard Air Interface for Fixed Broadband Wireless Access Systems. The document includes a flexible Media-Access Control (MAC) layer. The Physical Layer (PHY) is designed for 10–66 GHz. This latter layer is also called informally the Local Multipoint Distribution Service (LMDS) spectrum. At the time of writing, the standard is still under development, however, the draft has passed the Working Group's letter ballot, pending resolution of comments proposed to improve it and its publication is planned soon.

The Working Group is also developing amendments to the base IEEE 802.16 standard to accommodate lower frequencies. Amendment 802.16a will deal with the licensed bands from 2 to 11 GHz. The primary target in the United States is the Multichannel Multipoint Distribu-

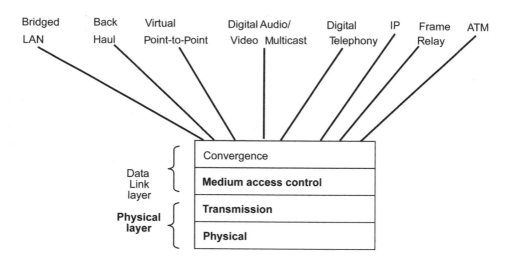

Figure 8.3   IEEE 802.16 protocol architecture

tion Service (MMDS) bands. The 802.16b amendment targets the needs of license-exempt applications around 5–6 GHz. The IEEE 802.16 committee maintains a close working relationship with standards bodies in the International Telecommunications Union (ITU) and the European Telecommunications Standards Institute (ETSI), especially in relation to the Hiperaccess and HiperMAN programs.

In the standards, the point-to-multipoint architecture assumes a time-division multiplexed downlink from the base station with subscriber stations in a given cell and sector sharing the uplink, typically by time-division multiple access. The uplink access is controlled by the base station, which has a set of scheduling schemes at its disposal in order to optimize the performance. The MAC protocol is connection-oriented and it is able to tunnel any protocol across the air interface with full QoS support. ATM and packet-based convergence layers provide the interface to higher protocols. However, the details of scheduling and reservation management are left unstandardized. There is a privacy sublayer that provides both encryption and authentication to secure access to these systems and protects them from hackers and unauthorized users. Figure 8.4 shows a wireless competitive local exchange carrier using ATM for distribution.

One important feature of the MAC layer is the option of granting bandwidth to a subscriber station rather than to the individual connections that it supports. This has the advantage of allowing a smart subscriber station to manage its bandwidth allocation among its users. Clearly, this has the potential of making more efficient allocation in multitenant, commercial or residential buildings. Moreover, efficiency is improved by the provision for header suppression, concatenation, fragmentation and packing.

As mentioned earlier, the 802.16 group has been developing a standard for 2–11 GHz BWA. In the United States, the primary targeted frequencies are in the MMDS bands, mostly from 2.5 to 2.7 GHz. In other parts of the world, 3.5 GHz and 10.5 GHz are likely applications. Due to the fact that non-line-of-sight operation is practical and because of the lower component costs, those bands are seen as good prospects for residential and small business services. The spectrum availability is suitable for those uses. The group has decided to

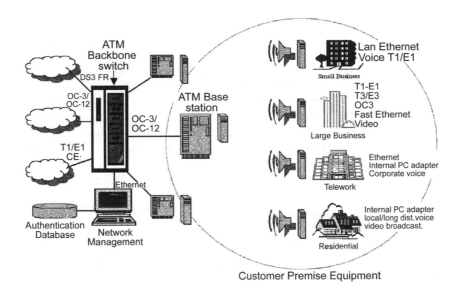

Figure 8.4 A wireless Competitive Local Exchange Carrier (CLEC) using Asynchronous Transfer Mode (ATM) for distribution [1–3,12].

support both single-carrier and multi-carrier PHY options. In the single-carrier proposal, submitted by representatives of 16 companies, frequency domain equalization is used. In the multicarrier proposal, submitted by representatives of 17 companies and an industry consortium, Orthogonal Frequency-Division Multiplexing (OFDM) and Orthogonal Frequency-Division Multiple Access (OFDMA) techniques are proposed. For more details see the proposals on the Web (http://ieee802.org/16/). As the time of writing, the MAC enhancements are about to be finalized. The MAC enhancements under development include optional mesh architecture in addition to the point-to-multipoint topology – testimony to the flexibility of the 802.16 MAC [11,12].

8.6 Summary

Broadband Wireless Technology (BWT) provides a cost-effective deployment plan with minimal dislocation for the community and the environment. Moreover, BWT can meet and may exceed the capabilities of other broadband alternatives. The development of wireless broadband access services can help reduce the congestion on the Public Switched Telephone Networks (PSTN), especially as related to Internet access. Wireless infrastructures can be divided into two main categories: mobile such as cellular and PCS networks and fixed such as LMDS and MMDS networks. The US Federal Communications Commission (FCC) has licensed wireless broadband services at four locations in the radio spectrum: the Multichannel Multipoint Distribution Service (MMDS), Digital Electronic Messaging Service (DEMS), Local Multipoint Distribution Service (LMDS), and Microwave Service. It is expected that over 4.4 million subscribers will choose fixed wireless broadband services by 2004. MMDS networks utilize a single omnidirectional central antenna that can provide MMDS service to an area faster and with a much smaller investment than other broadband services. One MMDS supercell can cover an area of about 9972 km^2. However, it is not easy to obtain line-of-sight, which may affect as many as 60% of households. Local Multipoint Distribution Service (LMDS) requires easy deployment. It was developed to provide a radio-based delivery service for a wide variety of broadband services. Due to the huge spectrum available, LMDS can provide high speed services with data rates reaching 155 Mbps. However, LMDS requires small cell sizes due to the high frequency at which they operate. Therefore, the average LMDS cell can cover between 32.6 and 73.3 km^2. The service provider can choose to launch its system at a pace to match its individual business plan without sacrificing QoS. Moreover, a LMDS subscriber will be able to utilize a rooftop or window-based antenna to receive signals from a radio base station.

References

[1] Rappaport T. S. *Wireless Communications: Principles and Practice*, Second Edition, Prentice Hall, Upper Saddle River, NJ, 2002.

[2] Stallings W. *Wireless Communications and Networks*, Prentice Hall, Upper Saddle River, NJ, 2002.

[3] Webb W. *An Introduction to Wireless Local Loop: Broadband and Narrowband*, Artec House, Boston, MA, 2000.

[4] The Insight Research Corporation, Wireless Broadband Access (WBA) Market Analysis: A White Paper, August, 1999, http://www.insight–corp.com

[5] Bolcske H., Paulraj A. J., Hari K. V. S., Nubar R. U. and Lu W., Fixed Broadband Wireless Access State of the Art: Challenges and Future Directions, *IEEE Communications Magazine*, January, 2001.

[6] Freeman R. *Radio System Design for Telecommunications*, John Wiley, New York, 1997.

[7] Correia A. J. and Prasad R. An Overview of Wireless Broadband Communications, *IEEE Communications Magazine*, January, 1997, 28–33.

[8] Xu H., Boyle R. J., Rappaport T. S. and Schaffner J. H. Measurement and Models for 38 GHz Point-to-Multipoint Radiowave Propagation, *IEEE Journal on Selected Areas in Communications: Wireless Communications Series*, 18(3), March, 2000, 310–321.

[9] Andrisano O., Trall V. and Verdone R. Millimeter Waves for Short–range Multimedia Communications Systems, *Proceedings of IEEE*, 86(July), 1998, 1383–1401.

[10] Nordbotten A. LMDS Systems and Their Applications, *IEEE Communications Magazine*, June, 2000.

[11] The IEEE 802.16 Working Group on Broadband Wireless Access Standards. http://grouper.ieee.org/groups/802/16/.

[12] http:// ieee802.org/16/.

9

Wireless Local Area Networks

9.1 Introduction

The growth of Wireless Local Area Network (WLANs) commenced in the mid-1980s and was triggered by the US Federal Communications Commission (FCC) decision to authorize the public use of the Industrial, Scientific and Medical (ISM) bands. This decision eliminated the need for companies and end users to obtain FCC licenses to operate their wireless products. Since then, there has been a substantial growth in the area of WLANs. Lack of standards, however, enabled the appearance of many proprietary products thus dividing the market into several, possibly incompatible parts. Consequently, the need for standardization in the area appeared.

The first attempt to define a standard was made in the late 1980s by IEEE Working Group 802.4, which was responsible for the development of the token-passing bus access method. The group decided that token passing was an inefficient method to control a wireless network and suggested the development of an alternative standard. As a result, the Executive Committee of IEEE Project 802 decided to establish Working Group IEEE 802.11 which has been responsible since then for the definition of physical and MAC sublayer standards for WLANs. The first 802.11 standard was finalized in 1997 and was developed by taking into consideration existing research efforts and market products, in an effort to address both technical and market issues. It offered data rates up to 2 Mbps using spread spectrum modulation in the ISM bands. In September 1999, two supplements to the original standard were approved by the IEEE Standards Board. The first standard, 802.11b, extends the performance of the existing 2.4 GHz physical layer, with potential data rates up to 11 Mbps. The second, 802.11a, aims to provide a new, higher data rate (from 20 up to 54 Mbps) physical layer in the 5 GHz band. The family of 802.11 standards is shown in Figure 9.1.

In addition to IEEE 802.11, another WLAN standard, High Performance European Radio LAN (HIPERLAN), was developed by group RES10 of the European Telecommunications Standards Institute (ETSI), as a Pan-European standard for high speed WLANs. The HIPERLAN 1 standard, like 802.11, covers the physical and MAC layers, offering data rates between 2 and 25 Mbps by using traditional radio modulation techniques in the 5.2 GHz band. Upon completion of the HIPERLAN 1 standard, ETSI decided to merge the work on Radio Local Loop and Radio LANs through the formation of Broadband Radio Access Networks (BRAN). This project aims to specify standards for Wireless ATM (HIPERLAN Types 2, 3, 4). The family of HIPERLAN standards is shown in Figure 9.2.

	802.11	802.11a	802.11b
Application	WLAN	WLAN	WLAN
Frequency band	2.4 GHz	2.4 GHz	5 GHz
Max. Data Rate	2 Mbps	54 Mbps	11 Mbps

Figure 9.1   The IEEE 802.11 family of standards

9.1.1 Benefits of Wireless LANs

The continual growth in the area of WLANs can be partly attributed to the need to support mobile networked applications. Many jobs nowadays require people to physically move while using an appliance, such as a hand-held PC, which exchanges information with other user appliances or a central computer. Examples of such jobs are healthcare workers, police officers and doctors. Wired networks require a physical connection between the communicating parties, a fact that poses great difficulties in the implementation of practical equipment. Thus, WLANs are the technology of choice for such applications.

Another benefit of using a WLAN is the reduction in infrastructure and operating costs. A wireless LAN needs no cabling infrastructure, significantly lowering its overall cost. Moreover, in situations where cabling installation is expensive or impossible (e.g. historic buildings, monuments or the battlefield) WLANs appear to be the only feasible means to implement networking. Lack of cabling also means reduced installation time, a fact that drives the overall network cost even lower.

A common fact in wired networks is the problems that arise from cable faults. Cable faults are responsible for most wired network failures. Moisture which causes erosion of the metallic conductors and accidental cable breaks can bring a wired network down. Therefore, the use of WLANs helps reduce the downtime of the network and eliminates the costs associated with cable replacement.

9.1.2 Wireless LAN Applications

The four major areas for WLAN applications [1] are LAN extension, cross-building interconnection, nomadic access and ad hoc networking. In the following sections we briefly examine each of these areas.

As mentioned, early WLAN products aimed to substitute wired LANs. A WLAN reduces installation costs by using less cable than a wired LAN. However, with advances in data transmission technology, companies continue to rely on wired LANs, especially those that use category 3 unshielded twisted pair cable. Most existing buildings are already wired with this type of cabling and new buildings are designed by taking into account the need for data

	HIPERLAN 1	HIPERLAN 2	HIPERLAN 3	HIPERLAN 4
Application	WLAN	WATM Indoor Access	Fixed Wireless Access-WATM Remote Access	Wireless Point to Point links-WATM interconnection
Frequency band	5 GHz	5 GHz	5 GHz	17 GHz
Max. Data Rate	23.5 Mbps	20 Mbps	20 Mbps	155 Mbps

Figure 9.2   The ETSI HIPERLAN family of standards

applications and are thus pre-wired. As a result, WLANs were not able to substitute their wired counterparts to any great extent. However, they were found to be suitable in cases were flexible extension of an existing network infrastructure was needed. Examples include manufacturing plants, warehouses, etc. Most of these organizations already have a wired LAN deployed to support servers and stationary workstations. For example, a manufacturing plant typically has a factory floor, where cabling is not present, which must be linked to the plant's offices. A WLAN can be used in this case to link devices that operate in the uncabled area to the organization's wired network. This application area of WLANs is referred to as LAN extension.

Another area of WLAN application is nomadic access. It provides wireless connectivity between a portable terminal and a LAN hub. One example of such a connection is the case of an employee transferring data from his portable PC to the server in his office upon returning from a trip or meeting. Another example of nomadic access is the case of a university campus, where students and working personnel access applications and information offered by the campus through their portable computers.

Ad hoc networking is another area of WLAN use. An ad hoc network is a peer-to-peer network that is set up in order to satisfy a temporary need. An example of this kind of application is a conference room or business meeting where the attendants use their portable computers in order to form a temporary network in order to share information during the meeting.

Another use of WLAN technology is to connect wired LANs located in nearby buildings. A point-to-point wireless link controlled by devices that usually incorporate a bridge or router functionality, connects the wired LANs. Although this kind of application is not really a LAN, it is often included in the area of WLANs.

9.1.3 Wireless LAN Concerns

The primary disadvantage of wireless medium transmission, compared to wired transmission, is its increased error rate. The wireless medium is characterized by Bit Error Rates (BERs) having an order of magnitude even up to ten times the order of magnitude of a LAN cable's BER. The primary reason for the increased BER is atmospheric noise, physical obstructions found in the signal's path, multipath propagation and interference from other systems. The latter takes either an inward or outward direction.

Inward interference comes from devices transmitting in the frequency spectrum used by the WLAN. However, most WLANs nowadays implement spread spectrum modulation, which operates over a wide amount of bandwidth. Narrowband interference only affects part of the signal, thus causing just a few errors, or no errors at all, to the spread spectrum signal. On the other hand, wideband interference, such as that caused by microwave ovens operating in the 2.4 GHz band, can have disastrous effects on any type of radio transmission. Interference is also caused by multipath fading of the WLAN signals, which results in random phase and amplitude fluctuations in the received signal. Thus, precautions must be taken in order to reduce inward interference in the operating area of a WLAN. A number of techniques that operate either on the physical or MAC layer (like alternative modulation techniques, antenna diversity and feedback equalization in the physical layer, Automatic Repeat Requests (ARQ), Forward Error Control (FEC) in the MAC sublayer) are often used in this direction. Outward interference occurs when the WLAN signals disrupt the operation of adjacent

(a)   (b)

Figure 9.3   Terminal scenarios: (a) 'hidden" and (b) 'exposed'

WLANs or radio devices, such as intensive care equipment or navigational systems. However, as most WLANs use spread spectrum technology, outward interference is considered insignificant most of the time.

A significant difference between wired and wireless LANs is the fact that, in general, a fully connected topology between the WLAN nodes cannot be assumed. This problem gives rise to the 'hidden' and 'exposed' terminal problems, depicted in Figure 9.3. The 'hidden' terminal problem describes the situation where a station A, not in the transmitting range of another station C, detects no carrier and initiates a transmission. If C was in the middle of a transmission, the two stations' packets would collide in all other stations (B) that can hear both A and C. The opposite of this problem is the 'exposed' terminal scenario. In this case, B defers transmission since it hears the carrier of A. However, the target of B, C, is out of A's range. In this case B's transmission could be successfully received by C, however, this does not happen since B defers due to A's transmission.

Another difference between wired and wireless LANs is the fact that collision detection is difficult to implement. This is due to the fact that a WLAN node cannot listen to the wireless channel while sending, because its own transmission would swamp out all other incoming signals. Therefore, use of protocols employing collision detection is not practical in WLANs.

Another issue of concern in WLANs is power management. A portable PC is usually powered by a battery having a finite time of operation. Therefore, specific measures have to be taken in the direction of minimizing energy consumption in the mobile nodes of the WLAN This fact may result in trade-offs between performance and power conservation.

The majority of today's applications communicate using protocols that were designed for wire-based networks. Most of these protocols degrade significantly when used over a wireless link. TCP for example was designed to provide reliable connections over wired networks. Its efficiency, however, substantially decreases over wireless connections, especially when the WLAN nodes operate in an area where interference exists. Interference causes TCP to lose connections thus degrading network performance.

Another difference between wired and wireless LANs has to do with installation. When preparing for a WLAN installation one must take into account the factors that affect signal propagation. In an ordinary building or even a small office, this task is very difficult, if not impossible. Omnidirectional antennas propagate a signal in all directions, provided that no obstacle exists in the signal's path. Walls, windows, furniture and even people can significantly affect the propagation pattern of WLAN signals causing undesired effects. MOST of the time, this problem is addressed by performing propagation tests prior to the installation of WLAN equipment.

Security is another area of concern in WLANs. Radio signals may propagate beyond the geographical area of an organization. All a potential intruder has to do is to approach the WLAN operating area and with a little bit of luck eavesdrop on the information being exchanged. Nevertheless, for this scenario to take place, the potential intruder needs to possess the network's access code in order to join the network. Encryption of traffic can be used to increase security, which, however, has the undesired effect of increased cost and overhead. WLANs are also susceptible to electronic sabotage. Most of them utilize CSMA-like protocols where all nodes are obliged to remain silent as long as they hear a transmission in progress. If someone sets a node within the WLAN area to endlessly transmit packets, all other nodes are prevented from transmitting, thus bringing the network down.

Finally, a popular issue that has to do not only with WLANs, but also with wireless communications in general, is human safety. Despite the fact that a final answer to this question has yet to be given, WLANs appear to be, in the worst case, just as safe as cellular phones. Radio-based WLAN components operate at power levels between 50 and 100 mW, which is substantially lower than the 600 mW to 3 W range of a common cellular phone. In infrared WLAN systems, the threat to human safety is even lower. Diffused Infrared (IR) WLANs offer no hazard under any circumstance.

9.1.4 Scope of the Chapter

The remainder of this chapter provides an overview of the WLAN area. In Section 9.2 the two types of WLAN topologies, infrastructure and ad hoc, are investigated. In Section 9.3 the requirements a WLAN is expected to meet are discussed. These requirements impact the implementation of physical and MAC layers for WLANs. In Section 9.4, physical layer matters are investigated and the five technology alternatives used today are presented. In Section 9.5 MAC sublayer issues are discussed and the two existing WLAN standards, IEEE 802.11 and HIPERLAN 1, are examined. Section 9.6 presents the latest developments in the WLAN area. The chapter ends with a brief summary in Section 9.7.

9.2 Wireless LAN Topologies

There arc two major WLAN topologies, ad hoc and infrastructure (Figure 9.4). An ad hoc WLAN is a peer-to-peer network that is set up in order to serve a temporary need. No networking infrastructure needs to be present, as the only things needed to set up the WLAN are the mobile nodes and use of a common protocol. No central coordination exists in this topology. As a result, ad hoc networks are required to use decentralized MAC proto-cols, such as CSMA/CA, with all nodes having the same functionality and thus implementa-tion complexity and cost. Moreover, there is no provision for access to wired network

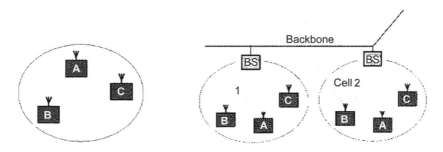

Figure 9.4   WLAN topologies: ad hoc and infrastructure

services that may be collocated in the geographical area in which the ad hoc WLAN operates. Another important aspect of ad hoc WLANs is the fact that fully connected network topologies cannot be assumed [2]. This is due to the fact that two mobile nodes may be temporarily out of transmission range of one another.

An infrastructure WLAN makes use of a higher speed wired or wireless backbone. In such a topology, mobile nodes access the wireless channel under the coordination of a Base Station (BS). As a result, infrastructure-based WLANs mostly use centralized MAC protocols like polling, although decentralized MAC protocols are also used (For example, the contention-based 802.11 can be implemented in an infrastructure topology). This approach shifts implementation complexity from the mobile nodes to the Access Point (AP), as most of the protocol procedures are performed by the AP thus leaving the mobile nodes to perform a small set of functions. The mobile nodes under the coverage of a BS, form this BS's cell. Although a fully connected network topology cannot be presumed in this case either, the fixed nature of the BS implies full coverage of its cell in most cases. Traffic that flows from the mobile nodes to the BS is called uplink traffic. When the flow of traffic follows the opposite direction, it is called downlink traffic.

Another use of the BS is to interface the mobile nodes to an existing wired network. When a BS performs this task as well, it is often referred to as an Access Point (AP). Despite the fact that it is not mandatory that the BS and AP be implemented in the same device, most of the time BSs also include AP functionality. Providing connectivity to wired network services is an important requirement, especially in cases where the mobile nodes use applications originally developed for wired networks.

The presence of many BSs and thus cells is common in infrastructure WLANs. Such multicell configurations can cover multiple-floor buildings and are employed when greater range than that offered by a single cell is needed. In this case, mobile nodes can move from cell to cell while maintaining their logical connections. This procedure is also known as roaming and implies that cells must properly overlap so that users do not experience connection losses. Furthermore, coordination among access points is needed in order for users to transparently roam from one cell to another. Roaming is implemented through handoff procedures. Handoff can be controlled either by a switching office in a centralized way, or by mobile nodes (decentralized handoff) and is implemented by monitoring the signal strengths of nodes. In centralized handoff, the BS monitors the signal strengths of the mobile nodes and reassigns them to cells accordingly. In decentralized handoff, a mobile node may decide to request association with a different cell after determining that link quality to that cell is superior to that of the previous one.

As far as the cell size is concerned, it is desirable to use small cells. Reduced cell sizes means shorter transmission ranges for the mobile nodes and thus less power consumption. Furthermore, small cell sizes enable frequency reuse schemes, which result in spectrum efficiency. The concept of frequency reuse is illustrated in Figure 9.5. In this example, nonadjacent cells can use the same frequency channels. If each cell uses a channel with bandwidth B, then with frequency reuse, a total of $3 \times B$ bandwidth is sufficient to cover the 16-cell region. Without frequency reuse, every cell would have to use a different frequency channel, a scheme that would demand a total $16 \times B$ of bandwidth.

The above strategy is also known as Fixed Channel Allocation (FCA). Using FCA, channels are assigned to cells and not to mobiles nodes. The problem with this strategy is that it does not take advantage of user distribution. A cell may contain a few, or no mobiles nodes at

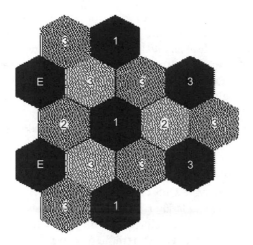

Figure 9.5   Example of frequency reuse

all and still use the same amount of bandwidth as a densely populated cell. Therefore, spectrum utilization is suboptimal. Dynamic channel allocation (DCA) [3–5], Power Control (PC) or integrated DCA and PC [6] techniques try to increase overall cellular capacity, reduce channel interference and conserve power at the mobile nodes. DCA places all available channels in a common pool and dynamically assigns them to cells depending on their current load. Furthermore, the mobile nodes notify BSs about experienced interference enabling channel reuse in a way that minimizes interference. PC schemes try to minimize interference in the system and conserve energy at the mobile nodes by varying transmission power. When increased interference is experienced within a cell, PC schemes try to increase the Signal to Interference noise Ratio (SIR) at the receivers by boosting transmission power at the sending nodes. When the interference experienced is low, sending nodes are allowed to lower their transmitting power in order to preserve energy.

Comparison of the above two WLAN topologies yields several differences [7]. However, most of these results stem from the assumption that ad hoc WLANs utilize contention MAC protocols (e.g. CSMA) whereas infrastructure networks use TDMA-based protocols. Based solely on topology, one can argue that the main advantage of infrastructure WLANs is their ability to provide access to wired network applications and services. On the other hand, ad hoc WLANs are easier to set up and require no infrastructure, thus having potentially lower costs.

9.3 Wireless LAN Requirements

A WLAN is expected to meet the same requirements as a traditional wired LAN, such as high capacity, robustness, broadcast and multicast capability, etc. However, due to the use of the wireless medium for data transmission, there are additional requirements to be met. Those requirements affect the implementation of the physical and MAC layers and are summarized below:

- *Throughput.* Although this is a general requirement for every network, it is an even more crucial aspect for WLANs. The issue of concern in this case is the system's operating throughput and not the maximum throughput it can achieve. In a wired 802.3 network, for example, although a peak throughput in the area of 8 Mbps is achievable, it is accompanied by great delay. Operating throughput in this case is measured to be around 4 Mbps, only 40% of the link's capacity. Such a scenario in today's WLANs with physical layers of a couple of Mbps, would be undesirable. Thus, MAC sublayers that shift operating throughput towards the theoretical figure are required.

- *Number of nodes.* WLANs often need to support tens or hundreds of nodes. Therefore the WLAN design should pose no limit to the network's maximum number of nodes.
- *Ability to serve multimedia, priority traffic and client server applications.* In order to serve today's multimedia applications, such as video conferencing and voice transmission, a WLAN must be able to provide QoS connections and support priority traffic among its nodes. Moreover, since many of today's WLAN applications use the client-server model, a WLAN is expected to support nonreciprocal traffic. Consequently, WLAN designs must take into consideration the fact that flow of traffic from the server to the clients can often be greater than the opposite.
- *Energy saving.* Mobile nodes are powered by batteries having a finite time of operation. A node consumes battery power for packet reception and transmission, handshakes with BSs and exchange of control information. Typically a mobile node may operate either in normal or sleep mode. In the latter case, however, a procedure that wakes up a transmission's destination node needs to be implemented. Alternatively, buffering can be used at the sender, posing the danger of buffer overflows and packet losses, however. The above discussion suggests that schemes resulting in efficient power use should be adopted.
- *Robustness and security.* As already mentioned, WLANs are more interference prone and more easily eavesdropped. The WLAN must be designed in a way that data transmission remains reliable even in noisy environments, so that service quality remains at a high level. Moreover, security schemes must be incorporated in WLAN designs to minimize the chances of unauthorized access or sabotage.
- *Collocated network operation.* With the increasing popularity of WLANs, another issue that surfaces is the ability for two or more WLANs to operate in the same geographical area or in regions that partly overlap. Collocated networks may cause interference with each other, which may result in performance degradation. One example of this case is neighboring CSMA WLANs. Suppose that two networks, A and B are located in adjacent buildings and that some of their nodes are able to sense transmissions originating from the other WLAN. Furthermore, assume that in a certain time period, no transmissions are in progress in WLAN A and a transmitting node exists in WLAN B. Nodes in A may sense B's traffic and falsely defer transmission, despite the fact that no transmissions are taking place in their own network.
- *Handoff – roaming support.* As mentioned earlier, in cell structured WLANs a user may move from one cell to another while maintaining all logical connections. Moreover, the presence of mobile multimedia applications that pose time bounds on the wireless traffic makes this issue of even greater importance. Mobile users using such applications must be able to roam from cell to cell without perceiving degradation in service quality or connection losses. Therefore, WLANs must be designed in a way that allows roaming to be implemented in a fast and reliable way.
- *Effect of propagation delay.* A typical coverage area for WLANs can be up to 150--300 m

in diameter. The effect of propagation delay can be significant, especially where a WLAN MAC demands precise synchronization among mobile nodes. For example, in cases where unslotted CSMA is used, increased propagation delays result in a rising number of collisions, reducing the WLANs performance. Thus, a WLAN MAC should not be heavily dependent on propagation delay.

- *Dynamic topology.* In a WLAN, fully connected topologies cannot be assumed, due to the presence of the 'hidden' and 'exposed' terminal problems. A good WLAN design should take this issue into consideration limiting its negative effect on network performance.
- *Compliance with standards.* As the WLAN market progressively matures, it is of significant importance to comply with existing standards. Design and product implementations based on new ideas are always welcome, provided, however, that they are optional extensions to a given standard. In this way, interoperability is achieved.

9.4 The Physical Layer

9.4.1 The Infrared Physical Layer

Infrared and visible light are of near wavelengths and thus behave similarly. Infrared light is absorbed by dark objects, reflected by light objects and cannot penetrate walls. Today's WLAN products that use IR transmission operate at wavelengths near 850 nm. This is because transmitter and receiver hardware implementation for these bands is cheaper and also because the air offers the least attenuation at that point of the IR spectrum. The IR signal is produced either by semiconductor laser diodes or LEDs with the former being preferable because their electrical to optical conversion behavior is more linear. However, the LED approach is cheaper and the IEEE 802.11 IR physical layer specifications can easily be met using LEDs for IR transmission.

Three different techniques are commonly used to operate an IR product. Diffused transmission that occurs from an omnidirectional transmitter, reflection of the transmitted signal on a ceiling and focused transmission. In the latter, the transmission range depends on the emitted beam's power and its degree of focusing and can be several kilometers. It is obvious that such ranges are not needed for most WLAN implementations. However, focused IR transmission is often used to connect LANs located in the same or different buildings where a clear LOS exists between the wireless IR bridges or routers.

In omnidirectional transmission, the mobile node's transmitter utilizes a set of lenses that converts the narrow optical laser beam to a wider one. The optical signal produced is then radiated in all directions thus providing coverage to the other WLAN nodes. In ceiling bounced transmission, the signal is aimed at a point on a diffusely reflective ceiling and is received in an omnidirectional way by the WLAN nodes. In cases where BSs are deployed, they are placed on the ceiling and the transmitted signal is aimed at the BS which acts as a repeater by radiating the received focused signal over a wider range. Ranges that rarely exceed 20 m characterize both this and the omnidirectional technique.

IR radiation offers significant advantages over other physical layer implementations. The infrared spectrum offers the ability to achieve very high data rates. Ref. [8] uses basic principles of information theory to prove that nondirected optical channels have very large

Shannon capacities and thus, transfer rates in the order of 1 Gbps are theoretically achievable. The IR spectrum is not regulated in any country, a fact that helps keep costs down.

Another strength of IR is the fact that in most cases transmitted IR signals are demodulated by detecting their amplitude, not their frequency or phase. This fact reduces the receiver complexity, since it does not need to include precision frequency conversion circuits and thus lowers overall system cost. IR radiation is immune to electromagnetic noise and cannot penetrate walls and opaque objects. The latter is of significant help in achieving WLAN security, since IR transmissions do not escape the geographical area of a building or closed office. Furthermore cochannel interference can potentially be eliminated if IR-impenetrable objects, such as walls, separate adjacent cells.

IR transmission also exhibits drawbacks. IR systems share a part of the spectrum that is also used by the Sun, thus making use of IR-based WLANs practical only for indoor application. Fluorescent lights also emit radiation in the IR spectrum causing SIR degradation at the IR receivers. A solution to this problem could be the use of high power transmitters, however, power consumption and eye safety issues limit the use of this approach. Limits in IR transmitted power levels and the presence of IR opaque objects lead to reduced transmission ranges which means that more BSs need to be installed in an infrastructure WLAN. Since BSs are connected with wire, the amount of wiring might not be significantly less than that of a wired LAN. Another disadvantage of IR transmission, especially in the diffused approach, is the increased occurrence of multipath propagation, which leads to ISI, effectively reducing transmission rates. Another drawback of IR WLANs is the fact that producers seem to be reluctant to implement IEEE 802.11 compliant products using IR technology. Furthermore, HIPERLAN does not address IR transmission at all.

The IEEE 802.11 physical layer specification uses Pulse Position Modulation (PPM) to transmit data using IR radiation. PPM varies the position of a pulse in order to transmit different binary symbols. Extensions 802.11a and 802.11b address only microwave transmission issues. Thus, the IR physical layer can be used to transmit information either at 1 or 2 Mbps. For transmission at 1 Mbps, 16 symbols are used to transmit 4 bits of information, whereas in the case of 2 Mbps transmission, 2 data bits are transmitted using four pulses. Figures 9.6 and 9.7 illustrate the use of 16 and 4 PPM. Notice that the data symbols follow the Gray code. This ensures that only a single bit error occurs when the pulse position is varied by one time slot due to ISI or noise.

Both the preamble and the header of an 802.11 frame transmitted over an IR link are

Data bits	1 Mbps PPM symbol	Data bits	1 Mbps PPM symbol
0000	0000000000000001	1100	0000000100000000
0001	0000000000000010	1101	0000001000000000
0011	0000000000000100	1111	0000010000000000
0010	0000000000001000	1110	0000100000000000
0110	0000000000010000	1010	0001000000000000
0111	0000000000100000	1011	0010000000000000
0101	0000000001000000	1001	0100000000000000
0100	0000000010000000	1000	1000000000000000

Figure 9.6   16-Pulse position modulation code

Data bits	2 Mbps PPM symbol
00	0001
01	0010
11	0100
10	1000

Figure 9.7   4-Pulse position modulation code

always transmitted at 1 Mbps. The higher rate of 2 Mbps, if employed, modulates only the sent MPDU. The following describes the frame fields:

- *SYNC*. Contains alternating pulses in consecutive time slots. It is used for receiver synchronization. The size of this field is between 57 and 73 bits.
- *Start frame delimiter*. A 4-bit field that defines the beginning of a frame. It takes the value 1001.
- *Data rate*. A 3-bit field that takes the values 000 and 001 for 1 and 2 Mbps, respectively.
- *DC level adjustment*. Consists of a 32-bit pattern that stabilizes the signal at the receiver.
- *Length*. A 16-bit field containing the length of the MPDU in milliseconds.
- *FCS*. A 16-bit frame check sequence used for error detection.
- *MPDU*. The 802.11 MAC protocol data unit to be sent. The size of this field ranges from 0 to 4096 octets.

9.4.2 Microwave-based Physical Layer Alternatives

The microwave radio portion of the electromagnetic spectrum spans from 10^7 to about 10^{11} MHz. Being of lower frequency, the Radio Frequency (RF) channel behaves significantly differently from that of IR. Radio transmission can penetrate walls and nonmetallic materials, providing both the advantage of greater coverage and the disadvantages of reduced security and increased cochannel interference. RF transmission is robust to fluorescent lights and outdoor operation thus being the only possible technology to serve outdoor applications. Nevertheless, RF equipment is subject to increased cochannel interference, atmospheric, galactic and man-made noise. There are also other sources of noise that affect operation of RF devices, like high current circuits and microwave ovens, making the RF bands a crowded part of the spectrum. However, careful system design and use of technologies such as spread spectrum modulation, significantly reduce interference effects in most cases.

RF equipment is generally more expensive than IR. This can be attributed to the fact that most of the time sophisticated modulation and transmission technologies, like spread spectrum, are employed. This means complex frequency or phase conversion circuits must be used, a fact that might make end products more expensive. However, the advances in fabrication of components promise even larger factors of integration and constantly lowering costs. Finally, as far as the WLAN area is concerned, RF technology has an additional advantage over IR, due to the large installed base of RF-WLAN products and the adoption of RF technology in current WLAN standards.

Microwave radio transmission was first used for long distance communications using very focused beams. However, in recent years, this part of the spectrum has experienced great popularity among electronic equipment manufacturers. As a result, cordless telephones, paging devices and WLAN products that use this band for transmission have appeared. When a company wants to deploy a product that uses a part of the microwave spectrum for transmission, licensing from the relevant authorities is needed. Such authorities are the Federal Communications Commission (FCC) in the United Stated and the Conference of European Postal and Telecommunications Administrations (CEPT) in the European Union.

Licensing poses both advantages and disadvantages. A significant advantage is that immunity to interference is guaranteed. If a product experiences performance degradation due to presence of interference, the corresponding authority will intervene and cease operation of the interfering source, since the latter is operating in a part of the spectrum licensed to another user. Disadvantages of licensing are the fact that the procedure can take a significant period of time and the electromagnetic spectrum is a scarce resource, so not everyone gets the desired bandwidth. The latter is true, especially in cases where the product is new and its market success not ensured. Such was the case for WLANs in the mid-1980s, when the licensing authorities seemed to be reluctant to authorize spectrum parts to WLAN vendors. This was due to the fact that the corresponding market was in a premature stage having no significant presence, while traditional voice oriented product vendors continued to demand more bandwidth. Thus, the need to satisfy the bandwidth needs of both the WLAN and existing product communities appeared.

The first step taken to resolve the problem was the authorization by FCC of license-free use of the Industrial, Scientific and Medical (ISM) bands (902–928 MHz, 2400–2483.6 MHz and 5725–5850 MHz) of the spectrum. This decision significantly boosted the WLAN industry in the United States. Since then, manufacturers and users do not need to license bandwidth to operate their products, a fact that lowers both the overall cost and the time needed for deployment and operation of a WLAN. However, to prevent excessive cochannel interference, certain specifications must be met for a product to use these bands, the most important of which is the mandatory use of spectrum spreading and low transmission power.

In 1993, CEPT announced bands at 5.2 and 17.1 GHz for HIPERLAN. One year later, the FCC released an additional 20 MHz of spectrum between licensed bands in the 1.9 GHz band after a request made by WINFORUM. The latter is an alliance between major computer and communication companies and its objective is to obtain and efficiently use license-free spectrum for data communication services. Another initiative started by WINFORUM led FCC to grant public use to 300 MHz of spectrum in the 5 GHz Unlicensed-National Information Infrastructure (U-NII) bands. This decision was taken in 1997 and is compatible with the European 5.2 GHz band allocation for HIPERLAN by CEPT. In these bands, FCC lifted the restriction of using only spread spectrum technology, thus providing the ability for higher data rates.

Today, the majority of WLAN products operate in the ISM bands. These bands are characterized by a number of significant differences. The most obvious is the fact that the higher bands, being wider, offer more bandwidth and thus higher potential transmission rates. Furthermore, the higher the band, the most challenging and expensive is the implementation of the corresponding RF equipment. The lower band, for example, can be supported with low-cost silicon-based devices. On the other hand, the upper band requires use of expensive gallium arsenide (GaAs) equipment. The middle band can be supported by both technologies and is thus characterized by a moderate cost.

However, the situation reverses when noise and interference are taken into account. From this point of view, the higher a band's frequency, the more appealing is its use, since at high frequencies less interference and noise exist. For example, the 902 MHz band is extremely crowded by devices such as cellular and cordless telephones, RF heating equipment, etc. The 2.4 GHz band experiences less interference with the exception of microwave ovens whose kilowatt level powers are concentrated towards the band's lower end. The 5.8 GHz band is even more interference-free. The same situation characterizes galactic, atmospheric and man-made noise [7]. The higher a band's frequency, the more noise-free the band is.

As far as transmission range is concerned, the lower the frequency of a band, the higher the achievable range. It is estimated [7] that the range in the 2.4 GHz band is around 5% less than that in the 902 MHz band. For the 5.8 GHz band, this number rises to 20%. As a rule of thumb, one can say that the properties of the three ISM bands vary monotonically with frequency. Both significant advantages or disadvantages characterize the high and low bands. The 2.4 GHz band stands in the middle, having the additional advantage of being the only one available worldwide.

Currently, the most popular WLANs use RF spread spectrum technology. The spread spectrum technique was developed initially for military applications. The idea is to spread the transmitted information over a wider bandwidth in order to make interception and jamming more difficult. In a spread spectrum system, the input data is fed into a channel encoder, which uses a carrier to produce a narrowband analog signal centered around a certain frequency. This signal is then spread in frequency by a modulator, which uses a sequence of pseudorandom numbers. In the receiving end, the same sequence is used to demodulate the spread signal and recover the original narrowband analog signal. The latter of course is fed into a channel decoder to recover the initial digital data. A random number generator, using an initial value called the seed, produces the pseudorandom sequence of numbers. Those numbers are not really random, since the generator algorithm is a deterministic one. A given seed always produces the same set of random numbers. However, a good random number generator produces number sequences that pass many tests of randomness, thus making interception of the spread signal practically possible only when the receiver possesses knowledge both of the algorithm and the seed used.

Among its other advantages, spread spectrum technology turns out to be quite successful in combating fading. As already mentioned, fading is frequency selective. Thus, since a spread spectrum signal is very wide in frequency, fading only affects a small part of it. In the following paragraphs, the two spread spectrum techniques, Frequency Hopping Spread Spectrum (FHSS) and Direct Sequence Spread Spectrum (DSSS) and their use as a physical layer for WLANs is presented. Next the alternatives of narrowband microwave transmission and orthogonal frequency division multiplexing physical layers are discussed.

9.4.2.1 The Frequency Hopping Spread Spectrum Physical Layer

Using this technique, the signal is broadcast over a seemingly random set of frequency channels, hopping from frequency to frequency at constant time intervals. The time spent on each channel is called a chip. The receiver executes the same hopping sequence while remaining in synchronization with the transmitter and thus receives the transmitted data. Any attempt to intercept the transmission would result in reception of only a few data bits. Attempts to jam the transmission succeed in erasing only a few random bits of the original message.

As mentioned in the previous paragraph, the hopping sequence is defined by the seed of the random number generator. The hopping rate, also known as chipping rate, defines the nature of the frequency hopping system. If set to a value greater than the transmission time of a single bit, multiple bits are transmitted over the same frequency channel. This technique is known as slow frequency hopping. If the hopping speed is set to a value less than the transmission time of a single bit, one bit is transmitted on more than one frequency. This technique is called fast frequency hopping. In both cases, when in a single channel, the actual transmitted signal is the result of modulation of the channel's center frequency with the original signal. FCC regulations state that each frequency channel is 0.5 MHz (902 MHz band) or 1 MHz (2.4 and 5.8 GHz bands) wide. In the 902 MHz bands, 52 FH channels exist of which, of which 50 must be used. In the middle band and upper bands, these channel numbers are 100 (83 in the United States), 75, 125 and 75, respectively. Furthermore, FCC rules state that the transmitters must not spend more than 0.4 s on any one channel every 20 s in the 902 MHz band and every 30 s in the upper bands. Since the peak transmission rate for a FHSS system is equal to a single channel's bandwidth, the two upper bands offer the highest peak transmission rate.

FHSS WLANs are very robust to narrowband interference due to the way they use the channel. Consider the case where a 2.4 GHz FHSS WLAN operates in the presence of 2 MHz narrowband interference. It is obvious that errors will occur only when the system hops to frequencies within the polluted 2 MHz. Since the 2.4 GHz band is 83.5 MHz wide, one concludes that the overall error rate will be very small. Furthermore, an intelligent FH system can replace the polluted channels with new ones. It can choose to use a new hop pattern that contains either a subset, or none, of the polluted channels. In this way, it can continue to operate in the presence of interference experiencing only small performance degradation.

Another advantage of FHSS WLANs is that they can operate simultaneously in the same geographical area. This is achieved by setting the WLANs to use orthogonal hopping sequences. Sets of such sequences can be defined, so that the members of each set present optimal cross-correlation properties. The orthogonality property ensures that any two patterns taken from the same set collide at most on a single frequency. As the pattern size can be set to be quite large, multiple FHSS WLANs can operate with acceptable performance in the same area.

The IEEE 802.11 FHSS physical layer specification calls for use of Gaussian Frequency Shift Keying (GFSK) to transmit data either at 1 or 2 Mbps in the 2.4 GHz band. The digital signal is fed into a GFSK modulator, which produces an analog signal centered on a certain frequency. The analog signal is then fed into a FH spreader, which makes use of a pseudorandom number sequence as an index into a table of frequencies. At each successive interval, the spreader selects a frequency, which is then modulated by the analog signal produced by the initial modulator. The result is a signal of the same shape bounded in the frequency channel chosen from the table. Repetition of this procedure produces the frequency-hopped signal. Transmission at 1 Mbps is implemented using two level GFSK, with a logical 0 transmitted at a frequency of $f_t - f_c$ and logical 1 at $f_t + f_c$. 2 Mbps data transmission is achieved using four level GFSK. The input to the modulator is a combination of two bits. Each of these 2-bit symbols is transmitted at 1 Mbps using the following frequency shifting scheme: logic 00 is transmitted at $f_t - 2f_c$, logic 01 at $f_t - f_c$, logic 11 at $f_t + f_c$ and logic 10 at $f_t + 2f_c$.

The 802.11 standard describes how to calculate optimal values for f_c. Furthermore, the

Data stream A	1			1			0			1			0		
Chip sequence B	1	0	0	1	0	1	1	0	1	0	0	1	0	1	0
Output signal C=A⊕B	0	1	1	0	1	0	1	0	1	1	1	0	0	1	0

Figure 9.8 DSSS modulation

standard defines three sets, each containing 26 hopping sequences designed to have minimal interference with one another within each set. Thus, BSs can be set to use sequences derived from the same set either to enable WLAN coexistence in the same area or to reduce cochannel interference.

Both the preamble and the header of an 802.11 frame transmitted over an FHSS link are always transmitted at 1 Mbps. The higher rate of 2 Mbps, if employed, modulates only the sent MPDU. The following describes the frame fields:

- *SYNC*. Consists of 80 alternating 0s and 1s used to synchronize the receiver.
- *Start frame delimiter*. A 16-bit field that takes the bit pattern 0000110010111101. It defines the start of a frame.
- *PLW*. A 12-bit field used to determine the end of the frame.
- *PSF*. A 4-bit field that takes the values 0000 and 0010 for 1 and 2 Mbps, respectively.
- *HEC*. A 16-bit field used for header error check.
- *Whitened MPDU*. The MPDU with special symbols stuffed every 4 bytes in order to minimize dc bias of the received signal. The size of this field ranges from 0 to 4096 octets.

9.4.2.2 The Direct Sequence Spread Spectrum Physical Layer

Using direct sequence spectrum spreading, each bit in the original signal is represented by a number of bits in the spread signal. This can be done by binary multiplication (XOR) of the data bits with a higher rate pseudorandom bit sequence, known as the chipping code. The resulting stream has a rate equal to that of the chipping code and is fed into a modulator, which converts it to analog form in order to be transmitted. The ratio between the chip and data rates is called the spreading factor and typically has values between 10 and 100 in modern commercial systems. This technique spreads the signal across a frequency band by a width proportional to the spreading factor. Figure 9.8 shows a binary data stream, a pseudorandom sequence having three times the rate of the data stream, and the resulting spread signal. Figure 9.9 depicts the demodulation of the spread signal at the receiver.

The actual data rate of the DS spread signal lowers with increasing spreading factor. FCC specifications state that in order for a DSSS product to operate in the ISM bands, a spreading factor of at least 10 must be used. For example, if a DSSS WLAN operates at a C MHz wide channel using a spreading factor of 10, the actual data rate cannot exceed $C/10$. On the other hand, a narrowband system can achieve data rates up to C. While seemingly wasteful of

Received signal C	0	1	1	0	1	0	1	0	1	1	1	0	0	1	0
Chip sequence B	1	0	0	1	0	1	1	0	1	0	0	1	0	1	0
Data stream A=C⊕B	1			1			0			1			0		

Figure 9.9 DSSS demodulation

bandwidth, DSSS has the significant ability to extract a signal from a background of narrow-band interference and noise, a fact that results in fewer retransmissions, thus enhancing throughput.

DSSS WLANs present a lower potential for interference cancellation than do FH ones. Returning to the example of the previous paragraph, we assume a DSSS WLAN operation occupying a 27 MHz wide channel. If the 2 MHz of noise are contiguous in the spectrum, the system can choose one of the other 27 MHz channels and continue to operate without experiencing interference. However, if the interfering source pollutes four nonadjacent 0.5 MHz channels, the DSSS WLAN cannot totally avoid interference in any case.

DSSS also has the ability to accommodate a number of simultaneous operating WLANs. Some DS WLANs may be designed to use less than the total available bandwidth. In such a case, additional WLANs using the remaining free channels can be admitted in the same geographical area. Nevertheless, as the number of DSSS subchannels is small, the number of collocated DSSS WLANs is generally smaller than in the FH case.

The IEEE 802.11 DSSS physical layer specification identifies the 2.4 GHz band for operation and divides the available bandwidth in 11 MHz wide subchannels using a chip sequence of rate 11 to spread each symbol. The specification uses Binary Phase Shift Keying (BPSK) to transmit the spread digital data stream at 1 Mbps. BPSK shifts the phase of the carrier frequency in order to represent different symbols. In the case of transmission at 2 Mbps, Quadrature Phase Shift Keying (QPSK) is used to transmit pairs of two bits at a rate of 1 Mbps thus achieving a data rate of 2 Mbps. Of course, since the specification calls for a chip rate of 11, the actual transmitted DSSS signal has a rate of 11 Mbps. Multiple networks can coexist in the same area provided they use subchannels with center frequencies separated by at least 30 MHz in order to avoid interference.

Extending the DSSS physical layer specification, the IEEE 802.11b standard supports 11 Mbps operation with fallback rates of 5.5 Mbps, 2 Mbps, and 1 Mbps, in the 2.4 GHz frequency band. The modulation technique used is Complementary Code Keying (CCK). CCK is the mandatory mode of operation for the standard, and is derived from the Direct Sequence Spread Spectrum (DSSS) technology. The extension is backward compatible with legacy 802.11 systems.

Both the preamble and the header of a frame transmitted over an 802.11b link are always transmitted at 1 Mbps. The higher rates, if employed, modulate only the sent MPDU. The following describes the frame fields:

- *SYNC*. Contains alternating pulses in consecutive time slots. It is used for receiver synchronization. The size of this field is 128 bits.
- *Start frame delimiter*. A 16-bit field defining the beginning of a frame.
- *Signal*. An 8-bit field that indicates 1, 2, 5.5, or 11 Mbps operation.
- *Service*. An 8-bit field reserved for future use.
- *Length*. A 16-bit field containing the length of the MPDU in milliseconds.
- *FCS*. An 8-bit frame check sequence used for error detection.
- *MPDU*. The 802.11 MAC protocol data unit to be sent. It has adjustable maximum length.

9.4.2.3 The Narrowband Microwave Physical Layer

An alternative to spread spectrum is narrowband modulation. Until recently, all narrowband

WLAN products had to use licensed parts of the radio spectrum. However, today's products can either use the newly released parts of the spectrum where licensing is not needed, or use the ISM bands without implementing spectrum spreading. The latter is permitted only if the narrowband transmission is of low power (0.5 W or less).

A narrowband WLAN has generally the opposite characteristics of a spread spectrum one. It is more vulnerable to fading. However, interference is not common in the case of WLANs that license their operating bandwidth. Licensing also ensures proper operation of collocated WLANs. Finally, the peak data rate of a narrowband WLAN operating in a channel of bandwidth C, is generally higher than that of a spread spectrum one. A DSSS WLAN achieves peak data rates of $C/10$ and a FHSS one has a peak data rate that equals its subchannel's bandwidth, while a narrowband WLAN can achieve a peak data rate of C.

HIPERLAN 1 uses narrowband modulation in the 5 GHz band. It divides the available bandwidth into five channels with center frequencies separated by 23.5 MHz. The standard defines two data rates. The lower one is at 1.47 Mbps and is used to transmit control information using Frequency Shift Keying (FSK) modulation. The higher data rate, at 23.4 Mbps, is used for data transmission and uses Gaussian Minimum Shift Keying (GMSK) modulation. The physical layer adds to the MPDU the lower data rate header, 450 high rate training bits used for channel equalization, $496 \times n$ high rate bits of payload and a variable number of padding bits. The equalization training bits are necessary in order to support the higher data rate in the presence of ISI. However, the standard does not define the equalizing technique leaving it to each implementation.

9.4.2.4 The Orthogonal Frequency Division Multiplexing (OFDM) Physical Layer

IEEE 802.11a operates in the in the 5 GH z bands and use Orthogonal Frequency Division Multiplexing (OFDM) to spread the transmitted signal over a wide bandwidth. OFDM is a form of multicarrier transmission and divides the available spectrum into many carriers, each one modulated by a low rate data stream using PSK. OFDM resembles FDMA in that the multiple user access is achieved by subdividing the available bandwidth into multiple channels, which are then allocated to users. However, OFDM uses the spectrum in a more efficient way by spacing the channels much closer. This is achieved by making all the carriers orthogonal to one another, preventing interference between the closely spaced carriers. Each carrier is of a very narrow bandwidth, which means that its data rate is slow. Figure 9.10 shows the spectrum for an OFDM transmission.

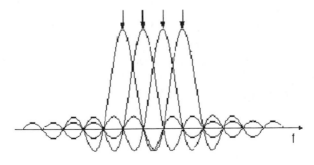

Figure 9.10   Detection of OFDM symbols

The spectrums of the subcarriers are not separated but partially overlap. However, the transmitted information can still be recovered due to the orthogonality relation, which gives the method its name. The spacing of the subcarriers is implicitly chosen in such a way that at the frequency where the received signal is evaluated (indicated as arrows), all other signals are zero. In order for the technique to work, however, perfect synchronization between the receiver and the transmitter is required.

OFDM effectively combats ISI. The OFDM symbols are artificially prolonged by periodically repeating the 'tail' of the symbol and precede the symbol with it. At the receiver, this so-called 'guard interval' is removed again. As long as the length of this interval is longer than the maximum channel delay all, reflections of previous symbols are removed and the orthogonality is preserved. However, by preceding the useful part of the length by the guard interval, we lose some parts of the signal that cannot be used for transmitting information.

In 802.11a multiple data rates are supported ranging from 6 to 54 Mbps. The mandatory data rates for 802.11a are 6, 12, and 24 Mbps. Depending upon the data rate, BPSK, QPSK, 16 QAM, or 64 QAM modulation is employed with OFDM in both standards.

9.5 The Medium Access Control (MAC) Layer

MAC protocols can be roughly divided into three categories: fixed assignment (e.g. TDMA, FDMA), random access (e.g. ALOHA, CSMA/CD, CSMA/CA) and demand assignment protocols (e.g. polling, token ring, PRMA). Fixed assignment protocols fail to adapt to changes in network topology and traffic and thus exhibit low performance in wireless data applications. Random access protocols, however, operate efficiently both without topology knowledge and under changing traffic characteristics. Nevertheless, their disadvantage is their nondeterministic behavior, a fact that causes problems in supporting QoS guarantees. Demand assignment protocols try to combine the advantages of fixed and random access protocols. However, knowledge of the network's logical topology is required in most cases. The latter, as mentioned, is hard to achieve in WLANs since fading and user mobility result in dynamically changing topologies. The token-based approach is generally thought to be inefficient. This is due to the fact that in a WLAN, token losses are much more likely to appear due to the increased BER of the wireless medium. Furthermore, in a token passing network, the token holder needs accurate information about its neighbors and thus of the network topology. In fact, the inefficiency of token passing was the reason the IEEE 802.4 Working Group, initially responsible for WLAN standardization, suggested the development of an alternative standard for WLANs. As a result, the IEEE 802.11 Working Group was formed in the late 1980s.

In the following paragraphs we examine the MAC sublayer of ETSI RES10 HIPERLAN 1 and IEEE 802.11. As mentioned earlier, collision detection is very difficult to implement in a WLAN receiver. Therefore, both of these standards employ CSMA/CA which reduces the probability of collisions. 802.11 includes an option that supports time-bounded applications. HIPERLAN 1 also supports time-bounded packet delivery by using an integrated priority mechanism. Issues like security, power saving and supported topologies are also discussed.

Lookup	Routing	Power saving	Priority mechanism
MAC			
Channel Access (EY-NPMA protocol)			
Physical Layer			

Figure 9.11   HIPERLAN 1 system architecture

9.5.1 The HIPERLAN 1 MAC Sublayer

The HIPERLAN 1 standard was released in 1995 aiming to define a WLAN technology of equal performance to that of traditional wired LANs and capable of supporting isochronous services. Unlike the IEEE 802.11 standard, the HIPERLAN committee was not driven by existing technologies and regulations. A set of requirements was set and the committee started working in order to satisfy them. The standard covers the physical and MAC layers of the OSI model.

The HIPERLAN 1 project, has defined the system architecture shown in Figure 9.11. It divides the functions of the Medium Access Control (MAC) into two subparts, which it refers to as Channel Access and Control (CAC) and MAC sublayers. The CAC layer defines how a given channel access attempt will be made depending on whether the channel is busy or idle, and at what priority level the attempt will be made, if contention is necessary. The HIPER-LAN MAC sublayer defines the various protocols which provide the HIPERLAN features of power conservation, lookup, security, and multihop routing, as well as the data transfer service to the upper layers of protocols. The routing mechanism supports the ability of HIPERLAN nodes to forward packets to stations out of their range with the help of inter-mediate forwarding stations. The lookup functionality enables collocated operation of more than one HIPERLAN network. Finally, the standard supports priorities, power conservation and support for encryption.

9.5.1.1 The Priority Mechanism and QoS Support

Although the HIPERLAN 1 standard does not define different priorities for the various traffic classes, like voice or multimedia, it tries to support time-bounded delivery of packets. HIPERLAN 1 dynamically assigns channel access priorities to packets by taking into account the packet's lifetime and its MAC priority. The MAC priority of a packet can be either normal or high, with normal being the default value. Every packet is generated with a specific lifetime ranging from 0 to 32767 ms, with the default value set at 500 ms. Packets that cannot be delivered within the allocated lifetime are dropped. The residual lifetime of a packet in combination with its priority define the packet's channel priority. Therefore, as time expires, the channel priority of each packet increases. Channel priority values range from 1 to 5, with priority p being higher than priority $p + 1$. This mechanism is used by HIPERLAN 1 to support time bounded applications.

9.5.1.2 The HIPERLAN 1 MAC Protocol

In HIPERLAN 1, a station can immediately commence transmission after sensing an idle medium for a duration of 1700 high rate bit times. However, even under moderate loads the above criterion is hardly ever fulfilled. When a station senses the medium busy, it waits until it becomes idle and then the Elimination Yield-Non-Preemptive Priority Multiple Access (EY-NPMA) protocol is applied. After the end of the detected transmission, all stations that want to transmit wait for another 256-bit period which is called a synchronization slot. Then, the EY-NPMA protocol is applied, which comprises the following phases:

- *The prioritization phase*. This phase is 1–5 slots long and each slot has a 256 high rate bit time duration. A station having to transmit a packet with channel priority p transmits a burst at slot $p + 1$, if it has not already sensed a higher priority burst from another station. Stations that sense higher priority bursts are dropped from contention and have to wait either for the next synchronization slot or for a 1700 bit idle period.
- *The elimination phase*. This phase consists of 1–13 slots each one being 256 high rate bits long. In this phase, stations that transmitted a burst during the previous phase, now contend for access to the medium. Each station transmits a burst for a geometrically distributed number of slots and then senses the medium for an additional slot. If it detects another burst during this slot, it stops contending for the channel, if not it proceeds to the next phase. Thus, stations that transmitted the longest burst and halted at the end of the same slot proceed to contend for access to the channel. The probability of a station's burst being of i slots, $(i < 12)$, is 0.5^{i+1}.
- *The yield phase*. This phase consists of 1–15 slots each one being 64 high rate bits long. Stations that make it to this phase defer for a geometrically distributed number of slots while sensing the channel. The probability of backing off by j slots is 0.1×0.9^j. The station that waits least seizes the channel and commences transmission. All other stations that made it to this phase sense the winner's transmission and wait until the next synchronization slot.

The purpose of the elimination phase is to reduce the contending stations and the yield phase tries to ensure that in the end, a single station gains access to the channel. According to the HIPERLAN 1 committee, the chances of two or more stations surviving all three phases (a fact that results in collision) are less than 3%. EY-NPMA simulation results in Ref. [9] show typical performance for a contention protocol:

- Performance increases for increasing packet sizes, since the larger the packet size, the less significant is the overhead added by the contention period.
- Decreasing throughput and increasing mean delay for an increasing number of stations.

Finally, overall throughput in HIPERLAN 1 is shown to be affected by the hidden terminal scenario, with increased intensity at high overall loads. The HIPERLAN 1 specification does not address this problem.

The combination of the EY-NPMA protocol and the priority mechanism supports time-bounded delivery of packets. It has to be noted, however, that time bounded does not mean QoS. HIPERLAN 1 just favors high priority packets, it cannot allocate a fixed portion of bandwidth to a particular application. From this point of view, it is just a best-effort network. Simulations in Ref. [9] show that for a small number of high priority stations increasing lower priority traffic does not affect the overall high priority throughput. However, increased numbers of high priority stations are likely to damage this good behavior, since with many high priority stations active, no mechanism for QoS establishment exists.

9.5.1.3 Supported Topologies and Multihop Routing

HIPERLAN 1 supports both infrastructure and ad hoc topologies. Furthermore, the standard supports multihop configurations, where a station can transmit a packet to another station out of its radio range without the need for additional infrastructure. This can be achieved with the help of intermediate stations that can forward packets destined for other stations. Each HIPERLAN station will select one and only one neighbor as its forwarder and transmit all packets destined for stations out of its range to the forwarder. Forwarded packets are relayed from forwarder to forwarder until they reach their destination. This means that a forwarder needs to know the network topology and maintain and dynamically update routing databases. However, it is optional for a station to forward packets. A station can announce its decision not to forward packets and become a nonforwarder. Nonforwarders are required to know only their direct neighbors.

Forwarding in a WLAN poses some problems. First, a forwarder needs to have a consistent image of the network topology at every moment. Since common routing algorithms are not designed for dynamically changing topologies, new algorithms need to be developed. Furthermore, maintenance of routing databases at a forwarder demands periodic exchange of information with its neighbors, a fact that limits the useful bandwidth of the channel.

Another problem arises due to the increased BER that characterizes wireless links. As a forwarded packet will travel over more than one such link, it is more likely to be corrupted or not arrive at all. Moreover, forwarding relies on the presence of stations willing to donate resources and processing power to serve other stations. Consequently, in a limited-resource HIPERLAN environment it is likely that forwarders are few. Simulations of forwarding topologies in HIPERLAN [9] depict decreased throughput performance when compared to a fully connected HIPERLAN topology.

9.5.1.4 Power Saving

The HIPERLAN 1 standard supports power saving by using both hardware-specific and protocol-based techniques. The first method relies on the existence of the two transmission speeds. As mentioned, the header of each packet is transmitted at the lower 1.47 Mbps rate. A node that hears a packet destined for another station can shut down the error correction, channel equalization and other receiver circuits until it receives a packet destined for itself.

Using the second power saving method, known as the p-saver method, a node can announce that it only powers up to receive incoming packets periodically. All other stations wishing to transmit to it, known as p-supporters, transmit to the p-saver only when it listens. A p-supporter may be an ordinary HIPERLAN device or a forwarder. As far as multicasts are concerned, p-supporters relaying multicasts announce their schedule for doing so, thus giving p-savers the option to power up in order to receive the multicast packets. P-saver schedules can be re-declared at any time in order to reflect new requirements.

9.5.1.5 Security

The MAC sublayer offers the ability to encrypt the transmitted MPDU. Each HIPERLAN packet carries a 2-bit field in the payload header that tells whether the payload is encrypted or not. If it is, the header identifies one of three possible keys. The standard defines a small set of keys, however, key distribution mechanisms are not defined.

The HIPERLAN 1 security algorithm operates as follows:

- At the transmitter, the key is XORed with a random bit sequence of equal length. Both are 30 bits. The resulting 30-bit value is used as a random number generator that outputs a bitstream of length equal to the MPDU length. The two bitstreams are again XORed to produce the encrypted data.
- The encrypted MPDU is encapsulated into a physical layer frame and transmitted to the destination. The key and the encrypted data are transmitted within the packet to the destination
- Upon extraction of the encrypted MPDU at the destination, the process is executed in reverse and the unencrypted data is obtained.

9.5.2 The IEEE 802.11 MAC Sublayer

The IEEE 802.11 standard covers the physical and MAC layers of the OSI model. It defines a single MAC sublayer for use with all the aforementioned 802.11 physical layers. There was considerable discussion within the committee before release of the final standard. The MAC protocol used is a CSMA/CA protocol called Distributed Foundation Wireless MAC (DFWMAC) and is very similar to the IEEE 802.3 Ethernet LAN line standard. DFWMAC, also referred to as the Distributed Coordination Function (DCF); it offers only a best-effort service. However, the 802.11 Working Group included optional support for time-bounded services through the use of a contention-free mechanism. This service is known as the Point Coordination Function (PCF) and is offered only in 802.11 infrastructure networks.

The 802.11 Working Group has defined the system architecture shown in Figure 9.12. DCF operates on top of the physical layer providing ordinary asynchronous traffic. PCF is built on top of the DCF and uses services offered by the DCF to provide contention-free traffic. The IEEE 802.11 MAC sublayer also offers mechanisms for authentication and privacy, encryption and power saving.

9.5.2.1 The 802.11 MAC Protocol

9.5.2.1.1 Distributed Coordination Function   The DCF sublayer uses a slotted CSMA/CA algorithm Thus, data transmissions can only start at the beginning of each slot. The IEEE

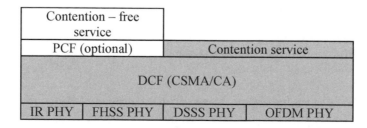

Figure 9.12   The IEEE 802.11 system architecture

802.11 standard utilizes a set of delays, known as Interframe Spaces (IFS). The steps taken for channel access are as follows:

- When a station has a packet to transmit, it first senses the medium. If the medium is sensed idle for an IFS, then the station can commence transmission immediately.
- If the medium is initially sensed busy, or becomes busy during the IFS, the station defers transmission and continues to monitor the medium until the current transmission is over.
- When the current transmission is over, the station waits for another IFS, while monitoring the medium. If it is still sensed idle, the station backs off a number of slots using a binary exponential backoff algorithm and again senses the medium. If it is still free, the station can commence transmission.

Of course, two or more stations can select the same slot to commence transmission, a fact that results in a collision. The actual size of the slot is physical layer dependent and is defined to be at least equal to the sum of the transmitter turn-on time plus busy medium detection time plus the maximum propagation delay between any two stations. This selection for the slot time ensures that collisions occur only when two or more stations select the same slot to transmit, as knowledge of a transmission commenced at slot k is propagated over the network before the start of slot $k + 1$. For the FHSS implementations, the slot time is 28 µs whereas in DSSS implementations it is 10 µs.

DCF uses three IFS values in order to enable priority access to the channel (Figure 9.13). These are, from the shortest to the longest, the Short IFS (SIFS), the Point Coordination Function IFS (PIFS) and the Distributed Coordination function IFS (DIFS). Their actual duration is defined by the slot duration and is thus physical layer dependent. Ref. [9] provides simulation results of the performance of the IEEE 802.11 DCF over three 802.11 physical layer specifications, concluding that end performance is highly dependent on the above two parameters. The Infrared (IR) physical layer shows better performance than the Direct Sequence Spread Spectrum (DSSS) layer, which in turn is proved to be superior to the Frequency Hopping Spread Spectrum (FHSS) physical layer.

DIFS is the minimum delay for asynchronous traffic contending for medium access. PIFS is used by the PCF portion of the MAC sublayer. Since it is shorter than DIFS it gives the Polling Coordinator (PC) the ability to lock out asynchronous traffic and allocated bandwidth for time bounded operations. The point coordination function is discussed later. SIFS is used in conjunction with the following 802.11 MAC operations:

- *MAC level acknowledgment (ACK).* When a station receives a frame destined only for itself it responds with an ACK frame after waiting only for a SIFS. Thus, a station acknowledging a received frame has to wait less time than stations trying to transmit

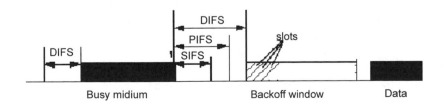

Figure 9.13   DCF operation

packets. As a result, the acknowledging station is favored to gain access to the medium. MAC level acknowledgment provides for efficient collision recovery, since collision detection is not implemented in IEEE 802.11. When an ACK is not received for a transmitted frame, the transmitting station assumes a collision occurred and re-contends for the channel.

- *Fragmentation*. MAC frames are passed down from the Logical Link Control (LLC) sublayer to the MAC sublayer. The MAC sublayer can choose to fragment unicast packets in order to increase transmission reliability. Unicast packets of size greater than the user manageable parameter *Fragmentation_Threshold*, are fragmented into multiple packets of size *Fragmentation_Threshold* and transmitted sequentially to the destination. Upon receipt of the first fragment, the destination waits for a SIFS and transmits an ACK. Upon receipt of the ACK, the source station immediately (after SIFS) sends the next fragment. As a result, the source station seizes the channel until all of the packet's fragments have been delivered.

- *RTS/CTS*. This mechanism enhances the two-way handshake CSMA/CA algorithm (DATA-ACK) to a four-way handshake algorithm (RTS-CTS-DATA-ACK). When a station wants to transmit a packet, it sends a small Request To Send (RTS) packet to the data packet destination. The latter, if ready to receive the data packet, responds after a SIFS with a Clear To Send (CTS) packet allowing the sending station to commence data transmission a SIFS after the CTS reception.

The RTS/CTS mechanism tries to combat the hidden terminal problem. The RTS and the CTS packets inform the neighbors of both communicating nodes about the length of the ongoing transmission. Stations hearing either the RTS or the CTS packet defer until the DATA and ACK transmissions are completed. RTS and CTS packets are very small (20 and 14 bytes, respectively) compared to the maximum 802.11 data frame (2346 bytes). As a result, when a collision between RTS or CTS packets occurs, less bandwidth is wasted when compared to collisions involving larger data frames. However, the use of the mechanism in a lightly loaded medium or in environments that are characterized by small data packets imposes additional delay due to the RTS/CTS overhead.

The use of the RTS/CTS mechanism is optional. RTS/CTS usage can be asymmetrical inside the same WLAN as only a subset of the WLAN nodes may decide to use the mechanism. A station can choose to never use RTS/CTS, use RTS/CTS when the data frame to be transmitted exceeds a certain user defined value (called the RTS threshold) or always use RTS/CTS. Simulations in Ref. [10] identify that the RTS threshold value that leads to optimal network performance is not constant, but depends on the length of the preamble added by the physical layer. The optimal value for the RTS threshold increases for increased preamble length.

The collision avoidance part of the protocol is implemented through a random backoff procedure. As mentioned, when a station senses a busy medium, it waits for an idle SIFS period and then computes a backoff value. This value consists of a number of slots. Initially, the station computes a backoff time ranging from 0 to 7 slots. When the medium becomes idle, the station decrements its backoff timer until it reaches zero, or the medium becomes busy again. In the latter case, the backoff timer freezes until the medium becomes idle again. When two or more station counters decrement to zero at the same time, a collision occurs. In this case, the stations compute a new backoff window given in slots by the formula

$[2^{2+i} \times \text{ranf}()] \times Slot\_time$, where i is the number of times the station attempts to send the current data frame, ranf() is a uniform variate in $(0, 1)$ and $[x]$ is the largest integer less than or equal to x. Successive collisions cause the size of the backoff window, also known as Contention Window (CW) to increase exponentially. When it reaches a certain maximum, which is a user defined parameter known as *CWMax*, i is reset to 1 and the size of the backoff window is reinitialized to 7. When a certain number of retransmissions occur for a specific frame, the frame is discarded.

However, consider the case of two stations, A and B, competing for access to the medium. A has either newly entered the competition or selects a backoff time due to a collision that occurred during its last transmission. Therefore, A selects a backoff value between 0 and CW. B, however, deferred a few slots ago and decrements its backoff timer when it senses the medium to be idle. Assume that B's backoff timer has decremented to a value of K slots $(0 < K < \text{CW})$ when A selects its backoff value. It is obvious that the slots between 0 and K have a higher probability of being chosen. This is due to the fact that although A uniformly selects slots between 0 and CW, the remaining backoff value for B can range only between slots 0 and K. Therefore, the backoff algorithm does not efficiently assign slots to competing stations and the increased selection likelihood of 'early' slots leads to increased collisions. This scenario is depicted in Figure 9.14. Ref. [10] proposes two algorithms that try to distribute contention slots to users in a uniform way. Stations that newly enter the competition select 'late' slots with higher probability, thus reducing collisions.

Simulation results [8,10–12] of DFWMAC reveal that under a fairly noiseless medium (BER $= 10^{-6}$) maximum throughput can reach satisfying percentage values, higher than those achieved for the same number of stations by HIPERLAN 1. However, the higher data rate offered by the physical layer of HIPERLAN 1 translates to higher transmission rates.

In a noiseless medium the use of large *Fragmentation_Threshold* values is preferable. This is due to the fact that for increased packet sizes, the resulting protocol overhead is not significant. Under harsh fading (BER $= 10^{-3}$) the protocol's performance drops sharply. Under such conditions, the use of small *Fragmentation_Threshold* values is preferable, as smaller packets are more likely to be transmitted without suffering errors. Being a random access protocol, DFWMAC peak performance decreases as the number of WLAN nodes increases. This is due to increased contention which leads to more collisions. Finally, the hidden terminal scenario greatly affects the performance of DFWMAC. Simulations in Ref.

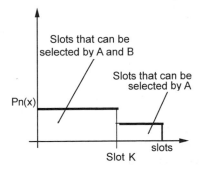

Figure 9.14   Slot selection probabilities. B is continuing a previous backoff and A newly enters the competition

[12] show that when the number of hidden pairs exceeds 10%, the protocol's performance drops sharply. However, significant performance improvements are achieved when using the RTS/CTS mechanism to reserve bandwidth for frame transmissions. Although the problem is not completely solved, 802.11 has an advantage over HIPERLAN 1 which does not address the hidden terminal problem at all.

9.5.2.1.2 Point Coordination Function PCF is an optional access method that supports isochronous, contention-free traffic and is built on top of the DCF. PCF is implemented only in infrastructure 802.11 WLANs. It operates by polling with a centralized polling master, known as the Point Coordinator (PC), which is usually the AP inside a cell. The PC makes use of the PIFS mentioned before. Since PIFS is smaller than DIFS, the PC can lock out all asynchronous traffic while it polls stations and receives responses. To avoid complete seizure of the medium by the PC, the 802.11 standard defines an interval known as the superframe. The first part of this interval serves contention-free traffic, while at the second part, the PC remains idle to give stations the chance to contend for medium access using DCF. During each PCF period, the PC polls stations demanding isochronous service. These stations are known as Contention Free Period (CFP) aware stations. A station that chooses not to participate in the CFP is called a non-CFP aware station. If at the end of the superframe the medium is busy, the PC has to wait until it becomes idle again in order to seize it. As a result, the next superframe is of reduced size.

Several user-definable parameters govern the joint operation of DCF and PCF. The Contention Free Period Repetition Interval (*CFP_Rate*) defines the nominal superframe length. The CFP maximum duration (*CFP_Max_Duration*) determines the maximum duration of the PCF. It can take a value no larger than the one that is required by the DCF in order to transmit a maximum size data frame successfully using the RTS–CTS–DATA–ACK mechanism. At the beginning of each superframe, the PC senses the medium. If the medium is idle for a PIFS period, the PC transmits these parameters using a Beacon frame. Stations that hear the Beacon frame defer until the CFP ends. The CFP can be terminated before the expiration time determined by *CFP_Max_Duration*. This can happen when all CFP–aware stations have transmitted their isochronous traffic. In this case, the PC terminates the CFP by transmitting a CFP-END frame.

The PC polls a CF aware station by sending it a CF-POLL frame. The station then responds by broadcasting either a CF-ACK frame, or a CF-ACK + DATA frame. In the first case, the PC receives a single acknowledgment of the CF-POLL receipt since the polled station does not have isochronous traffic buffered. Users who are idle repeatedly are removed from the poll cycle after k polls and are polled again at the beginning of the next CFP. In the second case, the PC receives a packet containing both the acknowledgment and data. In this case, the PC can resume polling by transmitting either a Data + CF-ACK + CF-POLL or a CF-ACK + CF-POLL frame. The CF-ACK part of the frame acknowledges the receipt of the previous data frame sent to the PC and the CF-POLL part is used to poll the next station. Of course, a CF aware station can also send isochronous data to stations other than the PC. In this case, the destination station transmits a DCF acknowledgment to the source and the PC resumes polling a PIFS interval after the receipt of the DCF ACK. The combined polling and data transmission mechanism reduces protocol overhead and increases CFP performance.

The PCF portion of 802.11 supports time-bounded applications better than HIPERLAN 1, as the polling mechanism guarantees transmission time to stations requesting it. However, when the number of stations requesting contention-free service increases, the polling algorithm must decide either to reduce the bandwidth offered to each station or deny contention-free service to some stations. The 802.11 standard, however, does not define the implementation of the polling algorithm and leaves it to the PC implementor. Joint simulations of the DCF and PCF in Ref. [11] reveal that setting k to 1 is optimal when all time-bounded data are voice data streams. This is explained by the fact that in relation to the duration of the CFP, voice streams are sent in slow on-off bursts.

9.5.2.2 Supported Topologies

The 802.11 standard supports both infrastructure and ad hoc network configurations. Infrastructure networks comprise one or more cells that contain mobile nodes. The mobile nodes access the backbone network, referred to as the Distribution System in 802.11 terminology, via APs. The set of stations associated with a given AP forms this AP's Basic Service Set (BSS). Two or more BSSs are interconnected using a Distribution System (DS). The 802.11 protocol does not define a specific DS. As a result, technologies like 802.x wired LANs, ATM or even another WLAN may be used as a DS.

The interconnection of multiple BSSs via the DS is called an Extended Service Set (ESS). Inside an ESS, data moves between BSSs through the DS. An ESS appears as a single logical WLAN to the LLC layer. An ad hoc network, having no AP, is called an Independent Basic Service Set (IBSS). The standard allows infrastructure and ad hoc topologies to coexist.

The BSSs inside an ESS can be disjoint, overlap or be physically collocated. Disjoint BSSs offer the advantage of reduced interference, paying the price of lack of continuous coverage, however. The reverse holds for BSSs that overlap. Finally, physically collocated BSSs can be used to form a higher performance WLAN. For example, multiple FHSS 802.11 WLANs using orthogonal hopping sequences might operate in the same geographical area to provide higher aggregate throughput.

The 802.11 standard identifies the following three mobility types: no transition, BSS transition, and ESS transition. The first type refers to nodes that either move inside a single BSS or do not move at all. The second type refers to nodes that roam from one BSS to another BSS while remaining in the same ESS. The third type refers to nodes that roam from a BSS in one ESS to a BSS in a different ESS. The 802.11 standard supports the first two types of mobility. However, it does not specify how roaming is performed, leaving this task to 802.11 product implementors.

Stations inside a BSS must remain synchronized in order for the MAC protocol to function properly. In infrastructure networks, the AP periodically transmits beacon frames that contain synchronization information, such as the hopping sequence that is used inside the BSS and timing information. In the case of IBSS networks, all stations periodically send beacon frames for synchronization purposes.

9.5.2.3 Security

The 802.11 standard defines two security procedures. The first allows for encrypted frame transmissions, in a way similar to that implemented by HIPERLAN 1. Encryption is implemented by using the Wired Equivalent Privacy (WEP) algorithm, which implements symmetric encryption. The WEP algorithm generates secret shared keys that can be used by both source and destination nodes to encrypt and decrypt data transmissions. However, the standard does not define the process of installing keys in stations.

The steps taken to encrypt a frame are the following:

- At the sending station, the WEP generates a 32-bit integrity value for the payload of the MAC frame. This value is used to alert the receiving station of possible data modification.
- A shared encryption key is used as an input to a pseudorandom number generator to produce a random bit sequence of length equal to the sum of the lengths of the MAC payload and the integrity value. Those fields are then encrypted by binary multiplication (XOR) with the bit sequence produced.
- The sending station places the encrypted MAC payload inside a MAC frame and hands it down to the physical layer for transmission.
- At the receiving station, the WEP algorithm uses the same key to decrypt the MAC payload and calculates an integrity value for the MAC payload. If the calculated value is the same as the one sent with the frame, it passes the MAC payload to the LLC.

The second security procedure concerns authentication between two communicating stations. Two authentication procedures are defined: Open System Authentication and Shared Key Authentication. The Open System Authentication procedure, is a two way handshake mechanism and is used when a high level of security is not required. Using this procedure, a station announces its desire to communicate with another station or AP by transmitting to it an authentication frame. The receiving station responds with another authentication frame that identifies success or failure of the authentication.

Shared Key Authentication, is a four-way handshake mechanism, which uses the WEP algorithm. The steps are as follows:

- The requesting station sends an authentication frame to another station.
- Upon receipt of an authentication frame, a station responds by transmitting another authentication frame containing a sequence of 128 bytes.
- The requesting station encrypts the received sequence using the WEP algorithm and sends it to the responding station.
- At the receiving station the bit sequence received is decrypted. If the decrypted sequence matches the one sent to the requesting station, the latter is informed of successful authentication.

9.5.2.4 Power Saving

The 802.11 standard supports power saving by buffering of traffic at the transmitting stations. When a mobile node is in sleep mode, all traffic destined to it is buffered until the node wakes up. In an infrastructure network, mobile nodes periodically wake up and listen to beacons sent by the access point. A station that hears a beacon indicating that the AP has buffered data for that station wakes up and requests reception of the data. In ad hoc networks, stations that implement power saving, wake up periodically to listen for incoming frames.

9.6 Latest Developments

9.6.1 802.11a

One of the latest developments is due to IEEE, which has developed 802.11a, a new specification for WLANs. The new specification enhances the physical layer of IEEE 802.11 while keeping upper layers intact. The advantages of IEEE 802.11a are its better interference immunity, and significantly higher speed, up to 54 Mbps. Devices utilizing 802.11a are required to support speeds of 6, 12, and 24 Mbps. Optional speeds go up to 54 Mbps, but will also typically include 48, 36, 18, and 9 Mbps. These differences are the result of implementing different modulation techniques and FEC levels. To achieve 54 Mbps, 64 QAM is used.

802.11a utilizes 300 MHz of bandwidth in the 5 GHz band. Although the lower 200 MHz is physically contiguous, the FCC has divided the total 300 MHz into three distinct 100 MHz domains, each with a different legal maximum power output. The 'low' band operates from 5.15 to 5.25 GHz with devices that use this band producing 50 mW output signals. The 'middle' band is in the 5.25–5.35 GHz band with a maximum of 250 mW. The 'high' band utilizes 5.725–5.825 GHz, with a maximum of 1 W. Because of the high power output, devices transmitting in the high band will tend to be building-to-building products. The low and medium bands are more suited to building into wireless products. One requirement specific to the low band is that all devices must use integrated antennas.

As mentioned earlier, 802.11a uses OFDM at the physical layer. The high data rate is accomplished by combining many lower-speed subcarriers to create one high-speed channel. 802.11a uses OFDM to define a total of eight nonoverlapping 20 MHz channels across the two lower bands; each of these channels is divided into 52 subcarriers, each approximately 300 kHz wide. The subcarriers are transmitted in parallel. Forward Error Correction (FEC) is also used in 802.11a (contrary to 802.11, which is described later). Because of the high data rates at the physical layer, 802.11a can accommodate FEC overhead with negligible impact on performance. Finally, OFDM specifies such a slower symbol rate to reduce the chance that a symbol will interfere with the following one, due to multipath propagation.

9.6.2 802.11b

Another recent development is the IEEE 802.11b specification. Like 802.11a, it is also a specification for the physical layer. It operates in the unlicensed 2.4 GHz band. The standard is backwards compatible to earlier IEEE 802.11 specifications (unlike 802.11a), allowing speeds of 1, 2, 5.5 and 11 Mbps on the same transmitters. The standard uses 14 channels, which are located in the 2.4 GHz band. Different channels are legal in different countries, and only channels 1, 6, and 11 have no overlap among them.

IEEE 802.11b achieves rates higher than its predecessor variants of 802.11 due to the use of Complementary Code Keying (CCK). CCK uses a series of codes called Complementary Sequences. Because there are 64 unique code words that can be used to encode the signal, up to 6 bits can be represented by any one particular code word. The CCK code word is then modulated with the QPSK technology used in 2-Mbps wireless DSSS radios. This allows for an additional 2 bits of information to be encoded in each symbol. Eight chips are sent for each 6 bits, but each symbol encodes 8 bits because of the QPSK modulation. The spectrum math for 1 Mbps transmission works out as 11 megachips per second times 2 MHz equals 22 MHz of spectrum. Likewise, at 2 Mbps, one is modulating 2 bits per symbol with QPSK, 11 megachips per second, and thus has 22 MHz of spectrum. To send 11 Mbps, one sends 11 million bits per second times 8 chips/8 bits, which equals 11 megachips per second times 2 MHz for QPSK-encoding, yielding 22 MHz of frequency spectrum.

9.6.3 802.11g

Task Group g of the 802.11 Working Group targets a system that will operate at the unlicensed 2.4 GHz band allowing for data rates up to 54 Mbps. 802.11g has two objectives:

- Mandatory use of OFDM as a part of the 802.11g standard, which will provide data rates up to 54 Mbps. Until very recently the FCC prohibited the use of OFDM in the 2.4-GHz band. Now, a common modulation format can be used in both the 2.4 GHz and 5 MHz bands.
- Backward compatibility with existing 802.11b systems. Thus, 802.11g will also support CCK.
- The optional elements included in the 802.11g standard are:
  - *CCK/OFDM.* This is a hybrid of CCK and OFDM designed to facilitate use of the OFDM waveform while supporting backward compatibility with existing CCK radios. CCK is used to transmit the packet preamble/header and OFDM is used to transmit the payload. CCK/OFDM supports data rates up to 54 Mbps. The CCK header alerts all legacy 802.11b devices that a transmission is beginning and informs those devices of the duration of that transmission. The payload can then be transmitted at a much higher rate using OFDM.
  - *Packet Binary Convolutional Coding (PBCC).* This is a 'single carrier' solution backed by Texas Instruments (TI). This waveform can also be described as a hybrid. It uses the CCK to transmit the header/preamble portion of each packet and PBCC to transmit the payload. PBCC supports data rates up to 33 Mbps. PBCC is actually a more complex signal constellation (8-PSK for PBCC vs. QPSK for CCK) and it employs a different code structure.

It is very likely that many IEEE 802.11g radios will implement only the mandatory modes.

9.6.4 Other Ongoing Activities within Working Group 802.11

At the time of writing, there are a number of ongoing standardization efforts within Working Group 802.11. They are all still ongoing. Their targets are summarized below [13].

9.6.4.1 Task Group d

The target of Task Group d is to add the requirements and definitions necessary to allow 802.11 WLAN equipment to operate in markets not served by the current standard.

9.6.4.2 Task Group e

The purpose of this task group is to enhance the current 802.11 MAC to expand support for LAN applications with QoS requirements, provide improvements in security, and in the capabilities and efficiency of the protocol. These enhancements, in combination with recent improvements in physical layer capabilities from 802.11a and 802.11b, will increase the performance of 802.11 networks.

9.6.4.3 Task Group f

This project proposes to specify the necessary information that needs to be exchanged between 802.11 Access Points (APs).

9.6.4.4 Task Group h

The aim of this group is to enhance the current 802.11 MAC and 802.11a PHY with network management and control extensions for spectrum and transmit power management in 5 GHz bands. Furthermore, it aims to provide improvements in channel energy measurement and reporting, channel coverage in many regulatory domains and provide Dynamic Channel Selection and Transmit Power Control mechanisms.

9.6.4.5 Task Group i

The target of this group is to enhance the current 802.11 MAC to provide security improvements.

9.7 Summary

In this chapter we cover the area of Wireless Local Area Networks by focusing on a number of issues:

- *WLAN topologies.* The two types of wireless LAN topologies used today are the infrastructure and ad hoc topologies. Ad hoc WLANs are preferable in cases where temporary and rapid deployment of a WLAN is demanded. On the other hand, infrastructure WLANs offer the ability to access data and services that are offered by collocated wired LANs. Access to those services is made through the use of Base Stations that implement Access Point functionality. Each base station forms its own cell and provides wired network access to all the nodes within its coverage. By employing Frequency Reuse schemes in cellular structures, the total available bandwidth of a system can significantly increase.
- *WLAN requirements.* The requirements expected to be met by a WLAN stem from the use of the wireless channel as a means of transmission. Wireless transmission is characterized by increased BER and interference, increased threat for unauthorized access and in most cases the need for spectrum licensing or use of spread spectrum techniques. Furthermore, the mobile nature of WLAN nodes results in dynamically changing, possibly not fully connected, network topologies where measures for power preservation at the mobile nodes must be taken. Those facts greatly affect the implementation of the protocol stack of a WLAN and should be taken into consideration when designing WLAN products.

- *Physical layer alternatives.* The five current physical layer alternatives are based either on infrared or microwave transmission. The IR-based physical layer poses the advantages of greater security and potentially higher data rates, however, not many IR-based products exist. Microwave alternatives include Frequency Hopping Spread Spectrum Modulation, Direct Sequence Spread Spectrum Modulation, Narrowband Modulation and Orthogonal Frequency Division Multiplexing. The Spread Spectrum and the OFDM approaches offer superior performance in the presence of fading which is the dominant propagation characteristic of wireless transmission. The Spread Spectrum techniques trade off bandwidth for this superiority, offering moderate data rates. Narrowband modulation on the other hand can potentially offer higher data rates than Spread Spectrum, being subject, however, to increased performance degradation due to fading. The OFDM approach is a form of multicarrier modulation that achieves relatively high data rates. The 802.11 standard supports all of the above alternatives, except for narrowband modulation, which is used by HIPERLAN 1.
- *MAC alternatives.* The two WLAN MAC standards available today, IEEE 802.11 and HIPERLAN 1, employ contention-based CSMA-like algorithms in order to access the wireless channel. The 802.11 MAC sublayer, when used in conjunction with the 802.11b and 802.11a physical layer extensions can offer data rates up to 11 and 54 Mbps, respectively. However, only 802.11b products are available at this time. HIPERLAN 1 offers data rates up to 24 Mbps; however, they have the disadvantage of incompatibility with 802.11 and the absence of an installed product base. The 802.11 MAC includes a mechanism that combats the hidden terminal problem whereas such a technique is not included in the HIPERLAN 1 standard. The latter includes a mechanism for multihop network support, effectively increasing the network's operating area. However, it pays the price of reduced overall performance compared to the single hop case. Both standards try to support time-bounded services, with 802.11 addressing it through the use of an optional contention-free mechanism. HIPERLAN 1 has an integrated priority mechanism that tries to support time-bounded applications, however QoS cannot be offered due to the absence of a mechanism that assigns a certain amount of bandwidth to a station. 802.11 can offer support for QoS applications, through bandwidth assignment to stations by the polling procedure, however, the latter is not defined in the standard.
- *Latest developments in the WLAN area.* These include the 802.11a and 802.11b standards, which are physical layer enhancements of 802.11 that provide high data rates. Furthermore, the aims of the ongoing work within Task Groups d, e, f, g, h, i of Working Group 802.11 have been reported.

Besides being a useful and profitable business, the WLAN area is also an extremely rich field for research due to the difficulties posed by the wireless medium and the increasing demand for better and cheaper services. It is very difficult to foresee the state of the area in the next decades or even years. However, the WLAN market is likely to increase in size and possibly integrate with other wireless technologies, in order to offer support for mobile computing applications, of perceived performance equal to those of wired communication networks.

WWW Resources

1. *www.standards.ieee.org*: the page of the Institute of Electrical and Electronics Engineers that is responsible for several standards in the area of WLANs, including the 802.11 standard.
2. *www.hiperlan.com*: the page of the HIPERLAN Alliance, a group promoting the ETSI HIPERLAN 1 standard.
3. *www.hiperlan2.com*: the page of the HIPERLAN 2 Global Forum, a group promoting the ETSI HIPERLAN 2 standard. It contains detailed information on HIPERLAN 2 in the form of white papers and presentations.
4. *www.wlana.org*: the page of the wireless LAN association is a nonprofit educational trade association maintained by WLAN vendors that aims to promote WLAN technology.
5. *www.wi-fi.com*: the page of the Wireless Ethernet Compatibility alliance, a vendor organization whose mission is to certify interoperability of 802.11-based products.
6. *www.etsi.org*: the page of the European Telecommunications Standards Institute (ETSI), a nonprofit organization that produces European standards in the telecommunications industry.
7. *http://www.lucent.com/minds/techjournal/findex.html*: the *Bell Labs Technical Journal* publishes several WLAN-related articles.
8. *http://grouper.ieee.org/groups/802/11/QuickGuide_IEEE_802_WG_and_Activities.htm*: this page provides an overview of the status of ongoing work within the 802.11 Working Group.

References

[1] Pahlavan K., Probert T. H. and Chase M. E. Trends in Local Wireless Networks, *IEEE Communication Magazine*, March, 1995, 88–95.
[2] Chen K.-C. Medium Access Control of Wireless LANs for Mobile Computing, *IEEE Network*, September/October, 1994, 50–63.
[3] Lagrange X. Multitier Cell Design, *IEEE Communications Magazine*, August, 1997, 60–64.
[4] Zander J. Radio Resource Management in Future Wireless Networks: Requirements and Limitations, *IEEE Communications Magazine*, August, 1997, 30–35.
[5] Nettleton R. W. and Schoemer G. R. Self Organizing Channel Assignment for Wireless Systems, *IEEE Communications Magazine*, August, 1997.
[6] Lozano A. and Cox D. C. Integrated Dynamic Channel Assignment and Power Control in TDMA Mobile Wireless Systems, *IEEE Journal on Selected Areas in Communications*, 17(November), 1999, 2031–2041.
[7] Bantz D. F. and Bauchot F. J. Wireless LAN Design Alternatives, *IEEE Network*, March/April, 1994, 43–53.
[8] Barry J. R., Kahn J. M., Lee E. A. and Messerschmitt D. G. HighSpeed Nondirective Optical Communication for Wireless Networks, *IEEE Network Magazine*, November, 1991, 3764–3776.
[9] Weinmiller J., Schlager M., Festag A. and Wolisz A. Performance study of Access Control in Wireless LANs IEEE 802.11 DFWMAC and ETSI RES 10 HIPERLAN, *ACM Mobile Networks and Applications Special Issue on Channel Access*, 2(1), 1997, 55–67.
[10] Weinmiller J., Woesner H., Ebert J. P. and Wolisz A. Analyzing and Tuning the Distributed Coordination Function in the IEEE 802.11 DFWMAC Draft Standard, in *Proceedings MASCOT '96, Analysis and Simulation of Computer and Telecommunication Systems*.
[11] Crow B. P. Performance Evaluation of the IEEE 802.11 Wireless Local Area Network Protocol, Masters Thesis, Department of Electrical and Computer Engineering, University of Arizona, 1996.

[12] Kahol A., Khurana S. and Jayasumana A. P. Effect of Hidden Terminals on the Performance of IEEE 802.11 MAC Protocol, *in Proceedings of IEEE LCN 98*, 1998, pp. 12–20.
[13] http://grouper.ieee.org/groups/802/11/QuickGuide_IEEE_802_WG_and_ Activities.htm

Further Reading

[1] Pahlavan K. and Levesque A. *Wireless Information Networks*, Wiley, 1995.
[2] Gilbert E. Capacity of a Burst Noise Channel, *Bell System Technology Journal*, 39(September), 1960, 1253–1265.
[3] Falconer D. D., Adachi F. and Gudmundson B. Time Division Multiple Access Methods for Wireless Personal Communications, *IEEE Communications Magazine*, January, 1995, 50–57.
[4] Andersen J. B., Rappaport T. S. and Yoshida S. Propagation Measurments and Models for Wireless Communication Channels, *IEEE Communications Magazine*, January, 1995, pp 42–49.
[5] Badra R. E. and Daneshrad B. Asymmetric Physical Layer Design for High-Speed Wireless Digital Communications, *IEEE Journal on Selected Areas in Communications*, October, 1999, 1712–1724.
[6] Taylor L. HIPERLAN Type 1 Technology Overview, TTP Communications Ltd. Revision 0.9 June, 1999.
[7] Broadband Radio Access Networks (BRAN). HIgh PErformance Radio Local Area Network (HIPERLAN), Type 1, Functional specification V1.2.1 July, 1998.
[8] Chen K.-C. and Lee C.-H. RAP - A Novel Medium Access Control Protocol for Wireless Data Networks, in *Proceedings of IEEE GLOBECOM*, 1993, pp. 1713–1717.
[9] Chen K.-C. and Lee C.-H. Group Randomly Access Polling for Wireless Data Networks, in *Proceedings of IEEE ICC*, 1994, pp. 913–917.
[10] Hayes V. Standardization Efforts for Wireless LANs, *IEEE Network Magazine*, November, 1991, 19–20.
[11] Stallings W. *Data and Computer Communications*, Fifth Edition, Prentice Hall, Upper Saddle River, NJ.
[12] Stallings W. *Local and Metropolitan Area Networks*, Fifth Edition, Upper Saddle River, NJ.
[13] Obaidat M. S. and Ahmed C. B. Schemes for Mobility Management of Wireless ATM Networks, *International Journal of Communication Systems*, May/June, 1999, 153–166.
[14] Pahlavan K. and Krishnamurthy P. Wideband Local Access: Wireless LAN and Wireless ATM, *IEEE Communications Magazine*, November, 1997, 34–40.
[15] Geier J. Wireless *LANs, Implementing Interoperable Networks*, Macmillan Network Architecture and Development Series.
[16] Tannenbaum A. *Computer Networks*, Third Edition, Prentice Hall, Upper Saddle River, NJ.

10

Wireless ATM and Ad Hoc Routing

10.1 Introduction

Recently, considerable research effort has been put into the direction of integrating the broadband wired ATM [1] and wireless technologies. In 1996 the ATM Forum approved a study group devoted to wireless ATM, WATM. WATM [2–4] aims to provide end-to-end ATM connectivity between mobile and stationary nodes. WATM can be viewed as a solution for next-generation personal communication networks, or a wireless extension of the B-ISDN networks, which will support guaranteed QoS integrated data transmission. WATM will combine the advantages of freedom of movement of wireless networks with the statistical multiplexing (flexible bandwidth allocation) and QoS guarantees supported by traditional ATM networks. The latter properties, which are needed in order to support multimedia applications over the wireless medium, are not supported in conventional LANs due to the fact that these were created for asynchronous data traffic.

10.1.1 ATM

In this section, a brief introduction to ATM is made in order prior to discussing Wireless ATM. ATM, also known as cell-relay for reasons that will be described later, is a technology capable of carrying any kind of traffic, ranging from circuit-switched voice to bursty data, at very high speeds. ATM possesses the ability to offer negotiable QoS. Thus, ATM is the technology of choice for multimedia networking applications that demand both large bandwidths and QoS guarantees since these properties cannot typically be offered by conventional networks such as Ethernet LANs.

ATM is a packet-switching technology that somewhat resembles frame relay. However, the main difference is the fact that ATM has minimal error and flow control capabilities in order to reduce control overhead and also that ATM utilizes fixed-size (53 bytes) packets known as cells instead of variable-sized packets as in frame relay. Fixed size packets enable fast speeds for ATM switches and together with the reduced overhead give rise to the very high data rates offered by ATM.

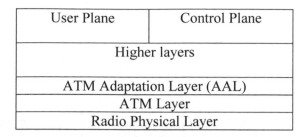

User Plane	Control Plane
Higher layers	
ATM Adaptation Layer (AAL)	
ATM Layer	
Radio Physical Layer	

Figure 10.1   ATM protocol architecture

The ATM protocol architecture is shown in Figure 10.1. Its main parts are:

- *Physical layer.* It involves the specification of the transmission medium and the signal encoding to be used. The two alternative speeds offered by the physical layer are 155 and 622 Mbps.
- *ATM layer.* This defines the transmission of ATM cells and the use of connections either between users, users and network entities or between network entities. These connections are referred to as Virtual Channel Connections (VCCs) and are analogous to the data link connections in frame relay. VCCs can carry both user traffic and signaling information. A collection of VCCs that share the same endpoints is known as a Virtual Path Connection (VPP).
- *The ATM Adaptation Layer (AAL).* This layer maps the cell format used by the ATM layer to the data format used by higher layers. Thus, at the transmitting side, AAL maps frames coming from higher layers to ATM cells and hands them over to the ATM layer for transmission. On the receiving side, ATM reassembles cells into the respective frames and passes frames to upper layers. A number of AALs exist, each of which corresponds to a specific traffic category. AAL0 is virtually empty and just provides direct access to the cell relay service. AAL1 supports services that demand a constant bit rate, which is agreed during connection establishment and must remain the same for the duration of the connection. This category of service is known as Constant Bit Rate (CBR) service with typical examples being voice and video traffic. AAL2 supports services that can tolerate a variable bit rate but pose limitations regarding cell delay. This category of service is known as Variable Bit Rate (VBR) service with typical examples being transmission of compressed (e.g. MPEG) video where bit rate can vary, however, delay guarantees are needed to avoid jerky motion. AAL3/4 and AAL5 support variable-rate traffic with no delay requirements. Such categories are VBR traffic with no delay bounds, Available Bit Rate (ABR), which is a best effort service that guarantees neither rate nor delay but only minimum and maximum rate and Unspecified Bit Rate (UBR) which is essentially ABR without a minimum rate guarantee.

The protocol architecture shown in Figure 10.1 also defines three separate planes. These are: (a) the user plane, which provides for transfer of user information and associated control information (e.g. FEC, ARQ); (b) the control plane, which performs call control and connection control; and (c) the management plane, which includes plane management for management of the whole system and coordination of the planes and layer management for management of functions relating to the operation of the various protocol entities.

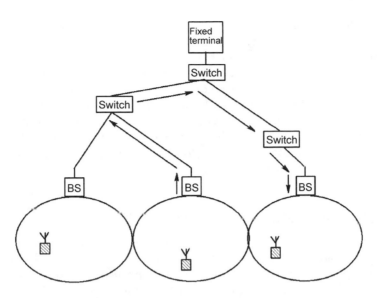

Figure 10.2   WATM network architecture

10.1.2 Wireless ATM

A simple network architecture for WATM is shown in Figure 10.2. It consists of a number of small cells, each of which contains a BS. The basic role of the base station is interconnection of the wireless and wired segment of the network. Each cell contains a number of mobile ATM-enhanced terminals. All terminals inside a cell communicate only with the cell's BS and not between each other. To support mobile terminals, BSs are connected to mobility-enhanced ATM switches. These in turn are interconnected by regular switches in the ATM backbone network. ATM switching is used for intercell traffic. Terminals are capable of roaming between cells and this gives rise to the need for techniques for efficient location management and efficient handoff.

There are proposals for two different scenarios [5] regarding the functionality of the BS in the above architecture. The first scenario calls for termination of the ATM Adaptation Layer (AAL) at the BS. In this case, the traffic transmitted over the wireless medium is not in the format of ATM cells. Rather a custom wireless MAC is used that encapsulates one or more ATM cell into a single packet. Using this grouping procedure, the overhead due to the header needed for wireless transmission is less than it would be for wireless transmission of single ATM cells. In the second, the BS relays ATM cells from the BS towards both the wired segment of the network and the mobile terminals.

ATM implementation over the wireless medium poses several design and implementation challenges that are summarized below:

- ATM was originally designed for a transmission medium whose BERs are very low (about 10^{-10}). However, wireless channels are characterized of low bandwidth and high BER values. It is questioned whether ATM will function properly over such noisy transmission channels.
- ATM calls for a high resource environment, in terms of transmission bandwidth. However,

as we have seen, the wireless medium is a scarce resource that calls for efficient use of medium. However, an ATM cell carries a header, which alone poses an overhead of about 10%. Such an overhead is undesirable in wireless data networks since it reduces overall performance. This problem can be alleviated by performing header compression. ATM was designed for stationary hosts. In the wireless case, users may roam from one cell to another thus causing frequent setup and release of virtual channels. Thus, fast and efficient mechanisms for switching of active VCs from the old wireless link to the new one are needed. When the handover occurs, the current QoS may not be supported by the new data path. In this case, a negotiation is required to set up new QoS. Handover algorithms should take those facts into consideration.

10.1.3 Scope of the Chapter

The remainder of this chapter discusses a number of issues relating to wireless ATM. It is assumed that the reader possesses basic knowledge on ATM. In Section 10.2, wireless ATM architecture is discussed covering issues related to the protocol stack of wireless ATM. This section also discusses location management and handoff in wireless ATM networks. Section 10.3 discusses HIPERLAN 2, an ATM-compatible WLAN developed by the European Tele-communications Standards Institute (ETSI). Contrary to WLAN protocols, HIPERLAN 2 is connection oriented and ATM-compatible. Slightly deviating from the contents of this chapter, Section 10.4 presents a number of routing protocols for multihop ad hoc wireless networks. Finally, the chapter ends with a brief summary of its contents in Section 10.5.

10.2 Wireless ATM Architecture

The protocol architecture currently proposed by the ATM Forum is shown in Figure 10.3.

The WATM items are divided into two parts: mobile ATM, which consists of a subpart of the control plane, and radio access layer (shaded items in the figure). Mobile ATM deals with the higher-layer control/signaling functions that support mobility. The radio access layer is responsible for the radio link protocols for wireless ATM access. Radio access layers consists of the physical layer, the media access layer, the data link layer, and the radio resource control. Up to now, only PHY and MAC are under consideration. The protocols and approaches for DLC and RRC have not been proposed yet. The physical, MAC and DLC layers for the radio access layer are briefly discussed below, while mobile ATM issues are discussed in later sections.

User Plane		Control Plane
ATM Adaptation Layer (AAL)		Wireless
ATM Layer		
DLC		
MAC		Control
Radio Physical Layer		

Figure 10.3 WATM protocol architecture

	Low-sped PHY	High-speed PHY
Band	5.15-5.35 GHz, 5.725-5.875 GHz	59-64 GHz
Cell Radius	80m	15m
Channel bandwidth	30MHz	150/700 MHz
Data Rate (PHY)	25 Mbps	155-622 Mbps
Modulation	16 – DQPSK	32-DQPSK
PHY packet length	PHY header + MAC header + 4 * ATM cells	

Figure 10.4   Physical layer requirements for WATM

10.2.1 The Radio Access Layer

10.2.1.1 Physical Layer (PHY)

Fixed ATM stations can typically achieve rates ranging from 25 to 155 Mbps at the PHY layer. However, due to the use of the wireless medium, such speeds are difficult to achieve in WATM. Thus, typical bit rates for WATM PHY are in the region of 25 Mbps, corresponding to the 25 Mbps UTP PHY option for wired ATM. Note that 25 Mbps is the speed at the physical layer. WATM VCs will typically enjoy bit rates ranging from 2 to 5 Mbps sustained and from 5 to 10 Mbps peak. Nevertheless, higher PHY speeds are possible and WATM projects under development such as the MEDIAN project succeeded in achieving data rates of 155 Mbps by employing OFDM transmission at 60 GHz. As far as hardware is concerned, WATM modems should be able to support burst operation with relatively small preambles in order to support transmission of small control packets and ATM cells and cope with delay spreads ranging from 100 to 500 ns.

The suggested physical layer requirements for WATM [6] are shown in Figure 10.4. Apart from the modulation techniques shown in the figure, a number of others have been proposed[3], such as equalized QPSK/GMSK, equalized QAM and multicarrier techniques such as OFDM. Of these, the most promising seem to be equalized QPSK/GMSK, which is simple to implement and can cope with moderate delay spreads (<250 ns) and OFDM which is more robust to interference and larger delay spreads. CDMA transmission, although efficient for frequency reuse and multiple access is not a potential candidate for WATM, because of the low DLC data rates it will offer due to spreading.

10.2.1.2 MAC Layer

A number of MAC protocols have been proposed for WATM [5,7]. Most of the proposals describe a form of a centralized TDMA system in which the frames are divided into two parts: one contention part, which is used by the mobiles to reserve bandwidth for transmission and one part in which information is transmitted.

Some general requirements for an efficient WATM MAC protocol are the following [5]:

- Allow for decreased complexity and energy consumption at the mobile nodes.
- Provide a means of supporting negotiated QoS under any load condition.
- Support the standard ATM services, such as UBR, ABR, VBR and CBR traffic classes.
- Provide adequate support for QoS-demanding traffic classes.
- Provide a low delay mechanism of channel assignment to connections.

- Support Peak Cell Rate (PCR), Sustainable Cell Rate (SCR), and Maximum Burst Size (MBS) requests.
- Support multiple physical layers. For example, the same MAC functionality should be able to operate over the 5 GHz and 60 GHz physical layer options.
- Efficiently manage and reroute ATM connections as users move while maintaining negotiated QoS levels.
- Provide efficient location management techniques in order to track mobiles and locate them prior to connection setup.

WATM, being a member of the ATM family, provides support for applications, like multimedia, which are characterized by stringent requirements, such as increased data rates, constant end-to-end delay and reduced jitter. Traditional WLANs cannot support those requirements, and have limited support for QoS applications, as we mentioned before. As a result, considerable research projects target the area of WLANs using ATM technology (WATM LANs). Such a project is HIPERLAN 2, a standard being developed by ETSI. The standard is described in later sections.

10.2.1.3 DLC Layer

The DLC layer interfaces the ATM layer to lower layers. Thus, in order to hide the deficiencies of the wireless medium from the ATM layer, DLC should implement error detection, retransmission and FEC. Different levels of coding redundancy might be used in order to support each ATM service class.

The DLC layer exchanges 53-byte ATM cells with the ATM layer above it. A DLC PDU is a packet that may consist of one or more cells. This packet is handed down to the physical layer for transmission as a single unit. The use of a multicell DLC packet reduces overhead but requires functionality to convert between the ATM cell format and the DLC packet format.

10.2.2 Mobile ATM

10.2.2.1 Location Management/Connection Establishment

Existing protocols for connection setup in ATM assume that the location of a terminal is fixed. Thus, the terminal's address can be used to identify its location, which is needed in processes such as call establishment. However, when terminals become mobile, this is no longer true and additional addressing schemes and protocols are needed to track the mobile ATM terminal.

Location management in a wireless ATM network can be either external to the connection procedure or integrated[3,8]. Here we describe the latter option. Each mobile terminal served by the network is associated with a 'home' BS or switch which provides it with a home ATM address. When the terminal moves to another cell, it is assigned a foreign address via this cell's BS. The home switch maintains a pointer from the permanent home address to the current foreign address of the mobile. This pointer maintenance is achieved by terminal transmission of address updates as they move to new cells. Connections to a mobile terminal are then established with a simple extension to the standard Q.2931 signaling procedure specified in existing ATM specifications. When a connection needs to be established to a specific terminal, a SETUP message is issued with the home address of the mobile as the destination. If the mobile is under coverage of its 'home' BS the connection is established. If the mobile has roamed to another cell, a RELEASE message is returned towards the source that requested the connection. The RELEASE message carries the foreign address of the terminal. Upon reception of the RELEASE message, the source can then issue a SETUP message with the terminal's foreign address as the destination. Thus, the connection with the roamed terminal will be set up.

10.2.2.2 Handover in Wireless ATM

The mobility nature of terminals in WATM networks means that the network must be able to dynamically switch ongoing connections of users that roam between cells. Handovers take place when mobiles move out of the coverage of a BS towards the coverage of a new one. In such a case the signal measurement at the mobile of the new BS gets stronger while that of the previous one weakens. Handoff can be network-controlled, mobile-assisted or mobile-controlled. In the first case, the mobile terminal is completely passive and all signal measurements and handoff initiations are a responsibility of the BS. In the second case, both the BS and the mobile terminal perform signal measurements, however, the handoff initiation is a responsibility of the BS. Finally, in the third case both the BS and the mobile terminal perform signal measurements and the handoff initiation is a responsibility of the mobile terminal.

A handover should be done in an efficient way such that the user does not notice performance degradation. Of course, there is a chance of the handoff being blocked. This means that the new BS is not able to serve the connections of the roaming user, either for reasons of bandwidth availability or due to the fact that it cannot guarantee the QoS of the user's connection. In the latter case, however, a renegotiation towards a lower level of QoS might be carried out in order for the connection to be kept alive.

A handoff generally involves switching the VCs of the roaming terminal from the former BS to the current one while maintaining route optimality and QoS to the maximum possible extent. A typical handoff in a wireless ATM network consists of the following phases[3]:

- The terminal initiates the handoff. This is done by sending a message to its current BS in order to initiate the procedure of moving the connection from the current BS to the new one.
- The network switches and BSs collectively determine the switch from which to reroute each VC. This switch is known as a 'crossover switch' (COS). When the handover occurs, the current QoS may not be supported by the new data path. In this case, a negotiation is required to set up new QoS. Handover algorithms should take those facts into consideration. Related to this fact is the identification of the optimal COS to be used in order to switch the connection. COs may be initiated either at the old or the new BS.
- Upon determination of the COS, the network routes a subpath from the COS to the new BS.
- Over the above path, the cell stream is switched to the new BS.
- The unused subpath from the COS to the old BS is released.
- Finally, the terminal drops its radio connection with the old BS, connects to the new one and confirms end-to-end handoff.

To minimize QoS disruption during the handoff, the network can perform a 'lossless handoff' [8] in order to maintain cell delivery in sequence without loss to the mobile terminal. This involves buffering of traffic in transit during the handoff process. Specifically, the COS sends a 'marker' ATM cell to the current BS before switching the terminal's connections to the new one. From that point onwards, when ATM cells are received at the new BS from the COS, these are buffered until the handoff is confirmed by the mobile terminal. Furthermore, the current BS buffers traffic received between the arrival of the marker cell and the arrival handoff confirmation. Upon this confirmation, the current BS forwards buffered traffic to the new BS. Finally, the new BS relays to the terminal the buffered cells from the current BS followed by the buffered cells from the new one. Thus, lossless, in-sequence delivery is achieved.

In order to support WATM handoff, a number of extensions to ATM signaling protocols have been proposed [3,8].

10.3 HIPERLAN 2: An ATM Compatible WLAN

HIPERLAN 2 [9–11] aims to provide high speed access (up to 54 Mbps at the physical layer) to a variety of networks including 3G networks, ATM and IP based networks and for private use as a wireless LAN system. Supported applications include data, voice and video, with specific QoS parameters taken into account. In contrast to the WLAN systems described in Chapter 9, HIPERLAN 2 is a connection-oriented system which uses fixed size packets. HIPERLAN 2 is compatible with ATM. Its connection-oriented nature makes support for QoS applications easy to implement. In the following subsections, we describe the main aspects of HIPERLAN 2.

10.3.1 Network Architecture

The HIPERLAN 2 standard adopts an infrastructure topology. As shown in Figure 10.5, the network coverage area comprises a number of cells, with traffic in each cell being controlled by an Access Point (AP). Mobile terminals within a cell communicate with the cell's AP through the HIPERLAN 2 air interface. Direct communication between two mobile terminals is also possible, however. this procedure is still in the development phase. Each mobile terminal can communicate only with one AP (that of the current cell). In order for such a communication to take place, an association procedure must first take place between the AP

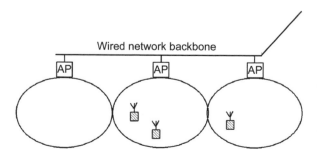

Figure 10.5   HIPERLAN 2 network architecture

and the mobile terminal. After the association takes place, mobile terminals can freely move within the coverage area of the HIPERLAN 2 network while maintaining their logical connections. Moving to another cell is made possible through a handover procedure. The APs automatically configure the network by taking into account changes in topology due to mobility. Association and handover are revisited later in this section.

Being compatible with ATM, HIPERLAN 2 is a connection-oriented network using fixed size packets. Signaling functions are used to establish connections between the mobile nodes and the AP in a cell and data is transmitted over these connections as soon as they are established, using a time division multiplexing technique. The standard supports two types of connections: bi-directional point-to-point connections between a mobile node and an AP, and unidirectional point-to-multipoint connections carrying traffic to the mobile nodes. Finally, there is a dedicated broadcast channel used by the AP to transmit data to all mobiles within its coverage.

The connection-oriented nature of HIPERLAN 2 makes support for QoS applications easy to implement. Each connection can be created so as to be characterized by certain quality requirements, like bounded delay, jitter and error rate. This support enables the HIPERLAN 2 network to support multimedia applications in a way similar to the ATM network.

HIPERLAN 2 also provides support for issues like encryption and security, power saving, dynamic channel allocation, radio cell handover, power control, etc. However, most of these issues are either not standardized yet or left to the vendors to implement.

10.3.2 The HIPERLAN 2 Protocol Stack

The protocol stack for the HIPERLAN 2 standard is shown in Figure 10.6. It comprises a control plane part and a user plane part following the semantics of ISDN functional partitioning. The user plane includes functionality for transmission of traffic over established connections, and the control plane provides procedures to control established connections. The protocol has three basic layers: the Physical Layer (PHY), the Data Link Control (DLC) layer, and the Convergence Layer (CL). At the moment, only the DLC includes control plane functionality. The various layers are discussed below.

Control Plane	User Plane
Higher Layers	
Convergence Layer	
Cell based	Packet based
DLC layer	
RRC ACF DCC	Error Control
RLC	
MAC protocol	
Physical layer	

Figure 10.6   The HIPERLAN 2 protocol stack

Mode	Modulation scheme	Code Rate	PHY speed (Mbps)
1	BPSK	½	6
2	BPSK	¾	9
3	QPSK	½	12
4	QPSK	¾	18
5	16 QAM	9/16	27
6	16 QAM	¾	36
7	64 QAM	¾	54

Figure 10.7   HIPERLAN 2 physical layer alternatives

10.3.2.1 HIPERLAN 2 Physical Layer

HIPERLAN 2 is characterized by high transmission rates at the physical layer, up to 54 Mbps. The use of OFDM in the physical layer effectively combats the increased fading occurrence experienced in indoor radio environments, such as offices, etc., where the transmitted radio signals are subject to reflection from a number of objects, thus leading to multipath propagation and consequently ISI. The channel spacing is 20 MHz with 52 subcarriers used for each channel. Of these, 48 subcarriers carry actual data and the remaining four are used as pilots in order to perform coherent demodulation.

HIPERLAN 2 is able to adapt to changing radio link quality through a Link Adaptation (LA) mechanism. Based on received signal quality which depends both on the AP-mobile terminal relative position and interference from nearby cells, LA dynamically selects the method of modulation and the Forward Error Correction (FEC) code to use in an effort to provide a robust physical layer. The alternative modulation methods are BPSK, QPSK, 16 QAM and 64 QAM. FEC is performed by a convolutional code with rate 1/2 and constraint length 7. The physical layer alternatives offered by LA are shown in Figure 10.7.

10.3.2.2 HIPERLAN 2 Data Link Control (DLC) Layer

The DLC layer is used to establish the logical links between APs and the MTs. The DLC layer comprises a number of sublayers providing medium access and connection handling services to upper layers. The DLC layer consists of three sublayers: the Medium Access Control (MAC) sublayer, the Error Control (EC) sublayer and the Radio Link Control (RLC) sublayer.

10.3.2.2.1 MAC Protocol and Channel Types   The MAC protocol used by HIPERLAN 2 is based on time-division duplex (TDD) and dynamic time-division multiple access (TDMA). MAC control is centralized and performed by each cell's AP. The wireless medium is shared in the time domain through the use of a circulating MAC frame containing slots dedicated either to uplink or downlink traffic. The length of the MAC frame is fixed at 2 ms and comprises a number of parts which are not fixed. Rather, their lengths are variable in nature and are determined by the AP. Uplink and downlink slots within a frame are allocated dynamically depending on the need for transmission resources. All data from both mobile terminals and APs is transmitted in dedicated time slots. For mobile terminal

| Broadcast | Downlink | Direct link | Uplink | Random Access |

Figure 10.8   Structure of the 2 ms MAC frame

transmission, slots are allocated after bandwidth requests made to the AP. The exact form of the MAC frame is shown in Figure 10.8, where one can see that apart from the parts dedicated to uplink and downlink traffic there are also broadcast, direct link and random access phases. The broadcast frame carries the broadcast control channel and the frame control channel (both are described below). The direct link phase enables exchange of user traffic between mobile terminals without intervention of the AP. As mentioned above, this is optional. Finally, the random access phase carries the random access channel (described below). This phase is used by mobile terminals either for purposes of association with an AP, for control signaling when the terminal has not been allocated uplink slots within the MAC frame and during handover to a new AP for the purpose of switching ongoing connections to the new AP.

The MAC frame consists of several transport channels:

- The Broadcast Channel (BCH) is a downlink channel used to convey to the mobiles control information regarding transmission power levels, wake-up indicators for nodes in power save mode, length of the FCH and the RCH channels (described below) and the means to identify the HIPERLAN 2 network and the AP to which the mobile belongs.
- The Frame Control Channel (FCH) is a downlink channel used to notify the mobile nodes about resource allocation within the current MAC frame both for uplink and downlink traffic and for the RCH.
- The Random Access Channel (RCH) is used in the uplink both in order to request transmission in the downlink and uplink portions of future MAC frames and to transmit signaling messages. The RCH comprises contention slots which are used by the mobiles to compete for reservations. Collisions may occur and the results from RCH access are reported back to the mobiles in the Access Feedback Channel (ACH). When the request for transmission resources from the MTs arise, the AP can allocate more resources for the RCH in order to serve the increased demand.
- The Access Feedback Channel (ACH) is used on the downlink to notify about previous access attempts made in the RCH.

The above transport channels are used as a means to support a number of logical HIPERLAN 2 channels. The mapping is shown in Figures 10.9 and 10.10. The logical channels are as follows:

- The Slow Broadcast Channel (SBCH). All nodes within a cell can access the SBCH. It is a

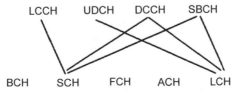

Figure 10.9   Mapping from logical to transport channels (downlink)

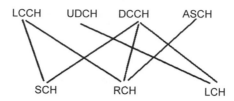

Figure 10.10   Mapping from logical to transport channels (uplink)

downlink channel that conveys broadcast control information concerning all the nodes within a cell. This transmission is initiated upon decision of the AP and may contain information regarding (a) the seed to be used for encryption, (b) handover acknowledgments, (c) MAC address assignments to non-associated mobile terminals, and (d) broadcast of RLC and CL information.

- The Dedicated Control Channel (DCCH) is of bidirectional nature and is implicitly established when a terminal associates with the AP within a cell. After association with an AP has taken place, a terminal has its dedicated DCCH which is used to convey control signaling. The DCCH is realized as a DLC connection upon which RLC messages regarding association and control of DLC connections are exchanged.

- The User Data Channel (UDCH) transports user data cells between a mobile node and an AP and vice versa. A UDCH for a specific mobile node is established through signaling transmitted over the node's DCCH. The UDCH establishment takes place after negotiation of certain quality parameters that characterize a connection. The DLC guarantees in-sequence delivery of the transmitted data cells to the convergence layer. The use of ARQ techniques is possible in UDCH operation, although there might be connections where ARQ is not used, such as multicasts and broadcasts. For uplink traffic, mobile requests for UDCH bandwidth are conveyed to the AP which then notifies the mobile whether or not it has been granted bandwidth through the FCH. For downlink traffic, the AP can reserve UDCH bandwidth without requests from mobiles.

- The Link Control Channel (LCCH) is a bidirectional channel used to exchange information regarding error control (EC) over a specific UDCH. The AP determines the necessary transmission slots for the LCCH in the uplink and the grant is announced in an upcoming FCH.

- The Association Control Channel (ASCH) is used by the mobile nodes either to request association or disassociation from a cell's AP. Such messages are exchanged only (a) when a mobile terminal de-associates with an AP and (b) when a handover takes place.

10.3.2.2.2 Error Control Protocol   The Error Control (EC) protocol of the HIPERLAN 2 protocol stack uses a selective repeat ARQ scheme in order to provide error-free, in-sequence data delivery to the convergence layer. Positive and negative acknowledgments are transmitted over the LCCH channel. In-sequence delivery is guaranteed by assigning proper sequence numbers to all frames of the connection. The number of retransmission attempts per frame is configurable. Furthermore, in an effort to support QoS for applications that are vulnerable to delay, the EC layer includes an out-of-date frame discard mechanism. If a data cell becomes obsolete, then the sender EC layer can decide to discard it together with frames in the same connection with lower sequence number. In such a case, the responsibility for dealing with the data loss belongs to upper layers.

10.3.2.2.3 Radio Link Control Protocol   The Radio Link Control (RLC) protocol provides services to the Association Control Function (ACF), Radio Resource Control function (RRC), and the DLC user Connection Control function (DCC). These signaling entities implement the DLC control plane functionality for exchange of control information between the AP and the mobile terminals.

The ACF is used by mobile nodes for purposes of:

- *Association.* In this case, a mobile node chooses among multiple APs the one with the best link quality. These measurements are made by listening to the BCH from the various APs, since the BCH provides a beacon signal to be used for this purpose. If association takes place, the AP grants the mobile terminal a unique MAC identity number. Then, the ASCH is used to exchange information with the AP regarding the capabilities of the DLC link to be established. For example, a mobile terminal may request from the AP information regarding the capabilities and characteristics of the links it can offer, such as the physical layer used, whether encryption is possible or not, supported authentication and encryption procedures and algorithms, supported convergence layers, etc. The AP replies with a set of supported PHY modes, a single convergence layer and a selected authentication and encryption procedure; an alternative is support for no authentication/encryption. Supported encryption algorithms are DES and 3-DES. The two alternatives for authentication are public key-based and pre-shared key authentication. If encryption is to be employed then the mobile terminals start a Diffie–Hellman key exchange procedure in order to determine the secret session key. This is used for encryption of all unicast traffic between the AP and the mobile terminal. Moreover, broadcast and multicast traffic can also be encrypted. This procedure takes place by using common keys at the mobile terminals and the AP (all mobile terminals under the same AP use the same common key). Common keys are distributed encrypted through the use of the unicast encryption key. Periodic changes of the various encryption keys increases system security. After the mobile node and the AP have associated, the AP can assign a DCCH to the mobile node which is used by the latter to establish one or more DLC connection, possibly each of different QoS.
- *Deassociation.* This can have either an explicit or an implicit form. In both cases the AP frees the resources which were allocated to the deassociated mobile terminal. In the first case, the AP is notified by the mobile terminal that the latter wants to deassociate (e.g. when the terminal is about to shut down). In the second case, the AP deassociates with a specific terminal, when the latter remains unreachable from the AP for a specific time period.

No user data traffic transmission can take place unless at least one DLC connection has been established between the mobile terminal and the AP. Thus, the DCC function is used to establish DLC user connections by transmitting signaling messages over the DCCH. The AP assigns a unique connection identifier to each DLC connection. The signaling scheme is quite straightforward, comprising a request for a specific QoS connection followed by an acknowledgment when the request can be fulfilled. There also exist connections that manage both DLC connection release and modification of the parameters that characterize an existing DLC connection.

The RRC function manages the following procedures:

- *Handover.* For a mobile terminal handover is initiated when the quality of the link between the terminal and the current AP is inferior to that of a link to another AP. There are two handover methods in HIPERLAN 2: reassociation and handover via signaling across the fixed network. The first method takes place when the mobile terminal deassociates with an AP and reinitiates association with another AP. The second method involves exchange of information regarding association and connection control between the old and new APs. This information transfer between the APs takes place across the fixed network. The method for making link quality measurements for handovers is not defined in HIPERLAN 2. Rather, each vendor is free to either base it on signal strength or another quality criterion.
- *Dynamic frequency selection.* This RRC function automatically allocates frequencies to the various APs of a HIPERLAN 2 network. Allocation is made in a way that avoids use of interfering frequencies through measurements made by APs and mobile terminals. The latter contribute to the procedure upon request of their AP to perform measurements regarding the radio signals received from nearby APs. Since the radio environment is due to be dynamic, APs are likely to change operating frequency while already involved in an ongoing connection. Thus, RRC also includes signaling functionality to inform mobile terminals of an upcoming change in the operating frequency of their AP.
- *'Mobile terminal alive'.* This procedure enables second case of deassociation mentioned above. When mobile terminals are idle, their AP tracks them by periodically transmitting 'alive' messages to these terminals. Alive messages are followed by responses from idle terminals and thus APs are able to supervise them. If an idle terminal does not respond to the 'APs' alive messages, it is deassociated from the AP. Alternatively APs do not transmit alive messages but rather monitor idle terminals for a specific time period. When this period has elapsed the terminal is deassociated.
- *Power saving.* This is a process controlled by the mobile terminal. A mobile terminal selects a sleeping time of duration N frames, with $2 \leq N \leq 216$. After these N frames have elapsed, the following scenarios are possible: (a) the AP wakes up the mobile terminal due to data pending for this terminal at the AP; (b) the mobile terminal wakes up due to data pending for transmission at this terminal; (c) the mobile terminal goes to sleep for another N frames; (d) the mobile terminal, after failing to receive the wake-up messages from the AP, wakes up after N frames and performs the 'mobile terminal alive' procedure.

10.3.2.3 HIPERLAN 2 Convergence Layer (CL)

The CL of the protocol stack carries out two functions. The first is to segment the higher layer PDUs into fixed size packets used by the DLC. The second is to adapt the services demanded by the higher layers to those offered by the DLC. This function requires reassembly of the fixed-size DLC packets to the original variable-size packets used by the higher layers. There are currently two different types of CLs defined:

- *Cell-based CL.* The cell-based CL serves interconnection to ATM networks and transpar-

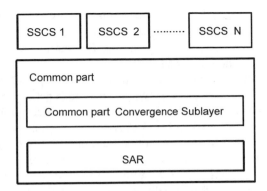

Figure 10.11   Structure of the packet based CL

ently integrates HIPERLAN 2 with ATM. In the cell-based CL, Segmentation and Reassembly (SAR) functionality is not included because ATM cells fit into the HIPERLAN 2 DLC PDU. Nevertheless, a compression of the ATM cell header is necessary, transmitting only its most important parts.

- *Packet-based CL.* The packet based CL is used to interconnect WATM mobiles to legacy wired LANs like Ethernet. The packet-based CL comprises a common part and several Service Specific parts (SSCS), as shown in Figure 10.11. SSCSs allow for easy interfacing with fixed networks. The common part has the responsibility of segmenting packets received from SSCSs before handing them down to lower layers. Similarly, it is a responsibility of the common part to reassemble segmented packets received from the DLC before these are handed to the appropriate SSCS. Furthermore, the common part is responsible for adding padding bytes in an effort to make common part Protocol Data Units (PDUs) and an integral number of DLC Service Data Units (SDUs).

The overall performance of a HIPERLAN 2 system depends on a number of factors, including available channel frequencies, propagation conditions and interference experienced. Tests have shown that, in most cases, data rates above 20 Mbps (at the DLC layer) are likely to be achieved.

10.4 Routing in Wireless Ad Hoc Networks

A brief introduction to packet routing in wireless ad hoc networks was made in Chapter 2. There, it was highlighted that the performance of such protocols largely depends on the efficiency of the routing protocol used. In wireless ad hoc networks, stations are free to move around. This, together with the fact that the transmission range of mobile terminals is fixed, results in a dynamically changing network topology: As stations move around, some network links are destroyed while the possibility of new links being established arises. It is obvious that such an environment cannot be served efficiently by routing protocols developed for wired networks. This is due to the fact that in such networks, the assumption of a static topology is made. Thus, new routing protocols are needed for the dynamically changing ad hoc wireless environment.

This section describes some representative routing protocols for ad hoc wireless networks.

In these protocols, it is assumed that all stations of the network have identical capabilities and employ the capability to perform routing-related tasks, such as route discovery/establishment to other nodes in the network and route maintenance. The routing protocols presented fall into two families: table-driven and on-demand [12].

Table-driven routing protocols aim to maintain consistent, up-to-date routing information from each node to all other nodes of the network. Thus, each network node maintains one or more routing table which is used to store the routes from this node to all other network nodes. This knowledge regarding every possible route needs to be present in every node irrespective of the fact that some of these routes may not be used by network connections. When topological changes occur, the relating information is relayed to all network nodes in an effort to provide the network nodes with up-to-date routing information.

On-demand routing protocols follow a different approach: a route is established only when required for a network connection. Thus, when a source node A needs to connect to a destination node B, then A invokes a routing discovery protocol to find a route connecting it to B. Upon route establishment, nodes A and B as well as intermediate nodes store the information regarding the route from A to B in their routing tables. The route is maintained until the destination is unreachable or the route is no longer needed.

Table-driven routing protocols obviously have the advantage of reduced end-to-end delay, since, upon generation of a network connection request, the route is already established. However, their disadvantage is the fact that routing information is disseminated to all network nodes leading to increased signaling traffic and power consumption. Thus, bandwidth for user traffic is reduced and the operating time of the battery-powered mobile nodes is reduced. On-demand routing protocols, on the other hand, have a lower power consumption and demand less control signaling; however, end-to-end connection delay is increased, since upon generation of a connection request between two nodes, the connection needs to wait some time for the link between the nodes to be established.

10.4.1 Table-driven Routing Protocols

10.4.1.1 Destination-Sequenced Distance-Vector (DSDV) Routing Protocol [13]

The DSDV routing protocol is an extension of the classical Bellman–Ford routing algorithm. The extensions incorporated in DSDV target freedom from loops in routing tables. In DSDV, each node maintains a routing table that contains information regarding all possible routes within the network, the number of hops of each route and the sequence number of each route. The latter is a number assigned by the destination of the route and shows how 'old' the route is. The lower the sequence number, the 'older' the route. When a node A needs to select a route to node B, it checks its routing table. If more than one such route is found, the newer one (the one with the largest sequence number) is used. If more than one route shares the same sequence number, then the shortest one (the one with the lower number of hops) is chosen.

Network nodes periodically broadcast their routing tables in order to propagate topology knowledge throughout the network. Apart from these periodic transmissions, a station can select to broadcast its routing table when significant topology changes have occurred. The propagation of routing tables obviously results in a large overhead. In an effort to alleviate this problem, two types of updates are defined: full-dump updates and incremental updates. In full-dump updates, stations transmit their entire routing table. Since routing tables are mostly quite large, a full-dump update typically involves more than one packet broadcast. This obviously consumes resources, so full dumps are transmitted infrequently. Incremental updates are transmitted between full dumps and convey only that information which has changed since the last update. Incremental updates thus consume less resources and are carried over a single packet. The relative frequency of full-dump and incremental updates depends on the nature of topological changes. In a network of a slowly changing topology, full dumps are rarely used since incremental dumps are able to convey the slow topological changes. On the other hand, in a network of fast changing topology, full dumps will be more frequent.

10.4.1.2 Clusterhead Gateway Switch Routing (CGSR) Protocol [14]

CGSR is a modification of DSDV. It is different from DSDV in that, while DSDV assumes a 'flat' network (which means that all nodes have identical responsibilities), CGSR partitions the network into a number of 'clusters'. Nodes inside a cluster are controlled by a node known as the clusterhead. Clusterheads are selected by the members of each cluster. It is obvious that as mobile nodes move, some clusters will disappear, new ones will be created and new nodes may be admitted into existing clusters. Thus, new clusterhead selections will appear from time to time. In an effort to reduce the overhead due to clusterhead selections, a Least Cluster Change (LCC) clusterhead selection algorithm is used. LCC states that clusterhead selections take place only when two clusterheads come into transmission range of one another or when a node moves out of the range of all the clusterheads.

CGSR uses a modification of DSDV as the routing scheme. Specifically, in CGSR all routes commencing from nodes inside a certain cluster pass through this cluster's clusterhead. If a route serves a connection between two nodes inside the same cluster, then the clusterhead routes packets of this connection to their destination. If the route serves a connection between nodes in different clusters, then the clusterhead routes this packet to a gateway node. These are the nodes that are within range of more than one clusterhead. Upon receipt of the packet by the gateway, this is routed to the clusterhead of the adjacent cluster. The procedure continues until the packet reaches the clusterhead of its destination. Then, it is routed to the destination station. An example of CGSR routing is shown in Figure 10.12.

In GGSR, nodes maintain two tables: The routing table and the 'cluster member table'. The 'cluster member table' contains the clusterhead of each node in the network. These tables are periodically transmitted by each node. Upon receipt of such a table from a neighbor, network nodes update their own 'cluster member table'. 'Cluster member tables' are needed for packet routing. Upon reception of a packet, a node will check its cluster member table to find the clusterhead of the next cluster along the route to the destination station. Then, it checks its routing table to find the next hop that should be selected to reach the next clusterhead and forwards the packet over this hop.

10.4.1.3 The Wireless Routing Protocol (WRP) [15]

In order for WRP to operate, each node must maintain four tables, a fact that can lead to substantial memory requirements, especially in the case of networks comprising many nodes. the four tables are the distance table, the routing table, the link-cost table and the Message Retransmission List (MRL) table. For a node A, the distance table of A contains the distance

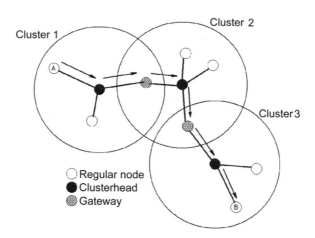

Figure 10.12   CGSR routing from node A to node B

to each destination node X via each neighbor Y of A. Moreover, each entry contains the downstream neighbor of Y through which the route from A to X traverses. The routing table of node A contains the distance to each destination node X, successor of A in this route and a flag that indicates whether this route is a simple one, or a loop. The link cost table of node A maintains the cost of the link from A to each neighbor Z and the number of timeouts since an error-free message was received from Z. Finally, the MRL contains entries regarding update messages sent from A. Such an entry comprises the sequence number of the update message, a retransmission counter, a flag indicating whether an acknowledgement is required from the neighbor for an update transmitted by A and a list of updates sent in the update message. Thus, the information in the MRL contains information regarding (a) neighboring nodes that have not acknowledged update messages from A, (b) when to retransmit update messages to these nodes.

In WRP, nodes exchange update messages with their neighbors both periodically and as a result of link changes. Such is the case of a link loss between two nodes, e.g. A and B. In such cases, A and B send update information to their neighbors, which in turn modify their tables and search for alternative routes that do not contain the link between A and B. Updates contain information regarding new route destinations (that may have been established by neighboring nodes and other nodes in the network), new distances of routes, the predecessor of each route's destination and a list of nodes that should acknowledge this update. When a node receives an update message from one of its neighbors, it modifies its distance table and checks for possible alternative paths for each route. In cases of slowly changing topologies, it is likely that the network topology around a certain node, e.g. A, might not have changed between two consecutive updates. In such a case, node A will not transmit an update message but only acknowledge its presence to its neighbors through transmission of a HELLO message. HELLO packets, although useful as described above, consume system bandwidth and prevent nodes form going to power-saving mode.

A unique feature of WRP is that it checks the consistency of all its neighbors upon detecting a change in link of any of its neighbors. This consistency check helps eliminate loop-free situations.

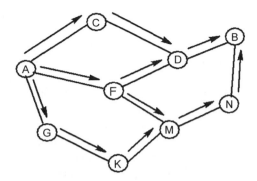

Figure 10.13 Propagation of the RREQ packet

10.4.2 On-demand Routing Protocols

10.4.2.1 Ad hoc On-demand Distance Vector (AODV) Routing [16]

The AODV algorithm is the on-demand counterpart of the table-based DSDV algorithm. Their primary difference lies in the fact that AODV creates routes on-demand while DSDV maintains the list of all the routes. In AODV, a route is created only when requested by a network connection and information regarding this route is stored only in the routing tables of those nodes that are present in the path of the route.

The procedure of route establishment is shown in Figures 10.13 and 10.14. In this example, we assume that node A wants to set up a connection with node B. In Figure 10.13, node A initiates a path discovery process in an effort to establish a route to node B, by broadcasting a Route Request (RREQ) packet to its immediate neighbors. Each RREQ packet is identified through a combination of the transmitting node's IP address and a broadcast ID. The latter is used to identify different RREQ broadcasts by the same node and is incremented for each RREQ broadcast. Furthermore, each RREQ packet carries a sequence number (similar to that of DSDV) which allows intermediate nodes to reply to route requests only with up-to-date route information. Upon reception of an RREQ packet by a node, this is forwarded to the immediate neighbors of the node and the procedure continues until the RREQ is received either by node B or by a node that has recently established a route to node B. If subsequent copies of the same RREQ are received by a node, these are discarded.

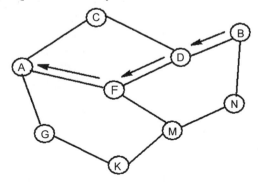

Figure 10.14   Propagation of the RREP packet

When a node forwards a RREQ packet to its neighbors, it records in its routing tables the address of the neighbor node where the first copy of the RREQ was received. This fact helps nodes to establish a reverse path, which will be used to carry the response to the RREQ. Returning to the previous example, we see in Figure 10.14 that when the RREQ has reached its destination, a route reply packet is sent back to A. Notice that the RREP follows the route B–D–F–A due to the fact that the first reception of the RREQ packet from B was due to node D and the first reception of the RREQ packet from D due to node F. As the RREP packet travels along the reverse path, the nodes that constitute the path (D, F, A) make appropriate changes in their routing tables (pointing to the next neighbor that is a part of this route) which identify the forward path from A to B. Due to the fact that the RREP packet travels along the reverse path traveled by the RREQ, AODV supports only the use of symmetric links. Support for asymmetric links is not provided. Upon establishment of a route, each route entry at each node is associated with a 'lifetime' value. A timer starts running when the route is not used. If the timer exceeds the value of the 'lifetime', then the route entry is deleted.

Routes may change due to the movement of a node (e.g. node X) within the path of the route. In such a case, the upstream neighbor of this node generates a 'link failure notification message' which notifies about the deletion of the part of the route and forwards this to its upstream neighbor. Upon reception of this message by a node, this is transmitted to the next upstream neighbor. The procedure continues until the source node is notified about the deletion of the route part caused by the movement of node X. Upon reception of the 'link failure notification message', the source node can reinitiate discovery of a route to the destination node.

10.4.2.2 Dynamic Source Routing (DSR) [17]

DSR uses source routing, rather than hop-by-hop routing. Thus, in DSR every packet to be routed carries in its header the ordered list of network nodes that constitute the route over which the packet will be relayed. Thus, intermediate nodes do not need to maintain routing information as the contents of the packet itself are sufficient to route the packet. This fact eliminates the need for the periodic route advertisement and neighbor detection packets that are employed in other protocols. On the other hand, the overhead in DSR is larger, since each packet must contain the whole sequence of nodes comprising the route. Therefore, DSR will be most efficient in cases of networks of small diameter.

DSR comprises the processes of route discovery and route maintenance. A source node wishing to set up a connection to another node initiates the route discovery process by broadcasting a ROUTE_REQUEST packet. This packet is received by neighboring nodes which in turn forward it to their own neighbors. A node forwards a ROUTE_REQUEST message only if it has not yet been seen by this node and if the node's address is not part of the route. The ROUTE_REQUEST packet initiates a ROUTE_REPLY upon reception of the ROUTE_REQUEST packet either by the destination node or by an intermediate node that knows a route to the destination. Upon arrival of the ROUTE_REQUEST message either to the destination or to an intermediate node that knows a route to the destination, the packet contains the sequence of nodes that constitute the route. This information is piggybacked on to the ROUTE_REPLY message and consequently made available at the source node. DSR supports both symmetric and asymmetric links. Thus, the ROUTE_REPLY message can be either carried over the same path with the original ROUTE_REQUEST, or the destination

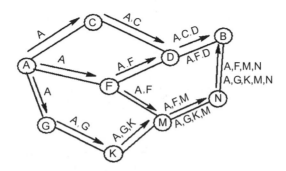

Figure 10.15   DSR route discovery

node might initiate its own route discovery towards the source node and piggyback the ROUTE_REPLY message in its ROUTE_REQUEST. Route discovery is shown schematically in Figure 10.15 for an example network.

In order to limit the overhead of this control messaging, each node maintains a cache comprising routes that were either used by this node or overheard. As a result of route request by a certain node, all the possible routes that are learned are stored in the cache. Thus, a ROUTE_REQUEST process may result in a number of routes being stored in the source node's cache.

Route maintenance is initiated by the source node upon detection of a change in network topology that prevents its packets from reaching the destination node. In such a case the source node can either attempt to use alternative routes to the destination node (if such routes reside in the source's cache) or reinitiate route discovery. Storing in the cache of alternative routes means that route discovery can be avoided when alternative routes for the broken one exist in the cache. Therefore route recovery in DSR can be faster than in other on-demand protocols.

Since route maintenance is initiated only upon link failure, DSR does not make use of periodic transmissions of routing information, resulting in less control signaling overhead and less power consumption at the mobile nodes.

10.4.2.3 Associativity Based Routing (ABR) [18]

The fundamental objective of ABR is to find longer-lived routes for ad hoc mobile networks. This obviously results in fewer route reconstructions and thus higher throughput. ABR defines a new routing metric, called 'degree of association'. This metric defines the level of association stability between neighboring nodes and is derived as follows: all nodes periodically generate and transmit beacons, in order to notify neighboring nodes of their existence. Beaconing intervals must be small enough to ensure accurate spatial and thus connectivity information. Whenever a node (e.g. A) receives such a beacon from a neighboring node (e.g. B), it updates its associativity table by incrementing a counter which signifies the degree of association between this node and the beaconing neighbor. Associativity values are reset when the neighbors of a node or the node itself move out of range. Thus, for two neighboring nodes A and B, the value of the association counter described above defines the degree of association stability between the two nodes. High values of the associativity counter for A and B indicate a low state of relative mobility, while a low value of the associativity counter may indicate a high state of node mobility.

ABR consists of three phases. These are described below:

- *Route discovery.* For purposes of route discovery, a node transmits a Broadcast Query (BQ) packet. This message contains the node's address and the values of the associativity counter with its neighbors. Upon reception of a BQ message, a node erases its upstream neighbor value of the associativity counters and maintains only the associativity counter concerning itself and its upstream node. Then, it forwards the message to its downstream neighbors. A node does not forward a BQ request more than once. Thus, as a BQ packet reaches the destination node, it will contain the values of the associativity counters along the route from the source to the destination. Upon receiving a number of BQ packets (each one corresponding to a different path), the destination will posses information regarding the overall degree of association stability for each route and can thus select the best route. If more than two routes have the same association stability, then the one having the minimum number of hops is selected. Upon selection of a route by the destination, a REPLY packet is sent back to the source along the path specified by the route. As the REPLY packet traverses the path, the corresponding route is marked as active, while the alternative routes remain inactive. The above procedure is known as the BQ-REPLY process.
- *Route reconstruction (RRC).* Depending on which node (or nodes) along the route move, RRC consists of partial route discovery, invalid route erasure, valid route updates, and new route discovery. When the source node moves, a new BQ-REPLY process is initiated and the old route is deleted. When the destination node moves, then its immediate upstream neighbor erases its route and checks if the destination is still accessible by performing a localized query process (LQ[H], where H stands for the number of hops from the upstream node to the destination node). If the destination node receives the LQ packet, it selects the best partial route and issues a reply message. Otherwise, the upstream neighbor of the destination node concludes that the latter is out of range and the next upstream neighbor is instructed to perform the LQ process. This procedure continues until either a new route has been established or the process has backtracked more than half the number of hops that constituted the route from the source to the destination. In the latter case, the procedure is aborted and a new BQ-REPLY process starts at the source node.
- *Route deletion (RD).* An RD broadcast is initiated when a route is no longer valid. Upon reception of an RD packet, all nodes along the route delete the corresponding entries from their routing tables. RD messages are propagated by a full broadcast because the source node may not be aware of any route node changes that occurred during RRCs.

10.4.2.4 Signal Stability Routing (SSR) [19]

SSR routes packets based on the signal strength between nodes and a node's location stability. Thus, SSR selects those routes having the strongest connectivity. This fact aims at fewer route reconstructions and thus higher throughput.

SSR comprises two cooperative protocols. These are the Dynamic Routing Protocol (DRP) and the Static Routing Protocol (SRP). DRP maintains the Signal Stability Table (SST) and the Routing Table (RT). SST is used to store the signal strength of neighboring nodes. The storage of these values in the SST is made possible by periodic link-layer beaconing of nodes in SSR. Based on the quality of the beacon signal, SST entries identify links as 'weak' or 'strong'. All packet transmissions are monitored by the DRP before being passed to the node's SRP which examines the packet in order to find out whether it is destined for this node or another one. In the first case, the packet is pushed up to higher protocol layers. In the second case the packet must be forwarded to its destination. Thus, the node searches in its RT for a route to the destination. If no route is found, then a route search process is initiated. The corresponding control packets are transmitted to the neighbors of the current node and the procedure continues until the destination has been reached. During this procedure, intermediate nodes are allowed to forward the control packet only if it (a) has not yet been received by the node and (b) it was received over a strong link. Upon arrival of the first control packet at the destination, the latter sends a reply message back to the initiator of the route search process. The reason for choosing the first control packet to arrive at the destination is that it is probable that this packet arrived over the shortest and/or least congested path. As the reply travels along the returning path, node DRPs update intermediate nodes' RTs correspondingly.

The fact that packets arriving over a weak channel are dropped at intermediate nodes means that route-search packets arriving at the destination have necessarily arrived on the path of strongest signal stability. Thus, the protocol routes packets over routes having the highest possible signal stability. However, under high BER conditions, few links may be classified as 'strong' In such cases, the route search process may not find a route to the destination. In such a case, the route search process initiator may chose to reinitiate the procedure indicating that weak links in the path of the route are acceptable.

When routes are 'broken' due to topological changes, SSR (and also AODV and DSR) initiates route discovery from the source node. Unlike ABR, partial route discovery (at intermediate nodes) is not performed. However, this is not necessarily a disadvantage since in some cases the failures of intermediate nodes to find a valid alternative route will again shift the process to the source node. Thus, an accompanying increase in delay of route construction compared to source-initiated route reconstruction might arise.

10.5 Summary

Recently, considerable research effort has been made on integrating broadband wired ATM and wireless technologies. WATM combines the advantages of wired ATM networks and wireless networks. These are the flexible bandwidth allocation offered through the statistical multiplexing capability of ATM and the freedom of terminal movement offered by wireless networks. This combination will enable implementation of QoS demanding applications over the wireless medium. This chapter covers a number of issues:

- The protocol stack for wireless ATM is presented and physical, MAC and DLC layers discussed. Furthermore, the issues of location management and handoff in wireless ATM networks are discussed.
- HIPERLAN 2, an ATM compatible WLAN standard developed by ETSI is presented. Contrary to WLAN protocols, HIPERLAN 2 is connection oriented and ATM compatible. HIPERLAN 2 will support speeds up to 25 Mbps at the DLC layer.

- Slightly deviating from the contents of this chapter, a number of routing protocols for multihop ad hoc wireless networks are presented.

WWW Resources

1. *www.atmforum.com*: the web location of the ATM forum, which promotes ATM technology.
2. *www.ittc.ukans.edu/~prasiths/wirelessatm/content.htm*: an on-line tutorial on WATM.
3. *www-vs.informatik.uni-ulm.de/projekte/wand/wand.html*: this web site contains information relating to the Magic WAND project. This is a European project aiming to develop a WATM system operating at the 5 GHz band at a transmission speed of 20 Mbps.
4. *www.imst.de/mobile/median/median.html*: this is the web site of the MEDIAN project, supporting wireless ATM extension at the 60 GHz band at a transmission speed of 155 Mbps.
5. *www.hiperlan2.com*: this is the web site of the HIPERLAN 2 Global forum which promotes the HIPERLAN 2 standard.

References

[1] Stallings W. *Data and Computer Communications*, Fifth Edition, Prentice Hall, Upper Saddle River, NJ.

[2] Awater G. A. and Kruys J. Wireless ATM - an Overview, *Mobile Networks and Applications*, 1, 235–243, 1996.

[3] Raychaudhuri D. Wireless ATM Networks: Technology Status and Future Directions, *Proceedings of the IEEE*, October, 1999, 1790–1806.

[4] Acampora A. Wireless ATM: a Perspective on Issues and Prospects, *IEEE Personal Communications,* August, 1996, 8–17.

[5] Kubbar O. and Mouftah H. T. Multiple Access Control Protocols for Wireless ATM: Problems Definition and Design Objectives, *IEEE Communications Magazine*, November, 1997, 93–99.

[6] Deane J. WATM PHY requirements ATM Forum/96-0785, June, 1996.

[7] Sanchez J., Martinez R. and Marcellin M. W. A Survey of MAC protocols proposed for Wireless ATM, *IEEE Network,* November/December, 1997, 52–62.

[8] Acharya A., Li J., Rajagopalan B. and Raychaudhuri D. Mobility Management in Wireless ATM Networks, *IEEE Communications Magazine*, November, 1997, 100–109.

[9] Johnsson M. HiperLAN/2 - The Broadband Radio Transmission Technology Operating in the 5 GHz Frequency Band, HiperLAN/2 Global Forum, 1999, Version 1.0.

[10] Broadband Radio Access Networks (BRAN); HIPERLAN Type 2; System Overview, ETSI Technical Report 101 683 V1.1.1.

[11] Jush J. K., Malmgren G., Schramm P. and Torsner J. HIPERLAN type 2 for broadband wireless communication, *Ericsson Review*, 2, 2000.

[12] Royer E. M. and Toh C.-K A Review of Routing Protocols for Ad-Hoc Mobile Data Networks, *IEEE Personal Communications,* April, 1999, 46–55.

[13] Perkins C. E. and Bwaghat P. Highly Dynamic Destination-Sequenced Distance-Vector Routing (DSDV) for Mobile Computers, *Computer communications Review*, October, 1994, 234–244.

[14] Chiang C. C. Routing in Clustered Multihop, Mobile Wireless Networks with Fading Channel, in *Proceedings of IEEE SICON*, 1997, pp. 197–211.

[15] Murthy S. and Garcia-Luna-Aceves J. J. An Efficient Routing Protocol for Wireless Networks, *ACM Mobile Networks and Applications Journal*, Special Issue on Routing in Mobile Communication Networks, October, 1996, 183–197.

[16] Perkins C. E. and Royer E. M. Ad Hoc On-Demand Distance Vector Routing, in *Proceedings of 2nd IEEE Workshop on Mobile Computer Systems and Applications*, February, 1999, pp. 90–100.

[17] Johnson D. B. and Maltz D. A. *The Dynamic Source Routing Protocols for Mobile Ad Hoc Networks*, IETF Draft, October, 1999.

[18] Toh C. K. A Novel Distributed Routing Protocol to Support Ad Hoc Mobile Computing, in *Proceedings of IPCCC '96*, 1996, pp. 480–486.

[19] Dube R., Rais C. D., Wang K.-Y. and Tripathi S. K. Signal Stability Based Adaptive Routing for Ad Hoc Mobile Networks, *IEEE Personal Communications*, February, 1997, 36–45.

11

Personal Area Networks (PANs)

11.1 Introduction to PAN Technology and Applications

11.1.1 Historical Overview

The concept of a Personal Area Network (PAN) differs from that of other types of data networks (e.g. LAN, MAN, WAN) in terms of size, performance and cost (Figure 11.1). PANs are the next step down from LANs and target applications that demand short-range communications inside the Personal Operating Space (POS) of a person or device. The term POS is used to define the space in the near vicinity of a person or device and can be thought of as a bubble that surrounds him. As the person goes through his regular daily activities, his POS changes to include a number of different devices (such as cellular phones, pagers, headphones, PC interfaces, etc.) with whom the ability for easy and transparent information exchange would be useful. PANs aim to provide such ability in an efficient manner.

There exist a number of different communication mediums to choose for implementing a PAN, such as electric and magnetic fields, radio and optical signal transmission. One of the first research concepts for PANs dates back to an IBM research project in 1996 and is known

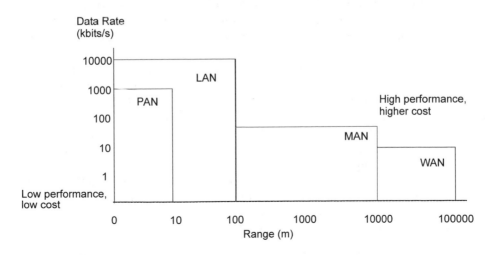

Figure 11.1 The various kinds of wireless networks

as 'Near-field Intra-body Communication PAN (NIC-PAN)' [1]. This approach uses the human body as the communication medium, which can conduct electricity due to its natural content of salt. According to this approach, NIC-PAN-compliant devices worn by a user can communicate with each other through the user's body, thus no wires are needed. Furthermore, a user wearing such a device can initiate communication with another device located on another user by means of a simple handshake. In order to transmit data between devices attached to the users' bodies or clothing, the NIC-PAN device transmitter charges and discharges the human body, thus resulting in an oscillating potential appearing between the body and the environment. These changes of potential are picked up by the receiving NIC-PAN device and thus a communication channel is established. The electrical current used in this approach is approximately 1 nA, which is much lower than the natural electrical current of the human body. In the context of this research, a small prototype was built, which achieved data rates around 2.4 kbps.

However, the NIC-PAN approach did not evolve into something more than a research project, whereas the true revolution in the area of PANs was brought about by the use of wireless transmission-based PANs (WPANs[1]). The first attempt to define a standard for PANs dates back to an Ericsson project in 1994, which aimed to find a solution for wireless communication between mobile phones and related accessories (e.g. hands-free kits). This project was named Bluetooth, after the king that united the Viking tribes. Bluetooth was a promising approach and as a result, Nokia, Intel, Toshiba and IBM joined Ericsson to form the Bluetooth Special Interest Group (SIG) in May 1998. The purpose of the Bluetooth SIG is to develop a de facto standard for PANs that meets the communication needs of all mobile computing and communication devices located in a reduced geographical space regardless of their size or power budget. The size of the SIG has grown from the initial five members to nearly 2000 others at the present day. Any interested company is allowed to join the SIG provided that it lets all other members of the SIG use its patents royalty-free, in an effort to keep the standard open. Version 1.0 of the Bluetooth specification was released by the SIG in 1999, followed by version 1.1 in 2001. Both these versions support 64 kbps voice channels and asynchronous data channels either asymmetric, with a maximum data rate of 721 kbps in one direction and 57.6 in the other or symmetric with a 432 kbps maximum rate in both directions. Bluetooth uses Frequency Hopping Spread Spectrum (FHSS) modulation in the 2.4 GHz ISM band. The supported range is 10 m with the possibility of extending this to 100 m [2].

Another initiative from industry members to develop a PAN standard was made in 1997 by the formation of the HomeRF Working Group. The primary goal of this group is to enable interoperable wireless voice and data networking within the home. Version 1.0 of HomeRF was published in 1999. It supported four 32 kbps voice connections and data rates up to 1.6 Mbps at ranges up to 50 m. Version 2.0 of HomeRF was released in 2001 and increased these numbers to eight channels and 10 Mbps, respectively, making HomeRF more suitable than Bluetooth for transmitting music, audio, video and other high data applications. However, Bluetooth seems to have more industry backing. Like Bluetooth, HomeRF also supports voice and asynchronous data channels using FHSS modulation in the 2.4 GHz ISM band.

After the appearance of the Bluetooth and HomeRF initiatives, IEEE also decided to join the area of developing specifications for PANs. Thus the 802.15 Working Group [2–5] was

---

[1] Since all PAN technology alternatives today employ wireless transmission, the terms PAN and WPAN are used in this book synonymously.

formed in March 1999, with the responsibility of defining physical and MAC layer specifications for PANs having low implementation complexity and low power consumption. Although Working Group 802.11, which deals with Wireless LANs, already existed, it was decided that a new Working Group was needed for PAN standardization. This is attributed to the fact that there is a much greater concern over power consumption, size, and product cost in PANs stemming from the demands for PAN devices compared to WLAN devices:

- Less size and weight, in order to be easily carried or worn for long periods of time. On the other hand, the size and weight of WLAN cards is a matter of secondary importance. This is because WLAN devices are typically either attached to portable computers, which of course are not carried by users all of the time, or to fixed desktop PCs.
- Lower cost, in order not to burden the total cost of the device. PAN devices aim to provide wireless connectivity to commercial electronic appliances. In order to enable small device size, PAN functionality will be integrated within the devices. To earn market acceptance, the cost burden of PAN functionality on the total cost of the device should be small. Users, on the other hand, can buy WLAN cards separately, thus the impact of their cost is less important.

There are four Task Groups (TGs) inside Working Group 802.15:

- *TG1*. This group is working on a PAN standard based on Bluetooth.
- *TG2*. This group aims to facilitate coexistence of PAN and WLAN networks.
- *TG3*. This group aims to produce a PAN standard with data rates exceeding 20 Mbps, while maintaining low cost, power consumption and interoperability with industry standards.
- *TG4*. This group aims to produce a PAN standard that will enable low-rate operation while achieving levels of power consumption so low, that a battery life of months or years will be possible.

Due to the fact that industry consortia initiatives for PAN standards development preceded the initiative of IEEE, a key mission of the 802.15 Working Group will be to work closely with such consortia, such as Bluetooth and HomeRF, in order to achieve interoperability for PANs coexisting in a shared wireless medium.

11.1.2 PAN Concerns

There are certain issues that need to be taken into account when designing a PAN. The most obvious one affects all types of wireless networks and concerns the increased Bit Error Rate (BER) of the wireless medium. As has been mentioned, the primary reason for the increased BER is atmospheric noise, physical obstructions found in the signal's path, multipath propagation and interference from other systems. The primary source of interference in the 2.4 GHz ISM band in which PANs operate, comes both from narrowband and wideband sources, such as microwave ovens. Apart from the good interference avoidance properties of SS modulation, which is employed for unlicensed transmission in the 2.4 GHz ISM band, Automatic Repeat Requests (ARQ) and Forward Error Control (FEC) techniques can also be used in this direction. Furthermore, PANs have to deal with interference from collocated PANs and WLANs, although this should not be a big problem due to the use of FHSS modulation.

PANs should provide full communication capability between all devices in the POS of a person. However, two PAN devices that initiate communication inside the POS should not interfere with other devices when this is not wanted. Consider for example the case of a person entering a conference room. After sitting in a chair, a PAN interface nearby could communicate with a handheld device of the person and deliver to him the information he wants. However, automatic initiation of communication with the devices of nearby conference attendants may not be desirable for privacy reasons.

A person carrying a PAN-enabled device could find him-/herself in a diverse range of situations, whether personal or professional, that demand information delivery through the PAN device. Therefore, PAN devices should be compatible to enable information exchange in all cases. Returning to the conference example, imagine two attendants wanting to exchange information, discovering that their systems are not compatible and thus they need to exchange the information on paper. In such case, the market community would see PANs as a waste of time and money. Compatibility not only involves following a specific PAN standard but also software compatibility. For example, the file formats on each of the above devices should be able to be read by both devices.

PAN-enabled devices will be typically be carried by people for long periods of time. Thus, they need to be small enough in order to be carried around without burdening. PAN devices should therefore be as small and light as possible. Their energy efficiency should be enough in order not to trouble the user with frequent recharging of batteries, while maintaining a low device weight. Therefore, both low power consumption devices and high capacity batteries are desirable. Furthermore, the efficiency in terms of size and weight should not come at an expense over the price of PAN devices, in order to enable market acceptance.

Security issues are also crucial in PANs. Communications should be secure and difficult to eavesdrop. Consider the case of a malicious user approaching pedestrians carrying PAN devices. The unpleasant 'Big Brother' scenario of Orwell's 1984 is obvious here and should be as difficult to achieve as possible. It should be very hard for the malicious user to obtain information regarding the unsuspected pedestrians, such as their names, home addresses, etc. Therefore, robust authentication and encryption schemes should be developed in an effort to prevent unauthorized initiation of communication and eavesdropping. These schemes should be developed while keeping in mind the relatively low processing and power capabilities of PAN devices which stem from the requirements for reduced cost, size and weight.

Finally, as in the case of all wireless networks human safety issues are of great concern. A PAN device will typically be very close to the user for long periods of time and therefore even small dangers could potentially have some impact on the user over time. The good thing here is that PAN (like WLAN) devices typically transmit at power levels up to 0.5 W. Despite the fact that a final answer to the question of radiation threats to human health has yet to be given, it is reassuring for the consumers to know that the operating power levels of a PAN device are substantially lower than the 600-mW to 3-W range of common cellular phones.

11.1.3 PAN Applications

The main goal of PANs is freedom from cables and easy sharing of information between all kinds of wireless devices. The number of different possible applications can be very large. In the following we outline a representative set:

- *Personal device synchronization.* Automatic data synchronization between mobile wireless equipment such as a mobile phone, notebook PC, etc. that execute similar applications.
- *Ad hoc connectivity.* Transferring files, and other information to another user's PAN-enabled device.
- *Cordless computer.* Wireless interfacing of devices like mice, keyboards, game pads to the computer.
- *Cordless peripherals.* Access to a variety of wireless peripherals including printers, scanners, fax, copier, storage systems, etc.
- *Localized wireless LAN access.* PAN-enabled devices can gain access to services offered by wired LANs through PAN-compatible Access Points (APs).
- *Internet access.* Downloads of email or browsing a web page using a PAN-enabled device, such as a mobile phone.
- *Wireless synchronization.* Synchronization of portable devices with the stationary servers via PAN APs.
- *Cordless telephony/headset.* A user selects a contact name from a handheld, the handheld wirelessly prompts the mobile phone in its proximity to dial the number and the audio from the call is wirelessly forwarded to the user's headset.
- *Home automation.* Seamless transfer of commands to PAN-enabled home devices. For example, automatic unlocking of the door upon the arrival of the user at his home, or automatic tuning of the television to the user's favorite channel upon his entrance to the room.
- *Electronic purchases/reservations.* PAN devices can be used to electronically book tickets. For example, the PAN device of a user can be programmed to instantly initiate a request for booking a ticket for a specific flight when the user enters the airport, thus avoiding the long queues and waiting times of the traditional booking procedures.
- *Emergency situations.* Medical devices with PAN interfaces can be used in order to increase the safety of patients. For example, pacemakers could be monitored and controlled remotely through PAN interfaces, or can be programmed to immediately call an ambulance while also transmitting the patient's medical condition in the case of a heart attack or other serious health problem.

11.1.4 Scope of the Chapter

The remainder of this chapter provides a detailed presentation of technological alternatives in the PAN area. Section 11.2 presents the Bluetooth specification and discusses, among others, the way Bluetooth devices establish connections and exchange data. Furthermore, Bluetooth's provisions on security and power management are discussed. Section 11.3 is a similar discussion on the HomeRF standard. The chapter ends with a brief summary Section 11.4.

11.2 Commercial Alternatives: Bluetooth

11.2.1 The Bluetooth Specification

The Bluetooth specification 1.1 [6–9] comprises two parts: core and profiles. The core

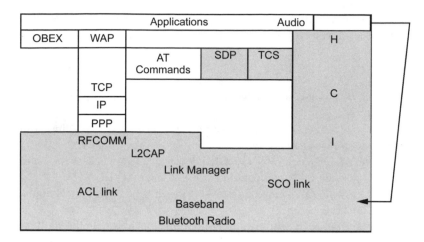

Figure 11.2 The Bluetooth protocol stack. Shaded layers implement Bluetooth-specific protocols

specification defines the layers of the Bluetooth Protocol stack. The aim of the stack (shown in Figure 11.2) is to provide a common data link and physical layer to applications and high-level protocols that communicate over the Bluetooth wireless link and maximize reuse of existing protocols at the higher layers. The protocols that run at different layers of the stack can be categorized into three groups: Bluetooth core, cable replacement protocols (RFCOMM), telephony control protocols (TCS and AT commands) and adopted protocols (such as IP, PPP, etc.). From these protocols, only the core protocols, RFCOMM and TCS are Bluetooth-specific protocols (those run at the shaded layers of Figure 11.2). The layers of the stack are summarized below:

- The radio layer provides the electrical specifications in order to send and receive bitstreams over the wireless channel. These specifications are discussed in Section 11.2.2.
- The baseband layer enables the operation of the Bluetooth links (described in Section 11.2.4) over the wireless medium. This layer is also responsible for framing, flow control and timing operations and it also manages the links between communicating Bluetooth devices.
- The Link Management (LM) layer runs the Link Management Protocol (LMP). This is an entity responsible for managing connection states, ensuring access fairness, performing power management and providing authentication and encryption services to upper layers. Power management issues are discussed in Section 11.2.7 while security is discussed in Section 11.2.8.
- The Logical Link And Adaptation Layer (L2CAP) provides connection-oriented and connectionless data services to upper layer protocols with protocol multiplexing capability, segmentation and reassembly (SAR) operations, and group abstractions. L2CAP permits higher-level protocols and applications to transmit and receive L2CAP packets up to 64 kilobytes in length. L2CAP supports only data traffic. As can be seen from Figure 11.2, audio data is not conveyed through L2CAP but is mapped directly to the baseband layer. Thus, data for audio connections is exchanged directly between the baseband layers of Bluetooth devices.
- The service discovery layer runs the Service Discovery Prototcol (SDP), which is used in

order for a Bluetooth device to learn about services on offer and neighboring device information. Using SDP, neighbors of a device can be queried and if some requirements are met, a connection can be established.

- The RFCOMM layer runs a serial line RS-232 control and data signal emulation protocol. It is used for cable replacement, offering transport capabilities over the wireless link to applications that use serial lines as a transport mechanism.
- The TCS layer defines call control signaling procedures for the establishment of voice and data calls between Bluetooth devices.
- The Host Controller Interface (HCI) is not a stack layer but an interface that provides the means for accessing the Bluetooth hardware capabilities.
- Layers that implement non-Bluetooth specific protocols (OBEX, WAP, etc.) are used to enable high-layer application functionality.

The profiles part of the specification is used to classify Bluetooth applications into nine application profiles, with each profile implementing only a certain set of the stack's protocols. This approach has received some criticism, which supports that the Bluetooth specification is essentially a set of nine standards instead of one, with the number likely to rise as new application profiles are added. However, the existence of application profiles aims to ensure interoperability between Bluetooth devices. In order for a device to be certified for a specific Bluetooth application, it has to follow the corresponding profile. Furthermore, the production of devices for a specific application means that the device could support only some of the application profiles, thus reducing its overall cost. Apart from the nine application profiles, version 1.1 of the Bluetooth specification also supports four system profiles, which include functionality common to one or more application profiles. The thirteen profiles are summarized below. Some of the profiles can exist only if they implement other profiles, as shown in Figure 11.3.

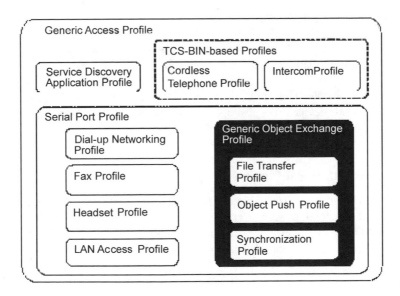

Figure 11.3   The Bluetooth profiles

- *Generic access profile*. This system profile is responsible for link maintenance between devices. This profile is not useful for supporting any useful application by itself, however, it needs to be supported in every Bluetooth device since it includes the functionality needed to use all the other system profiles.
- *Service discovery application profile*. This is another system profile that enables users to access the Service Discovery Protocol (SDP) in order to find out which applications are supported by a specific device. The support of the service discovery application profile is optional. If, however, this system profile is not supported, only applications can access the SDP, not users.
- *Intercom profile*. This application profile supports direct voice communication between two Bluetooth devices within range of each other.
- *Cordless telephony profile*. This application profile is designed in order to support the '3-in-1 phone' concept, meaning that a Bluetooth-compliant telephone can be used either as an intercom (communicating directly with another Bluetooth device), cordless (fixed-location) or mobile phone.
- *Serial port profile*. This system profile emulates RS232 and USB serial ports in order to allow applications to exchange data over a serial link.
- *Headset profile*. This application profile uses the serial port profile to provide connections between Bluetooth-enabled computers or mobile phones and Bluetooth-enabled wireless headset microphones.
- *Dial-up networking profile*. This application profile uses the serial port profile to provide dial-up connections via Bluetooth-enabled cellular phones.
- *Fax profile*. This application profile uses the serial port profile to enable computers to send a fax via a Bluetooth-enabled cellular phone.
- *LAN access profile*. This application profile enables Bluetooth devices either to form small IP networks among themselves or connect to traditional LANs through Access Points (APs).
- *Generic object exchange profile*. This system profile defines the functionality needed for Bluetooth devices to support object exchanges.
- *Object push profile*. This application profile defines the functionality needed to support 'pushed' data. Examples of such information are advertisements and news distribution.
- *File transfer profile*. This application profile enables file transfers between Bluetooth devices.
- *Synchronization profile*. This application profile enables automatic data synchronization between Bluetooth devices. For example, it can be used to synchronize address books between a desktop computer and a portable.

11.2.2 The Bluetooth Radio Channel

The Bluetooth radio channel [7], enabled by the radio layer, provides the electrical interface for the transfer of Bluetooth packets over the wireless medium. The radio channel operates at the 2.4 GHz ISM band by performing frequency hopping through a set of 79 (US and Europe) or 23 (Spain, France and Japan) RF channels spaced 1 MHz apart. The wireless link comprises time slots of 0.625 ms length each, with each slot corresponding to a hop frequency. The nominal hop rate is 1600 hops/s. At each hop, the transmitted signal is modulated using GFSK with a binary one being represented by a positive frequency shift

Power class	Max. transmit power (mW)	Min. transmit power (mW)
1	100	1
2	2.5	0.25
3	1	N/A

Figure 11.4   Power classes for Bluetooth devices

and a binary zero by a negative frequency shift. Despite the fact that this configuration provides link speeds up to 1 Mbps, the effective data transfer speeds offered are lower. This is because different protocol layers use parts of the packet data payload to add header information for purposes of communication with peer layers. More efficient modulation schemes could obviously achieve higher speeds, however, the use of GFSK is preferred as it allows for low-cost device implementations.

Depending on the transmitted power, Bluetooth devices can be classified into three classes, as shown in Figure 11.4. Power control mechanisms can be used to optimize the transmitter's power output. This is done by measuring the received signal strength and relaying LMP commands to the respective transmitter indicating whether power should be increased or decreased. Power control is required for class 1 equipment, whereas it is optional for equipment of classes 2 and 3. Using power class 1 a range of 100 m can be achieved in a Bluetooth system [2].

11.2.3 Piconets and Scatternets

When two Bluetooth devices want to connect, the one requesting the connection, is known as the master, whereas the other is known as the slave. The master always controls the link created between the two devices. The master–slave relationship is good only for a specific link establishment, since any Bluetooth unit can be either master or slave. A given master can maintain up to seven connections to active slaves. As a result, several very small networks, called piconets, can be established. However, if the master wants to open connections to more than seven slaves it can instruct one or more active slaves to 'sleep' for a specified period of time by putting them in the low-power PARK mode (described in Section 11.2.7). Then the master can admit new slaves into the piconet for the period the old set of slaves were put to sleep. Thus, piconets of many devices can be formed.

A piconet is shown in Figure 11.5. Devices inside a piconet hop together according to the master's clock value and its 48-bit device ID. The way the hopping sequence is used and the starting point within that sequence is selected is shown in Figure 11.6. The first parameter defines the hopping sequence used for FHSS transmission inside the piconet. The second parameter is derived from the native clock of the master and defines the phase within that sequence. Slave to slave transmission is not supported inside a piconet. If two slaves need to communicate, they either have to form a separate piconet in which one of them is the master, or use a higher layer protocol, such as IP in order to relay messages through the piconet's master.

A number of piconets can coexist in the same area. This coexistence is enabled due to the use of FHSS transmission. Since the hopping sequences used in Bluetooth are pseudorandom, different coexisting piconets will use different hopping sequences, resulting in a low prob-

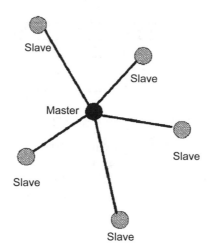

Figure 11.5   A piconet formed by six devices

ability of hop collisions. Still, if such a case occurs, a Bluetooth device can recognize and thus ignore packets originating from collocated networks by checking the access code field on the received packets which is different for every piconet, as we explain later. However, when a large number of piconets coexist in the same area, this probability rises and the performance of each piconet degrades.

A collection of overlapping piconets is called a scatternet. A scatternet will typically contain devices that participate in one or more piconets. Devices participating in two or more piconets are known as bridge devices and participate in each piconet in a time-sharing manner. After spending its time inside a piconet, a bridge device will change its hopping sequence to that of the new piconet in order to join it. Furthermore, it will select the starting point within that sequence based on the clock value of the master of the new piconet. By alternating among several piconets and buffering packets, bridge nodes can forward packets from one piconet to another. A bridge node that participates in several piconets can be either:

- *Slave in all the piconets.* In this case, when leaving the old piconet, the slave has to inform the master for the duration of its absence.
- *Master in one piconet and slave in all others.* In this case, all traffic in the old piconet is

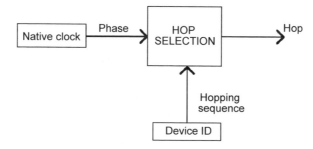

Figure 11.6   Combination of the master's device ID and native clock values to select the hopping sequence to be used and its starting point

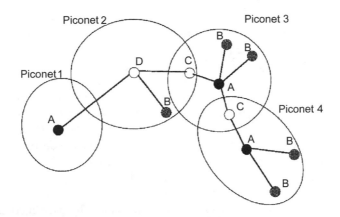

Figure 11.7   A scatternet formed by four piconets

suspended until the master returns to the piconet. The suspension of traffic is achieved by putting the piconet's slaves into the low-power HOLD mode (described in Section 11.2.7).

A bridge node cannot obviously be master in more than one piconet, since this would make these piconets use the same hopping sequence and thus collide.

Figure 11.7 shows a scatternet, where the participating nodes fall into four categories. Nodes in category A are masters within a single piconet, nodes in category B are slaves within a single piconet, nodes in category C are participating in two piconets as slaves and nodes in category D are participating in two piconets as slaves in the one and master in the other. Obviously, nodes in categories C and D are bridge nodes between the piconets in which they participate. Techniques for efficient interpiconet communication and efficient formation of scatternets from overlapping piconets are under development [10–12].

11.2.4 Inquiry, Paging and Link Establishment

Bluetooth devices can communicate as soon as they are within range of one another. However, since the in-range neighbors of a Bluetooth device change with time, a procedure that informs a device about its neighbors is needed. This procedure is carried out by issuing inquiries. Although an inquiry is a fairly simple procedure, it becomes complicated due to the use of FHSS at the physical layer. Assume that device A is within range of device B and wants to acquire knowledge about its neighbors. Since the hopping sequence of a link is defined by the master's device ID and clock values, A and B cannot exchange messages until they agree to a common hopping sequence as well as a common phase within that sequence. In order for two devices to exchange messages during the inquiry procedure, Bluetooth defines a specific hopping sequence to be used for inquiries by all devices. Furthermore, since there may be a phase uncertainty between A and B since those devices may not be completely synchronized (meaning that they start from a different hop inside the hopping sequence due to the differing Clock values of A and B), the sender hops faster than the listener, by transmitting a signal on each hop and listening between successive transmissions for an answer. The term Frequency Synchronization delay (or FS delay) refers to the time elapsed until the sender (A in this case) transmits at the frequency the receiver (B) is currently listening on. Upon reception of the inquiry, B waits for a randomly distributed period of time, called the random backoff delay. After the random backoff delay has elapsed, B replies to the next inquiry received by A by sending a FHSS packet containing its 48-bit device ID and clock values. The RB delay obviously aims to reduce the chance of collisions between two or more devices that are listening for inquiries at the same time.

The inquiry procedure provides a means for a master to gather neighborhood information. Two Bluetooth devices wanting to connect defines the paging procedure. Assuming that A is the master, B is the slave and that A does not participate in a piconet, the paging procedure is as follows: A is supposed to possess B's device ID and an estimate of its clock parameter from the inquiry procedure. To connect to B, A pages B using the hop sequence defined by B's device ID. However, since some time has elapsed from the inquiry procedure, A cannot possess the exact value of the native clock of B and thus the phase within the hopping sequence defined by B's device ID. Therefore, A transmits the page message not only in the hop in which it expects B to be, but also in neighboring hops, for a period of 10 ms. Upon reception of the page request, B responds by sending its device ID to A. Finally, A transmits to B a packet containing the master's device ID and clock values. Upon receiving this information, B creates a variable that contains its clock value, adds an offset to this variable in order to synchronize with A's clock and connects to A as slave. Information exchange between A and B can now initiate.

The above procedure becomes slightly more complicated when the master has already formed a piconet. In order for new slaves to join the piconet, the master needs to periodically suspend the traffic inside the piconet in order to scan for new slaves or accept slave requests. This traffic suspension obviously results in capacity reduction within the piconet. Therefore, when selecting the time period a master will spend searching for new slaves, a tradeoff has to be made between the latency of accepting a new slave to the piconet and overall piconet capacity.

11.2.5 Packet Format

As mentioned, each slot in Bluetooth corresponds to a single hop and lasts 0.625 ms. For a pair of communicating Bluetooth devices, the master always starts transmission at even numbered slots while slave transmission is set to be initiated only at odd numbered slots. Inside each piconet one packet can be transmitted in each slot. The format of packets exchanged over a Bluetooth link is shown in Figure 11.8. It comprises the following parts:

- A 72-bit access code, which is defined by the master and is unique for the piconet. This part is used by Bluetooth devices to identify incoming packets. If a Bluetooth device concludes that the access code of an incoming packet does not match the code of its piconet, the packet is discarded. Furthermore, this field is used by receiving units for synchronization purposes.
- A 54-bit packet header, which contains information related to MAC addressing, packet type, flow control, Automatic Repeat Request (ARQ) and Header Error Correction (HEC).

72 bits	54 bits	Variable-length
Access code	Packet header	Payload

Figure 11.8 The format of a transmitted Bluetooth packet

3 bits	4 bits	1 bit	1 bit	1 bit	8 bits
MAC address	Packet type	Flow control	AR Q	Sequence number	HEC

Figure 11.9 The fields of a Bluetooth packet header

MAC addresses are 3-bit fields, with address 000 specifying a broadcast packet. The 3-bit MAC field is the reason for the fact that up to seven active slaves are supported inside the same piconet and that further slaves can be admitted only if some others enter the PARK mode. The fields of a Bluetooth packet header are shown in Figure 11.9.

- A variable length payload may trail the header. Since the hop duration is 0.625 ms, 625 bits can be transmitted at a single hop over a 1 Mbps link. However, the Bluetooth specification states that the length of a packet transmitted at a single hop is 366 bits, in order to give transmitters and receivers enough time to hop to the next frequency and stabilize. Thus, the effective payload of a packet transmitted over a single hop equals 366 bits minus the access code and packet header size (126 bits), which results in 240 bits (30 bytes). The multislot packets, described later, can have a larger payload.

The 4-bit packet type field in the packet header defines 16 types of packets. Of these, four are control packets:

- The ID packet, which is used for signaling purposes.
- The NULL packet, which only contains the access code and packet header. This is used in case no payload has to be transmitted, but information in the packet header is needed for link management.
- The POLL packet, which is used by the master to poll slaves in an ACL link.
- The FHSS packet, which is used to exchange synchronization information between units, such as clock values of Bluetooth devices.

The remaining 12 types codes are divided into segments that define kinds of data packets. In an effort to improve efficiency, Bluetooth supports multislot packets, either three or five slots long, which are transmitted in consecutive slots. Thus, segment 1 specifies single-slot packets, segment 2 specifies three-slot packets and segment 3 specifies five-slot packets. Multislot packets are always sent on a single frequency, which is that used at the beginning of the multislot transmission. The transmission following that of the multislot packet occurs at the hop that would be used when the multislot packet was replaced by single hop packets. For example, consider four consecutive slots $k, k + 1, k + 2, k + 3$. In the case of three single hop transmissions, these would take place in frequencies f_k, f_{k+1}, f_{k+2}. In the case of a three-slot packet, the entire packet would be transmitted at frequency f_k with the next transmission on the channel beginning in slot $k + 3$ and using frequency f_{k+3}. The fact that multislot packets are transmitted on the same frequency results in a capacity increase, which comes, however, at an expense over the hopping rate of the system and thus lowering the system's interference avoidance.

11.2.6 Link Types

The baseband layer provides two types of links for Bluetooth: Synchronous Connection-

Oriented (SCO) and Asynchronous Connection-Less (ACL) links. A SCO link is a symmetric point-to-point link supporting circuit-switched traffic between a master and a slave. The master maintains SCO links using polling. Polling occurs at reserved slots at regular time intervals and only supports single-slot packets. SCO links are mainly used to convey voice information and do not support packet retransmission.

SCO links support three types of single-slot packets, HV1, HV2, HV3, each of which carry voice packets at a rate of 64 kbps. HV3 packets carry voice information without coding or protection. HV2 packets carry voice information with a 2/3 Forward Error Correction (FEC) code. Finally, HV1 packets offer the higher level of immunity to errors, voice information with a 1/3 FEC code. The speech coder in a Bluetooth unit generates 10 bytes every 1.25 ms. Thus, in the case of HV3 packets, one such slot in each direction of the SCO link will be needed every 3.75 ms (or every sixth slot), since a baseband packet carries 30 bytes of payload. This means that in the best case, up to three voice connections can be supported at the same time within a piconet. However, the use of a 2/3 FEC code in HV2 packets reduces the voice information carried in the SCO packets to 20 bits. Twenty voice bits are produced every 2.5 ms resulting in the need for SCO slot reservation in each direction every 20 ms (or every fourth slot). Finally, in the case of HV1 packets, a SCO slot reservation every 10 ms (or every second slot) will be needed in each direction, meaning that the entire piconet bandwidth will be allocated to the SCO link.

Asynchronous Connection-Less (ACL) links are point-to-multipoint links that support packet-switching between the master and all the slaves inside a piconet. An ACL link supports both single and multislot packets. An ACL link is used to carry data traffic and is maintained by the master through a polling mechanism. A slave is permitted to send packets to the master only when it has been polled by the master in the preceding master-to-slave slot. A packet transmission from the master to the slave implicitly polls the slave. However, when the master does not have a data packet to send to the slave, it polls the slave using a POLL control packet. This polling procedure makes sure that no collisions among ACL packets of slaves take place. Furthermore, the master can satisfy QoS requirements of slaves for an ACL link by polling the slave more frequently, changing the packet size, or performing both these operations. ACL links also support broadcast messages to all the slaves inside a piconet. Broadcasting is implemented by setting the MAC address of the packet to all zeroes (000).

ACL links support both single and multislot packets, all of which can be transmitted either with a 2/3 FEC code or without error correction at all. Link adaptation can be used for managing the ACL link by changing the packet length and protection level according to the link state. While the capacity of a SCO link is always 64 kbps, the capacity of an ACL link varies according to the number of SCO links and the amount of FEC protection used on SCO and ACL packets. Figure 11.10 summarizes the maximum throughput of an ACL link inside a piconet with no SCO links present for ACL packets with different sizes and FEC coding. It can be seen that the maximum capacity offered by an ACL link, using the most efficient five-slot packet and no FEC coding, is 432 kbps for a symmetric ACL link and 721.0/57.6 kbps for an asymmetric link. Furthermore, the use of FEC significantly reduces the maximum capacity of an ACL link.

Contrary to SCO links, ACL links support packet retransmission through a fast ARQ scheme. The payload of ACL packets is checked for errors using a CRC mechanism and the sender is notified about the status of its transmission in the slot following its own transmission. Assuming device A transmitted a packet to device B, the resulting ACK-

Packet type	Capacity (kbps)		
	Asymmetric, link 1	Asymmetric, link 2	Symmetric
One-slot, no FEC	172.8	172.8	172.8
Three-slot, no FEC	576	86.4	384
Five-slot, no FEC	721	57.6	432.6
One-slot, 2/3 FEC coded	108.8	108.8	108.8
Three-slot, 2/3 FEC coded	384	54.4	256
Five-slot, 2/3 FEC coded	477.8	36.3	286.7

Figure 11.10   Capacity scenarios on the ACL link

NAK notifications are piggybacked on the header of the packet transmitted by B, indicating successful or erroneous reception of A's transmission. Thus, A is informed whether retransmission is needed or not. If B does not have a packet to transmit in the slot following A's transmission, it will either have to transmit a NULL packet in order to relay the ACK message to A, or transmit no packet at all to notify A of erroneous reception. Thus, the absence of transmission in this case implies a NAK packet. This fast ARQ scheme is similar to the stop-and-wait ARQ scheme, however it has less delay due to the incorporation of the ACK-NAK messages on packet transmissions. The fast ARQ scheme is summarized in Figure 11.11.

SCO and ACL links may be time-multiplexed over the same wireless link, however, only a single ACL link is permitted to exist at any given time. Figure 11.12 shows communication inside a piconet of three units, in which the master maintains a SCO link with both slaves.

11.2.7 Power Management

Bluetooth devices should have an adequate level of energy efficiency so as not to trouble the user with frequent recharging of batteries. Thus, special attention has been paid to reducing the energy consumption of Bluetooth devices. Whenever a device is not part of a piconet, it enters a mode in which it only waits up periodically (every T seconds, with T ranging between 1.28 and 3.84 s) in order to send or receive inquiry messages for about 10 ms.

There also exist techniques for the reduction of power consumption of a Bluetooth slave

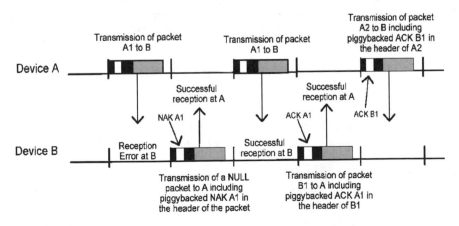

Figure 11.11   The fast ARQ scheme

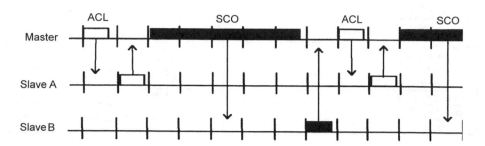

Figure 11.12 Time-multiplexing of SCO and ACL links inside a piconet of three devices

device operating inside a piconet. This is made possible with the definition of the following three low-power modes:

- *HOLD mode.* The master can instruct one or more active slaves to 'sleep' for a specified period of time by putting them in this mode. This is useful in cases where the master wants to suspend transfers in the piconet in order to perform inquiry and paging. In the HOLD mode, traffic on the ACL link is suspended while the operation of SCO links is maintained. The active transfers on the ACL link are resumed only when the slave exits this mode.
- *SNIFF mode.* In this mode, slaves listen to the piconet's master at a reduced rate. This means that the slave does not listen to the channel at every master-to-slave slot, but over larger time intervals.
- *PARK mode.* This is the mode that achieves less power consumption. In this mode, slaves stay synchronized to the master without actively participating in the piconet. As mentioned, this mode can be used to form piconets with more than seven slaves. Upon entering the PARK mode, a slave frees its 3-bit MAC address and is given two 8-bit addresses: A parked member address and an access request address. The former is used by the master to unpark the slave while the latter is used by the slave to request the master to unpark it. The master and the slave can still communicate by broadcasts on the ACL link.

Furthermore, power consumption of active slaves that receive a packet destined for another slave can be further reduced. This is made possible by exploiting the information in the packet header, which defines the packet type, the recipient and the packet's duration. Thus, upon reception of a packet not destined for itself a slave knows how much time it can spend in low-power mode.

11.2.8 Security

Acknowledging the fact that wireless transmission is subject to security errors, Bluetooth provides a number of security features. Apart from the inherent security of FHSS transmission, Bluetooth provides an authentication process to prevent unauthorized access and a mechanism for encryption of exchanged packet payloads to prevent eavesdropping of ongoing transmissions. Both these procedures are initiated at the LMP layer. In order for both authentication and encryption to take place, the same common secret link key must be present in both devices. This can be done by the user through typing of a randomly chosen PIN number on both devices. If, however, one or both devices do not have a keypad (e.g.

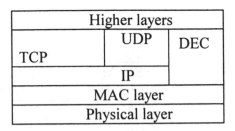

Figure 11.13 The HomeRF protocol stack

wireless headsets), the level of security decreases due to the fact that the chosen PIN, which is either entered by the user on one device or calculated by the device itself, is transmitted to the other device in clear text.[2] The Bluetooth specification provides a mechanism called pairing which is used for authentication of two certain devices for the first time. After pairing occurs between two devices, these can store the common secret link key in order to prevent future insecure PIN transmissions in the case of future connection establishments between them.

The authentication process uses a four-way handshake mechanism somewhat similar to that of IEEE 802.11. The device wanting to gain authentication (the claimant) transmits its 48-bit device ID to the device it wants to connect to (verifier). Upon receipt of this number, the verifier returns a 128-bit random address to the claimant. The claimant uses this number, a 128-bit common secret link key (described above) and the claimant's device ID to produce a 32-bit Signed RESponse (SRES). The SRES is then transmitted to the verifier, which compares it with its own SRES and notifies the claimant whether or not authentication was successful. The verifier permits connection establishment with the claimant only if their SRES numbers are the same.

SRES is also used for packet payload encryption. However, encryption also uses a 96-bit Authenticated Cipher Offset (ACO). The payload bits are modulo-2 added to a binary keystream. This keystream is produced in the following manner. Upon initiation of encryption over a Bluetooth link the master sends a random number RAND to the slave. The master's device ID, RAND, an encryption key and the current hop number produce the binary keystream. Using the current hop number in the production of the binary keystream leads to different keystreams for each transmitted packet. This is because packet transmission will typically occur on different hops. In the above procedure, the encryption key is produced from the 128-bit common secret link key (described above), RAND and ACO.

11.3 Commercial Alternatives: HomeRF

The HomeRF specification, like most networking interface standards, defines a protocol stack describing the functionality corresponding to the physical and MAC layer of the OSI model. The HomeRF protocol stack is shown in Figure 11.13. Above the physical and MAC layers the well known TCP-UPD/IP and DECT protocols are used. TCP and UDP deal with asynchronous transmissions and streaming media support while DECT employs the functionality needed for audio telephony transmissions. Version 2.0 of the HomeRF specification [13,14]

---

[2] As seen shortly, the common secret link key is the only real secret exchanged by the devices. If a malicious user manages to posses the PIN and thus this key, security is compromised.

	HomeRF Version 1.2	HomeRF Version 2.0
Frequency	2.4 GHz ISM	2.4 GHz ISM
Modulation	FHSS, 1 MHz channels	FHSS, 1 MHz and 5 MHz channels
Max speed	1.6 Mbps	10 Mbps
Hop rate	50 hops/s	50 hops/s 100 hops/s
Transmitter power	100 mW	100 mW, 500 mW
Roaming	No	Yes
MAC-level encryption	40-bit	128-bit

Figure 11.14 Main differences between HomeRF versions 1.2 and 2.0

was released in 2001. This version offers significantly higher data rates (up to 10 Mbps) than the previous version 1.2 [15,16] and is backwards compatible.

The next sections introduce the network topology of HomeRF and then a discussion on the physical and MAC layers of the system is made. This is accompanied by a discussion on synchronization, power management, security and compression services provided by the MAC layer. Due to the fact that at the time of writing the vast majority of HomeRF products are compliant with version 1.2 of the specification, we make a separate presentation of the different characteristics specified by this versions, in whichever situation such a differentiation occurs. Figure 11.14 presents the main differences between versions 1.2 and 2.0 of the HomeRF standard.

11.3.1 HomeRF Network Topology

11.3.1.1 HomeRF Device and Network Types

There are four device types in a HomeRF network [15]. These can be categorized as follows:

- *Connection Points (CPs).* A CP is the device that connects the HomeRF network to the Packet Switched Telephone Network (PSTN) and the Internet via a personal computer. CPs can be either separate devices that are connected to computers (typically through a USB port) or can be integrated into the computer. The presence of a CP inside a HomeRF network is optional.
- *Isochronous nodes (I-nodes).* I-nodes are the nodes that demand support for isochronous data delivery. The vast majority of I-nodes are concerned with voice data delivery. A typical example of an I-node is a cordless phone.
- *Asynchronous nodes (A-nodes).* A-nodes are the nodes that run applications exchanging data in an asynchronous manner. Typical examples of A-nodes are mobile personal computers and Personal Digital Assistants (PDAs).

- *Combined asynchronous and isochronous nodes (AI-nodes).*

As in the case of IEEE 802.11, the way a HomeRF network is controlled depends on the presence or not of a CP. When a CP is not present, the HomeRF system operates in an ad hoc manner and supports only asynchronous data transfers. In the ad hoc mode, network control is of course distributed. Therefore all A-nodes share responsibility for ensuring synchronization. Synchronization in an ad hoc HomeRF system is explained in Section 11.3.3.2.

With a CP present, the network becomes a centralized one (or managed, in HomeRF terminology). In such a case, where the control of the network is a responsibility of the CP, the HomeRF system can also provide support for voice data transfers in a contention-free manner. Arbitration of bandwidth for voice data transfer between HomeRF nodes is made by the CP using TDMA. Another responsibility of the CP is to provide synchronization information to A-nodes and I-nodes by transmitting beacon frames (explained in Section 11.3.3). Furthermore, in a managed network, the CP can be configured to provide power management services to the nodes of its network. Power management for a managed network is described in Section 11.3.3.3.

HomeRF provides methods for adapting both to CP malfunctions and to topology changes due to CP mobility. Within the same managed network, two or more CPs can exist, but only one can be active at a specific time. However, a passive CP can be used as a backup device. Whenever a passive CP does not hear 50 consecutive beacon transmissions (described in Section 11.3.3) from the active CP, it assumes the active CP is either malfunctioning or has moved out of range. Thus, the passive CP assumes the role of an active CP inside the HomeRF network. Furthermore, when the only CP of a HomeRF network moves out of range or is malfunctioning, the network's A-nodes are allowed to form their own ad hoc network after missing 100 consecutive receptions of the CP's beacons.

11.3.1.2 Network Initialization

Whenever a HomeRF device is switched on, it enters a scanning mode for a certain amount of time (equal to the superframe described in Section 11.3.3). In the scanning mode the node is searching for nearby CP or ad hoc network transmissions. Scanning is performed by employing a hopping sequence that equally covers all available hops. Whenever a packet is received, the HomeRF station reads the packet's header and records the packet's Network Identifier, which is a 24-bit number that uniquely characterizes each HomeRF network. Furthermore, it records information relating to the hopping sequence of the network as well as parameters for synchronizing with that network. Then the unit can decide either to join the network defined by the received NWID or keep searching for another network. If the unit decides to join the network, it alters its hopping sequence to the one used by the new network and then synchronizes with it. If the unit decides not to enter the network, it continues to operate in the scanning mode. When the unit has spent its time in the scanning mode and has not yet entered a network, it either gives up or forms its own network. However, not all device types can form the same network types: CPs are allowed to form a managed network and A-nodes are allowed to form ad hoc networks, whereas I-nodes are not allowed to form any kind of network on their own.

Moreover, since the HomeRF specification supports collocated network operation, a HomeRF unit can decide to form its own network even though it is within range of another ad hoc or managed network. In this case, the collocated networks can function without interfering with one another due to the use of FHSS at the physical layer. Therefore, the probability for collocated networks to transmit on the same hop at the same time is small. Still, if such a case occurs, a device can recognize and thus ignore packets originating from collocated networks by checking the NWID field on the received packets.

11.3.2 The HomeRF Physical Layer

The HomeRF physical layer operates in the 2.4 GHz ISM band. Since FHSS is needed for unlicensed transmission in this band, HomeRF employs this technique by hopping among a set of 79 (US and Europe) or 23 (Spain, France and Japan) RF channels spaced 1 MHz apart. This configuration produces a slotted channel as in the case of Bluetooth. However, in HomeRF the packets transmitted are significantly shorter than the hop duration. Thus, compared with Bluetooth, HomeRF can be thought of as a slow frequency-hopping system since in each hop, more than one packet can be transmitted.

Version 1.2 of the HomeRF specification defines a hopping rate of 50 hops/s, thus creating slot lengths of 20 ms. The speeds offered at the physical layer of this version are 1 Mbps using binary FSK and 2 Mbps using quadrature FSK over the 1 MHz channels used by the FSSS technique. Of course, the actual link speeds seen by higher layers are lower, 1.6 and 0.8 Mbps, respectively, due to the fact that protocol layers use parts of the packets' payload for purposes of communication with their peers. The transmission power specified in this version is 100 mW.

Version 2.0 of the specification uses a hopping rate of 50 or 100 hops/s, thus creating slot lengths that are either 20 or 10 ms long, respectively. The 10 ms slot length is used in cases of active voice calls whereas 20 ms is used when only asynchronous data traffic is exchanged. The link speed for data services offered by version 2.0 reaches 10 Mbps with fallback modes of 5, 1.6 and 0.8 Mbps, while voice transmission continues to use the 1.6 and 0.8 Mbps modes. As in version 1.2, the modulation scheme is binary FSK for the lower rate with quadrature FSK for the 1.6 Mbps rate. These speeds provide compatibility with older genera-tion products. The new rates of 5 and 10 Mbps for data transmission are achieved by dividing the available spectrum in the ISM band into 15 channels, 5 MHz each (in US and Europe). The higher capability is obviously due to the use of wider channels. These 15 5-MHz channels define the set among which 5 and 10 Mbps HomeRF systems hop. Finally, as far as transmission power is concerned, devices built using HomeRF version 2.0 will transmit signals with power levels up to 500 mW.

11.3.3 The HomeRF MAC Layer

11.3.3.1 The HomeRF MAC Protocol

The HomeRF MAC layer was designed to support both isochronous (e.g. voice) and asyn-chronous traffic. The MAC layer, which uses 48-bit device addressing, provides the function-ality for HomeRF devices to interoperate with the PSTN using a subset of the DECT standard. In order to support both asynchronous and isochronous data delivery, the HomeRF MAC protocol uses a superframe structure. Each superframe starts at a new hop and lasts one hop, which lasts either 10 ms or 20 ms, depending on the presence of isochronous connections. The

Parameter	Typical Value
SIFS size	142 μs
DIFS size	309 μs
Slot size	167 μs
Maximum contention window size	8 slots
Maximum contention window size (*CWmax*)	64 slots

Figure 11.15   CSMA/CA parameter values in HomeRF

superframe comprises two Contention-Free Periods (CFPs) and one contention period. The access methods used for isochronous (e.g. voice) and asynchronous traffic are TDMA and CSMA/CA, respectively. Thus, the MAC protocol of HomeRF can be referred to as a hybrid protocol.

11.3.3.1.1 Contention Period Operation   Inside the contention period, asynchronous traffic is served with CSMA/CA being the medium access mechanism. CSMA/CA in HomeRF behaves the same way as within the IEEE 802.11 MAC layer. Furthermore, as in the case of 802.11, the distributed nature of CSMA/CA enables HomeRF A-nodes to form ad hoc networks without the help of a central coordinator as in the CP. However, in such a case, CFP duration is zero. As a result, isochronous (such as voice) services for I-nodes will not be supported and an entire 20 ms superframe will be dedicated to asynchronous data transfer. Figure 11.15 displays typical values for the parameters used by the CSMA/CA part of the MAC part of HomeRF [15].

Due to the similarity of the operation of CSMA/CA inside the contention periods of HomeRF and IEEE 802.11 the operation of CSMA/CA for HomeRF is not discussed here in detail. However, we note the following difference in the contention period of HomeRF 2.0 compared with that of IEEE 802.11. Streaming multimedia is more delay-tolerant than voice data and obviously less compared with normal asynchronous traffic. In order to provide efficient support to streaming multimedia services within the contention period, version 2.0 includes a priority asynchronous data mechanism [14]. According to this mechanism a streaming multimedia session is assigned its access number upon establishment. The higher the access number, the lower the stream's priority. The contention-based data protocol is then adjusted so that the random backoffs by A-nodes do not include the numbers already assigned to the streams. HomeRF supports the assignment of up to eight simultaneous streams.

11.3.3.1.2 Contention-free Period Operation   As mentioned earlier, when a CP exists in a HomeRF network, isochronous services can be supported. Since voice calls are the primary isochronous application inside a HomeRF network, the discussion on CFP operation assumes that voice traffic is being served. The purpose and functionality of the CP is the same as that of the Point Coordinator (PC) in 802.11 networks: Inside the two CFPs of the HomeRF superframe, it grants permission to transmit to those I-nodes that want to transmit voice packets. Permission is granted using TDMA arbitration of the active I-nodes. As

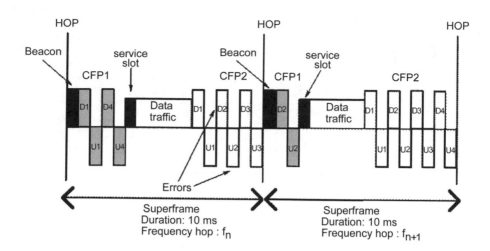

Figure 11.16 HomeRF superframe structure with support for voice traffic. For versions 1.x, the superframe has a fixed duration of 20 ms

mentioned above, when demand for voice connections arises, the HomeRF superframe changes its duration from 20 ms to 10 ms. The shorter duration, which is supported only in version 2.0 of the specification, provides decreased latency and increased interference immunity, thus enhancing the QoS demanded by applications such as voice calls.

As mentioned above, two CFPs exist per superframe. Each CFP comprises a number of slot pairs with the first slot corresponding to reception and the second one to transmission of voice packets. Since all voice calls are routed through the CP [16], those slots can also be referred to as downlink and uplink slots, respectively. The CFP at the start of the superframe (CFP1) can be thought of as the period for optional retransmission of voice data that was unable to be delivered by the CFP trailing the preceding superframe (CFP2). Thus, for voice connections, the initial transmissions are always made inside CFP2, while CFP1 is used for optional retransmissions made in CFP2 of the previous superframe. In order for the CP to determine the undelivered packets, and consequently determine whether voice retransmissions are needed, each voice packet transmitted by an I-node carries in its header an ACK message, acknowledging the last voice packet received by that node. Whenever the CP decides that a certain I-node will retransmit a voice packet, it announces this decision inside the beacon message at the start of the next superframe. A voice packet can be retransmitted only once.

Figure 11.16 shows two consecutive superframes in a HomeRF system providing support both for voice and data traffic. In this figure, shaded frames in CFP1 are retransmissions of damaged voice transmissions in CFP2 of the preceding superframe. The figure also shows the service slot, which enables I-nodes to make requests for slot assignment in the CFP2 of the next superframe. Due to the fact that there is only a single service slot, it is possible for two or more I-nodes to transmit requests at the same time, which of course will collide. The CP notifies I-nodes on the successful delivery of their connection requests by piggybacking an ACK message at the beacon transmitted at the start of the superframe. When such an ACK is not received, I-nodes assume a collision and perform a random backoff procedure to calculate the time to resend the request.

The voice coder of a HomeRF unit produces 32 kbps ADPCM voice data. Using the CFPs within the superframe structure, up to eight simultaneous voice calls can be supported in a HomeRF system [16], while still providing support for asynchronous data traffic.

11.3.3.1.3 Interference Reduction   HomeRF tries to provide support for reduction of the interference that characterizes wireless communications. Such a reduction is more crucial for voice connections, which are usually not delay-tolerant. The facts that:

- packets of adjacent superframes are transmitted on different frequencies, since with FHSS transmission each superframe is transmitted on a different frequency and
- CFP1 is associated with retransmissions concerning CFP2 of the preceding superframe

provide both frequency and time diversity to HomeRF CFP. Thus, the resistance of HomeRF to interference is enhanced. Furthermore, HomeRF provides an adaptive hopping mechanism [14] that aims to reduce the effect of static interference both on asynchronous and voice traffic. Adaptive hopping tries to ensure that two adjacent hops will not be inside a frequency range subject to interference. When a hop associated with interference is visited, adaptive hopping examines the hopping sequence used in order to determine whether any two consecutive hops are both within the range. When such a pair of hops is located, an attempt is made to replace a pair of the hops with one that is outside the interference range. In the presence of an interference range of up to 31 MHz, adaptive hopping leads to a selection of hops with no consecutive hops within that range.

11.3.3.2 Synchronization

In a HomeRF network, synchronization is performed by using the beacon frame. The beacon frame includes information on the hopping sequence used by the network and the length of the superframe. Inside a managed network, synchronization is a responsibility of the CP. During each superframe, the CP relays synchronization information to A- and I-nodes by including this information to the beacon frame it transmits.

In an ad hoc network, synchronization is performed collectively by all A-nodes. At the beginning of each superframe, A-nodes back off for a random time period from the scheduled ad hoc beacon transmission time in order to avoid simultaneous beacon transmissions. Using this technique, A-nodes manage to reduce beacon collisions and thus effectively relay synchronization information in a distributed manner.

11.3.3.3 Power Management

HomeRF provides a mechanism for reduction of power consumption inside a managed network. In such a case, the CP of the network must be configured to perform power management operations. Such a CP records the MAC addresses of the stations to which it provides power-management services. A HomeRF network can contain both Power Saving (PS) and non-PS nodes at the same time. Power management operations differ for I-nodes and A-nodes. Furthermore, power management operations for A-nodes differ for unicast and broadcast messages. These matters are discussed below.

11.3.3.3.1 Power Management for I-nodes   For I-nodes, reduction of power consumption is a straightforward operation. I-nodes will typically be in the sleep mode, which is characterized by reduced power consumption. Inside each superframe, I-nodes will only wake up to receive the beacon in order to receive slot assignment information for the superframe's CFPs. Then they can switch back to sleep mode and power on only when it is time to transmit/receive at their CFP slot. When an I-node is inactive, meaning that it does not have an ongoing voice connection, it wakes up periodically (every N superframes) to check for slot assignment in the CP's beacon transmission, which carries a list of I-nodes with pending voice traffic. By varying the value of N, a tradeoff can be made between the latency of establishing a voice connection and I-node power consumption.

11.3.3.3.2 Power Management for A-nodes Receiving Unicast Messages   A-nodes that want to reduce power consumption can spend some time in the sleep mode. In order to check for pending incoming unicast packets, A-nodes wake up from time to time to check for such an indication in the beacon frame of the CP. If traffic is pending, the station wakes up and stays powered on for the superframe in order to receive/send unicast packets. When the superframe ends, the station can re-enter sleep mode in order to conserve power.

11.3.3.3.3 Power Management for A-nodes Receiving Broadcast Messages   For broadcast messages, the above procedure for A-nodes is slightly modified. In this case, the CP broadcasts a parameter B within its beacon that defines the time when power-saving nodes should wake up to receive pending broadcasts. When an A-node receives a beacon, it knows it can sleep for B superframes before waking up to check for broadcasts. In the meantime, all broadcasts are buffered at the CP and are relayed to A-nodes when these exit the sleep mode. Parameter B provides a means for trading off latency, CP buffer size and battery life.

   In order to provide efficiency to the power management procedure, PS A-nodes periodically re-request power management services from the CP in an effort to check for an out of range CP. If the CP does not receive such a request for a specific A-node, it assumes the A-node has moved out of the network and frees the resources associated with this A-node's power management.

11.3.3.4 Security

The HomeRF MAC layer provides both an authentication mechanism in order to prevent unauthorized access and an encryption mechanism for secure delivery of exchanged packets. HomeRF performs authentication according to the DECT security model. As far as encryption is concerned, HomeRF provides support for a symmetric encryption scheme, both for asynchronous and isochronous services. Upon initiation of a connection between two nodes, the sender asks the receiver whether or not the latter can support encryption. If encryption is not supported, the connection will either not take place or take place in an unencrypted manner. However, if encryption is supported by the receiver, a shared-secret key procedure takes place. The secret key is shared by all nodes inside the same HomeRF network. The key may be installed in a node through incorporation to the node's Management Information Base (MIB), which is a database that includes parameters needed for the node's operation. In order to exchange an encrypted packet, the following events take place:

- For each packet, the sender produces a 32-bit Initialization Vector (IV) from the sequence number of the packet and a hash function of the sender's 48-bit MAC address.

- The sender uses a 56-bit (128-bit in HomeRF version 2.0) secret key and the IV to convert the transmitted message into an encrypted one.
- Upon receipt of the message, the receiver uses the same secret key and the sequence number of the packet in order to perform decryption.

The above discussion concerns only unicast traffic. Multicast and broadcast transmissions are not encrypted since it is not certain that all A-nodes within the same HomeRF network support encryption.

11.3.3.5 Compression

HomeRF includes support for optional packet compression, through a lossless combination of LZ77 and Huffman coding. Compression provides a tradeoff between bandwidth and power consumption and should of course be supported by both pairs of a HomeRF link. As with security, compression is also not performed for multicast and broadcast packets.

Although compressed transmissions help reserve bandwidth, they consume more power than ordinary transmissions, since extra power is needed for the compression procedure. Furthermore, compression reduces redundant symbols within the packet, thus making error correction more difficult to achieve. Therefore activation of compression should be decided by the system designer.

11.4 Summary

The concept of a PAN differs from that of other types of data networks in terms of size, performance and cost. PANs target applications that demand short-range communications inside the Personal Operating Space (POS) of a person or device, aiming to provide such capability in an efficient manner. This chapter provides a detailed presentation of technological alternatives in the PAN area. After a brief introduction to PANs and ongoing PAN-related activities by IEEE Working Group 802.15, it focuses on the two current technological alternatives, Bluetooth and HomeRF. The main issues presented in this chapter are summarized below, while a comparison of the characteristics of Bluetooth and HomeRF is given in Figure 11.17.

- Bluetooth, which is an industry initiative to develop a de facto standard for PANs that meets the communication needs of all mobile computing and communication devices located in a reduced geographical space, ranging up to 100 m. The Bluetooth protocol stack and profiles are presented. The Bluetooth radio channel which operates in the 2.4 GHz ISM band using FHSS modulation is discussed, followed by a discussion on the way Bluetooth devices connect to form small networks, known as piconets and piconet interconnections, known as scatternets. Bluetooth supports both voice and data transfers through the SCO and ACL links, respectively. A SCO link supports 64 kbps voice channels. An ACL link supports asynchronous data channels either asymmetrically, with a maximum data rate of 721 kbps in one direction and 57.6 in the other or symmetrically with a 432 kbps maximum rate in both directions. The maximum speeds achieved (as those seen by upper layers) are obviously lower than the actual speed at the physical layer of Bluetooth, due to overheads added by the various layers of the stack. Furthermore, power management and security services of Bluetooth are presented.

	Bluetooth Version 1.1	HomeRF Version 1.2	HomeRF Version 2.0
Frequency	2.4 GHz	2.4 GHz	2.4 GHz
Hopping rate	1600 hops/s	50 hops/s	50, 100 hops/s
Packets/hop	1 (fast frequency hopper)	>1 (slow frequency hopper)	>1 (slow frequency hopper)
Range	10, 100 meters	~50 meters	~50 meters
Topology	Ad-hoc	Ad-hoc: Supports data traffic Centralized: Supports data & voice traffic	Ad-hoc: Supports data traffic Centralized: Supports data & voice traffic
Voice support	Up to three 64 Kbps voice connections inside a single piconet	Up to eight 32 Kbps ADPCM voice channels	Up to eight 32 Kbps ADPCM voice channels
Bandwidth allocation method for voice traffic	Polling	TDMA	TDMA
Data support	Up to 723 Kbps	Up to 1.6 Mbps	Up to 10 Mbps
Bandwidth allocation method for data traffic	Polling	CSMA/CA	CSMA/CA
Security key length	128 bits	56 bits	128 bits

Figure 11.17   Comparison of Bluetooth and HomeRF characteristics

- HomeRF aims to enable interoperable wireless voice and data networking within the home at ranges a bit higher than those of Bluetooth. Version 1.2 of HomeRF supported speeds at upper layers of 1.6 and 0.8 Mbps, a little higher than the Bluetooth rates. However, version 2.0 provides for rates up to 10 Mbps by using wider (5 MHz) channels in the ISM band through FHSS, making it more suitable than Bluetooth for transmitting music, audio, video and other high data applications. However, Bluetooth seems to have more industry backing. Furthermore, HomeRF, due to its complexity (hybrid MAC, using CSMA/CA, higher capability PHY), is more expensive to implement than Bluetooth. Compared to Bluetooth, HomeRF can be thought of as a slow frequency-hopping system since at each hop more than one packet can be transmitted. The operation of the HomeRF MAC layer resembles that of IEEE 802.11, since it also defines a superframe structure comprising both contention and Contention-Free Periods (CFP), with the operation inside the contention period being almost identical to that of IEEE 802.11. Furthermore, control of CFP operation is a centralized procedure, as in IEEE 802.11. Finally, issues regarding system synchronization, power management and security are also discussed.

WWW Resources

1. *www.ieee802.org/15*: the web site of working Group 802.15 which deals with PAN standardization.
2. *www.bluetooth.com*: the Bluetooth SIG web site includes detailed technical information on the Bluetooth specification. It also contains information on Bluetooth products and copies of the SIG's newsletter.
3. *www.homerf.org*: the web site of the HomeRF initiative.
4. *www.mot.com/bluetooth/*: the Motorola web site contains information on Bluetooth products, links to other Bluetooth-related sites, news relating to the development of the standard and a list of Frequently Asked Questions (FAQ).

References

[1] Zimmerman T. G. Personal Area Networks: Near-Field Intrabody Communication, *IBM Systems Journal,* 35, 1996, 609–617.
[2] Schneiderman R. Bluetooth's Slow Dawn, *IEEE Spectrum*, November, 2000, 61–65.
[3] Siep T. M., Gifford I. C., Braley R. C. and Heile R. F. Paving the Way for Personal Area Network Standards: an Overview of the IEEE P802.15 Working Group for Wireless Personal Area Networks, *IEEE Personal Communications*, February, 2000, 37–43.
[4] IEEE Project 802.15. http://www.ieee802.org/15.
[5] Heile B., Gifford I. and Siep T. IEEE 802 Perspectives, The IEEE P802.15 Working Group for Wireless Personal Area Networks, *IEEE Network*, July, 1999.
[6] AU-system, Bluetooth Whitepaper 1.1, January, 2000.
[7] The Bluetooth Specification Version 1.0 B, Part A: Radio Specification, 1999.
[8] Bhagwat P. Bluetooth: Technology for Short-Range Wireless Apps, *IEEE Internet Computing*, May/June, 2001, 96–103.
[9] Haartsen J. The Bluetooth Radio System, *IEEE Personal Communications*, February, 2000, 28–36.
[10] Miklos G., Racz A., Turanyi Z., Valko A. and Johansson P. Performance Aspects of Bluetooth Scatternet Formation, in *Proceedings of Mobile and Ad Hoc Networking and Computing (MobiHOC)*, 2000, pp. 147–148.
[11] Kalia M., Garg S. and Shorey R. Scatternet Structure and Inter–Piconet Communication in the Bluetooth System, in *Proceedings of IEEE National Conference on Communications*, 2000, New Delhi.
[12] Salonidis T., Bhagwat P., Tassiulas L. and LaMaire R. Distributed Topology Construction for Bluetooth Personal Area Networks, in *Proceedings of IEEE INFOCOM*, 2001, pp. 1577–1586.
[13] The HomeRF Working Group, Wireless Networking Choices for the Broadband Internet Home, 2001.
[14] The HomeRF Working Group, Quality of Service in the Home Networking Model, 2001.
[15] Lansford J. and Bahl P. The Design and Implementation of HomeRF: A Radio Frequency Wireless Networking Standard for the Connected Home, in *Proceedings of the IEEE*, October, 2000, pp. 1662–1676.
[16] Lansford J. HomeRF: A Wireless Voice and Data System for the Home, in *Proceedings of IEEE ICASSP, 2000*, pp. 3718–3721.

Further reading

[1] Wireless Technologies, Dell White Paper.
[2] Haartsen J., Naghshineh M., Inouye J., Joeressen O. J. and Allen W. Bluetooth: Vision, Goals and Architecture, *ACM Mobile Computing and Communications Review*, October, 1998, 38–45.
[3] Albrecht M., Frank M., Martini P., Schetelig M., Vilavaara A. and Wenzel A. IP Services over Bluetooth: Leading the Way to a New Mobility, in *Proceedings of IEEE LCN*, 1998.
[4] Haartsen J. Bluetooth - The Universal Radio Interface For Ad–Hoc, Wireless Connectivity, *Ericsson Review, 3*, 1998, 110–117.
[5] Negus K., Stephens A. P. and Lansford J. HomeRF: Wireless Networking for the Connected Home, *IEEE Personal Communications*, February, 2000, 20–27.

12

Security Issues in Wireless Systems

The issue of security of computer systems and networks, especially security of wireless networks and systems has become essential, given the dependence of people on these systems in their daily life. This chapter presents the main issues for wireless networks and the need to secure access to such systems; any breach to such systems may entail loss of money, loss of national security information, or leak of such information and secrets to unwanted parties including competitors and enemies (see Section 12.1). Then, in Section 12.2, we review the types of attacks on wireless networks. Section 12.3 presents the classes of services of any reliable security system including confidentiality, nonrepudiation, authentication, access control, integrity, and availability. Section 12.4 presents the main aspects of the Wired Equivalent Privacy (WEP) Protocol. Section 12.5 introduces the security aspects of mobile IP. Section 12.6 investigates the main weakness of the WEP protocol. Then Section 12.7 presents virtual private network services as a cost-effective and secure scheme. Finally, we conclude by highlighting the main ideas presented in the chapter.

12.1 The Need for Wireless Network Security

A wireless local area network is a flexible data communication system implemented as an extension to or as an alternative to the wired local area network. Wireless LANs transmit and receive the data over the air using the radio frequency technology, thus minimizing wired connections. Thus, wireless LANs combine data connectivity with user mobility. Wireless LANs have gained strong popularity in a number of vertical markets and these industries have profited from the productivity gains of using hand held terminals and notebook computers to transmit real-time information to centralized hosts for processing. Today, wireless LANs are becoming more widely recognized as a general-purpose connectivity alternative for a broad range of business customers. But one of the scariest revelations is that wireless LANs are insecure and the data sent over the them can be easily broken and compromised. The security issue in wireless networks is much more critical than in wired networks. Data sent on a wireless system is quite literally broadcast for the entire world to hear. Therefore, unless some serious countermeasures are taken, wireless systems should not be used in situations where critical data is sent over the airwaves. Any computer network, wireless or wireline, is subject to substantial security risks. The major issues are [1–3]: (a) threats to the physical security of the network; (b) unauthorized access by unwanted parties; and (c) privacy.

A certain level of security is a must in almost all local area networks, regardless of whether they are wireless or wireline-based. There is no LAN owner who wants to risk having the LAN data exposed to unauthorized users or malicious attackers. If the data carried in the networks are sensitive, such as that found on the networks of financial institutions and banks, and e-commerce, e-government, and military networks, then extra measures must be taken to ensure confidentiality and privacy.

This chapter deals with various security issues related to wireless LANs including those that have been implemented in the IEEE 802.11 standard.

12.2 Attacks on Wireless Networks

The dependence of people on computer networks including wireless networks has increased tremendously in recent years and many corporations and businesses rely heavily on the effective, proper and secure operation of these networks. The total number of computer networks installed in most organizations has increased at a phenomenal rate. Corporations store sensitive and confidential information on marketing, credit records, income tax, trade secrets, national security data, and classified military data, among others. The access of such data by unauthorized users may entail loss of money or release of confidential information to competitors or enemies [2].

Attacks on computer systems and networks can be divided into passive and active attacks [1–3]. Active attacks involve altering data or creating fraudulent streams. These types of attacks can be divided into the following subclasses: (a) masquerade; (b) reply; (c) modification of messages; and (d) denial of service. A masquerade occurs when one entity pretends to be a different entity. For example, authentication can be collected and replayed after a valid authentication sequence has taken place. Reply involves the passive capture of a data unit and its subsequent retransmission to construct unwanted access. Modification of messages means that some portion of a genuine message is changed or that messages are delayed or recorded to produce an unauthorized result.

Passive attacks are inherently eavesdropping or snooping on transmission. The attacker tries to access information that is being transmitted. There are two subclasses: release of message contents, and traffic analysis. In the first type, the attacker reaches the e-mail messages or a file being transferred. In traffic analysis type of attack, the attacker could discover the location and identity of communicating hosts and could observe the frequency and length of encrypted messages being exchanged. Such information could be useful to the attacker as it can reveal useful information in guessing the nature of the information being exchanged [2,3].

In general, passive attacks are difficult to detect, however, there are measures that can be used to avoid them. On the other hand, it is difficult to prevent active attacks.

The main categories of attack on wireless computer networks are [2,5,6]:

- *Interruption of service*. Here, the resources of the system are destroyed or become unavailable.
- *Modification*. This is an attack on the integrity of the system. In this case, the attacker not

only gains access to the network, but tampers with data such as changing the values in a database, altering a program so that it does different tasks.

- *Fabrication*. This is an attack on the authenticity of the network. Here the attacker inserts counterfeit objects such as inserting a record in a file.
- *Interception*. This is an attack on the confidentiality of the network such as wiretapping or eavesdropping to capture data in a network. Eavesdropping is easy in a wireless network environment since when one sends a message over a radio path, everyone equipped with the proper transceiver equipment in the range of transmission can eavesdrop the data. These kinds of devices are usually inexpensive. The sender or intended receiver may not be able to find out whether their messages have been eavesdropped or not. Moreover, if there is no special electromagnetic shielding, the traffic of a wireless network can be eavesdropped from outside the building where the network is operating. In most wireless networks, there is a kind of link level ciphering done by the MAC entities.
- *Jamming*. Interruption of service attacks is also easily applied to wireless networks. In such a case, the legitimate traffic cannot reach clients or access points due to the fact that illegitimate traffic overwhelms the frequencies. An attacker can use special equipment to flood the 2.4 GHz frequency band. Such a denial of service can originate from outside the service area of the access point, or from other wireless devices installed in other work areas that degrade the overall strength of the signal.
- *Client-to-client attacks*. Wireless network users need to defend clients not just against an external threat, but also against each other. Wireless clients that run TCP/IP protocols such as file sharing are vulnerable to the same misconfigurations as wired networks. Also, duplication of IP or MAC addresses whether its intentional or accidental, may cause disruption of service.
- *Attacks against encryption*. The IEEE 802.11b standard uses an encryption scheme called Wired Equivalent Privacy (WEP) which has proven to have some weaknesses. Sophisticated attacker can break the WEP scheme.
- *Misconfiguration*. In order to have ease and rapid deployment, the majority of access points have an unsecured configuration. This means that unless the network administrator configures each access point properly, these access points remain at high risk of being accessed by unauthorized parties or hackers.
- *Brute force attacks against passwords of access points*. The majority of access points use a single password or key, which is shared by all connecting wireless clients. Attackers can attempt to compromise this password or key by trying all possibilities. Once the attacker guesses the key or the password, he/she can gain access to the access point and compromise the security of the system. Moreover, not changing the passwords or keys on a regular basis may put the network system at great risk especially if employees leave the company. On the other hand, managing a large number of access points and clients complicates the security system.
- *Insertion attacks*. This type of attack is based on deploying a new wireless network without following security procedure. Also, it may be due to installation of an unauthorized device without proper security review. For example, a company may not know that some of its employees have deployed wireless facilities on its network. Using such a rogue access point, the database of the company will be compromised. Clearly, there is a need to implement a policy to secure the configuration of all access points, in addition to a routine process by which the network is scanned for unauthorized devices in its wireless portion.

Another example is that an attacker may connect a laptop or a PDA to an access point without the authorization of the owner of the wireless network. If the attacker was able to gain access by getting a password or if there is no password or key requirement, then the attacker/intruder will be able to connect to the internal network.

Any network security system should maintain the following characteristics [2–4,6–12]:

- *Integrity*. This requirement means that operations such as substitution, insertion or deletion of data can only be performed by authorized users using authorized methods. Three aspects of integrity are commonly recognized: authorized actions, protection of resources, and error detection and correction.
- *Confidentiality*. This means that the network system can only be accessed by authorized users. The type of access can be read-only access. Another is privileged access where viewing, printing, or even knowing the existence of an object is permitted.
- *Denial of service*. This term is also known by its opposite, availability. An authorized individual should not be prevented or denied access to objects to which he has legitimate access. This access applies to both service and data. Denning [6] states that the effectiveness of access control is based on two ideas: (a) user identification and (b) protecting the access right of users.

Computer networks, in general, have security problems due to:

- *Sharing*. Since network resources are shared, more users have the potential to access networked systems rather than just a single computer node.
- *Complexity*. Due to the complexity of computer networks of all types, reliable and secure operation is a challenge. Moreover, computer networks may have dissimilar nodes with different operating systems, which makes security more challenging.
- *Anonymity*. A hacker or intruder can attack a network system from hundreds of miles away and thus never have to touch the network or even come into contact with any of its users or administrators.
- *Multiple point of attack*. When a file exists physically on a remote host, it may pass many nodes in the network before reaching the user.
- *Unknown path*. In computer networks, routes taken to route a packet are seldom known ahead of time by the network user. Also these users have no control of the routes taken by their own packets. Routes taken depend on many factors such as traffic patterns, load condition, and cost.

12.3 Security Services

Security services can be classified as follows [2,7–12]:

- *Confidentiality*. This service means the protection of data being carried by the network from passive attacks. The broadcast service should protect data sent by users. Other forms of this service include the protection of a single message or a specific field of a message. Another aspect of confidentiality is the protection of traffic from a hacker who attempts to analyze it. In other words, there must be some measures that deny the hackers from observing the frequency and length of use, as well as other traffic characteristics in the network.

- *Nonrepudiation.* This service prevents the sending or receiving party from denying the sent or received message. This means that when a message is received, the sender can confirm that the message was in fact received by the assumed receiver.
- *Authentication.* The authentication service is to ensure that the message is from an authentic source. In other words, it ensures that each communicating party is the entity that it claims to be. Also, this service must ensure that the connection is not interfered with in a way that a third party impersonates one of the authorized parties.
- *Access control.* This service must be accurate and intelligent enough so that only authorized parties can use the system. Also, this accuracy should not deny authorized parties from using the network system.
- *Integrity.* In this context, we differentiate between connection-oriented and connection-based integrity services. The connection-oriented integrity service deals with a stream of messages, and ensures that the messages are sent properly without duplication, modification, reordering or reply. Moreover, the denial of service aspect is covered under the connection-oriented service. The connectionless integrity service deals only with the protection against message modification. A hybrid type of integrity service was proposed to deal with the applications that require protection against replay and reordering, but do need strict sequencing [2–4]. A good security system should be able to detect any integrity problem and if a violation of integrity is reported, then the service should report this problem. A software mechanism or human intervention should resolve this problem. The software approach is supposed to resolve the problem automatically without human intervention.
- *Availability.* Some attacks may result in loss or reduction of availability of the system. Automated schemes can resolve some of these problems while others require some type of physical procedures.

12.4 Wired Equivalent Privacy (WEP) Protocol

The name, wired equivalent privacy (WEP), implies that the goal of WEP is to provide the level of privacy that is equivalent to that of a wired LAN. This was designed to provide confidentiality for network traffic using wireless protocols. WEP was intended to provide a similar level of privacy over wireless networks that one may get from a wired network. The WEP algorithm is used to protect wireless networks from eavesdropping. It is also meant to prevent unauthorized access to wireless networks. The scheme relies on a secret key that is shared between a wireless node and an access point. The secret key is used to encrypt data packets before sending them. The IEEE 802.11 standard does not specify how the standard key is established and most implementations use a single key that is shared between all mobiles and access points.

WEP relies on a default set of keys, which are shared between wireless LAN adapters and access points [13].

The IEEE 802.11 committee has established standards for wireless LANs and several companies have designed wireless LAN products that are compatible with these universal standards. Wireless networks users are primarily concerned that an intruder should not be able to: (a) access the network by using similar wireless LAN equipment; and (b) capture wireless LAN traffic by eavesdropping or other methods for further analysis [14].

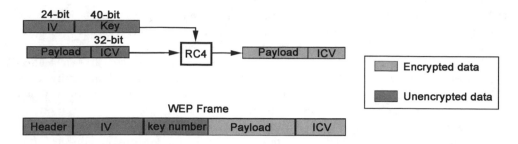

Figure 12.1   An authenticated frame [14,20]

In IEEE 802.11 networks, access to network resources is denied for any user who does not prove knowledge of the current key. Eavesdropping is prevented by using the WEP scheme whereby a pseudorandom number generator is initialized by a shared secret key. Based on the Rivest–Shamir–Adelman (RSA) RC4 algorithm, this simple WEP algorithm has the following properties: (a) *reasonably strong* – a brute force attack on this algorithm is difficult because every frame is sent with an initialization vector, which restarts the PseudoRandom Number Generator (PRNG) for each frame; (b) *self-synchronizing* – since just like in any LAN, the wireless LAN stations work in a connectionless environment where packets may get lost, the WEP algorithm resynchronizes at each message [13–23]. Figure 12.1 shows an authenticated frame.

The WEP algorithm uses the RC4 encryption scheme which is often called the stream cipher. RC4 is a stream cipher similar to the encryption scheme used in the Secure Socket Layer (SSL) to secure access to web sites. It works fine when used with SSL. This is because each transaction is assigned a unique 128-bit key. The WEP algorithm is part of the IEEE 802.11 standard and it defines how encryption must support the authentication, integrity, and confidentiality of packets sent using wireless systems. The standard committee selected RC4, a proven encryption scheme, to be used for wireless security and all wireless system manufacturers support IEEE 802.11. Designing systems that use cryptographic tools is a challenging task.

The open system authentication is the default authentication for the 802.11 standard. This scheme authenticates everyone that requests authentication. It relies on the default set of keys that are shared between the wireless devices and the wireless access points. Only a client with the correct key can communicate with any access point on the network. If a client without the correct key requests connection, then the request is rejected. The data is encrypted before transmitting, and an integrity check is performed to make sure that the packets are not modified in transit. Only a client with the correct key can decrypt the transmitted data preventing unauthenticated users from accessing the information.

The access control list can provide a minimal level of security. In order that vendors can provide security, they often use this mechanism by using the access control list, which is based on the Ethernet MAC addresses of the clients. This list consists of the MAC addresses of all of its clients and only the clients whose MAC addresses are listed can access the network. If the address is not listed, access is not granted. Figure 12.2 depicts WEP based security with the access control list [13–15].

The IEEE 802.11 standard specifies two methods for using the WEP. The first method provides a window of four keys. A station or an access point can decrypt packets enciphered

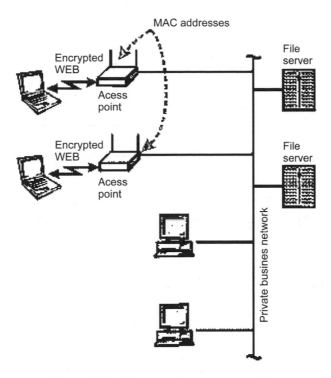

Figure 12.2   Security with access control list

with any of the four keys. The transmission is limited to any one of the four manually entered keys, which is known as the default key. The second method is called the key-mapping table where each unique MAC address can have separate keys. The use of a separate key for each client mitigates the cryptographic attacks found by others. The disadvantage is that all of these keys should be configured manually on each device or access point.

In the shared key authentication method, the station wishing to authenticate (initiator) sends an authentication request management frame indicating that it wishes to use the shared key authentication. The responder responds by sending the challenge text, which is the authentication management frame to the initiator. The PRNG with the shared secret and the random initialization vector generates this challenge text. After the initiator receives the challenge management frame from the responder, it copies the contents of the challenge text into the new management frame body. The new management frame body is then encrypted using the shared secret along with the new Initiating Vector (IV) selected by the initiator. This frame is then sent to the responder. The latter decrypts the received frame and verifies that the Cyclic Redundancy Code (CRC) Integrity Check Value (ICV) is valid, and that the challenge text matches the one that is sent in the first message. If they do, then the authentication is successful and the initiator and the responder switch roles and repeat the process to ensure mutual authentication.

Figure 12.3 shows what the authentication management frame looks like. The value is set to zero if successful and is set to an error value if unsuccessful. The element identifier

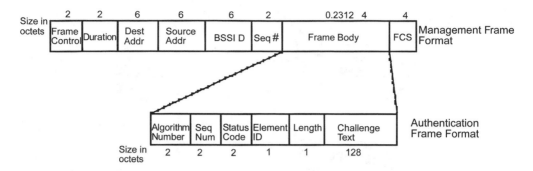

Figure 12.3   Authentication management frame [20]

identifies if the challenge text is included. The length field identifies the length of the challenge text, which includes a random challenge string [14–16].

12.5 Mobile IP

Mobile IP was developed in response to the increasing use of mobile computers in order to enable computers to maintain Internet connection during their movement from one Internet access point to another. It is important to note that the term mobile implies that the user is connected to one or more application across the Internet and the access point changes dynamically. Clearly, this is different from when a traveler uses his ISP account to access the Internet from different locations during his trip [17–20].

Mobile IP is the modification to the standard IP so that it allows the client to send and receive datagrams no matter where it is attached to the network. The only security problem using this mechanism is redirection attacks. A redirection attack occurs when a malicious client gives false information to the home agent in the mobile IP network. The home agent is informed that the client has a new care of address. So all IP datagrams addressed to the actual client are redirected to the malicious client.

Mobile IP is designed to resist two kinds of attacks: (a) a malicious agent that may reply to old registration messages and cut the node from its network, and (b) a node that may pretend to be a foreign agent and send a registration request to a home agent in order to divert traffic that is intended for a mobile node to itself. Message authentication and proper use of the identification field of the registration request and reply messages are often used to protect mobile IPs from these kinds of attack. In order to protect against such attacks, the use of message authentication and proper use of the identification field of the registration request and reply messages is supposed to be effective [20]. Each registration request and reply contains an authentication extension that has the following fields:

- *Type*. This is an 8 bit field that designates the type of authentication extension.
- *Length*. This is an 8 bit field that identifies the number of bytes in the authenticator.
- *Security Parameter Index*. This field has 4 bytes and is used to identify the security context between a pair of nodes. The configuration of the security context is made so that the two nodes share the same secret key and parameters relevant to the authentication scheme.
- *Authenticator*. This field has a code that is inserted by the sender into the message using a

shared secret key. The receiver uses the same code to make sure that the message has not been modified. The default authentication scheme is the keyed-MD5 (Message Digest 5) which produces a 128-bit message digest. MD5 was developed in 1994 as a one-way hash algorithm which takes any length of data and produces a 128-bit 'fingerprint' or 'message digest.' It is computationally not feasible to determine the original message based on the fingerprint.

12.6 Weaknesses in the WEP Scheme

The weakness of the WEP protocol involves the RC4 encryption algorithm and the Initialization Vector (IV). The RC4 takes an encryption key and generates a pseudorandom stream of bytes called the keystream [23]. The latter is pseudorandom as every key is guaranteed to produce a different keystream. Many researchers have found a number of flaws in WEP that seriously undermine the security of the system. WEP is weak against the following attacks: (a) active attacks that inject new traffic from unauthorized mobile stations; (b) active attacks to decrypt traffic based on fooling the access point; (c) passive attacks to decrypt traffic based on statistical analysis; and (d) dictionary-building attacks which allow real-time automated decryption of traffic after some analysis [23].

Active attack to inject traffic is due to the situation where an attacker knows the exact plain text for one encrypted message. By using this knowledge, the attacker can construct correct encrypted packets. This involves constructing a new message, calculating the CRC-32, and performing bit flips on the original encrypted message. This packet can now be sent to the access point or to a mobile node and accepted as a valid packet.

Another type of active attack is based on decryption traffic which is based on fooling the access point. Here, the attacker makes a guess about the header of the packet; not the packet's content. Basically, all that is needed to guess is the destination IP address. The attacker can then flip specific bits to transform the destination IP address to transmit the packet to a node under his control, and to transmit it using a rogue mobile station. Keep in mind that almost all wireless installations have Internet connection. The packet will be decrypted by the access point and forwarded unencrypted using routers to the attacker's machine, reporting the plain text. It is possible to change the destination port on the packet to port 80. This will allow the packet to be forwarded through most firewalls.

In the passive attack that is based on traffic decryption, an eavesdropper can intercept all wireless traffic until an IV collision occurs. The attacker can obtain the XOR of two plain text messages by XORing two packets which use the same Initialization Vector (IV). This result can be used to interpret data about the two messages. IP traffic is often predictable and has redundancy that can be used to eliminate many possibilities for the content of messages. Advanced guesses about the content of one or both of the messages can be obtained by using statistical analysis techniques to determine the exact content. Once it is possible to detect the entire plain text for one message, it is possible to detect the plain texts of all other messages with the same IV. Another scenario of this attack occurs when the attacker uses a host on the Internet to send traffic from outside to a host on the wireless system facilities. The attacker will be able to know the content of such traffic, hence, the plain text will be known. If the attacker intercepts the encrypted version of the message sent over an IEEE 802.11 system, he will be able to decrypt packets that use the same IV [14].

In table-based attack, the small space of possible initialization vectors allows an attacker to build an encryption table. Once the plain text for the packet is known, the attacker can compare the RC4 key stream generated by the IV. The latter can be used to decrypt all other packets that utilize the same IV. Clearly, the attacker can build up a table of inclusion vectors and the corresponding key streams over time. Once such a table, which requires small memory, is built, the attacker can decrypt all packets sent over that wireless link.

Such attacks can be implemented using inexpensive equipment. Therefore, it is highly recommended not to rely completely on WEP and consider using additional security techniques [14].

Although it is not easy to decode a 2.4 GHz digital signal, off-the-shelf hardware devices that can monitor IEEE 802.11 signals are available to attackers. Many IEEE 802.11 devices are available with programmable firmware that can be reverse-engineered in order to inject traffic. Hackers can distribute this firmware and sell it at high prices to interested parties including competitors and enemies.

12.7 Virtual Private Network (VPN)

A Virtual Private Network (VPN) connects the components and resources of one network over another network. VPNs accomplish this by allowing the user to tunnel through the wireless network or other public network in such a way that the tunnel participants enjoy at least the same level of confidentiality and features as when they are attached to a private wired network. A VPN is a group of two or more computer systems connected to a private network, which is built and maintained by the organization for its own use with limited public network access. A VPN solution for wireless access is currently the most suitable alternative to WEP. It is already widely deployed to provide remote workers with secure access to the networks via the Internet. In the remote user application, a VPN provides a secure, dedicated path called a tunnel over an untrusted network. A comprehensive VPN requires three main technology components: security, traffic control, and enterprise management [21].

VPNs provide the following main advantages [21,22]:

- *Security*. By using advanced encryption and authentication schemes, VPNs can secure data from being accessed by hackers and unauthorized users.
- *Scalability*. They enable organizations to use the Internet infrastructure within ISPs and devices in an easy and cost-effective manner. This will enable organizations to add large amounts of capacity without the need to add new significant infrastructure.
- *Compatibility with broadband technology*. VPN technology allows mobile users and telecommuters to benefit from the high-speed access techniques such as DSL and cable modem, to get access to their organization networks. This provides users with significant flexibility and efficiency. Moreover, such high-speed broadband connections provide a cost-effective solution for connecting remote offices.
- They are currently deployed on many enterprise networks
- They have low administration requirements.
- The traffic to the internal network is isolated until VPN authentication is performed.
- WEP key and MAC address list management become optional since the security measures are created by the VPN channel itself.

The main drawbacks of the current VPNs as applied to WLANs are [20,21]:

- Lack of support for multicasting and roaming between the wireless networks.
- They are not completely transparent since users receive a login dialog when roaming between VPN servers on the network or when a client system resumes from standby mode.

Various tunneling protocols, which are discussed below, are used to ensure security [20,21].

12.7.1 Point-to-Point Tunneling Protocol (PPTP)

This protocol is built on the Internet communications protocol called Point to Point Protocol (PPP) and the TCP/IP protocol. PPP offers authentication as well as methods of privacy and compression of data. PPTP allows the PPP session to be tunneled through an existing IP connection. The existing connection can be treated as if it were a telephone line. Therefore, a private network can run over a public network. Tunneling is achieved because PPTP provides encapsulation by wrapping packets of information within IP packets for transmission through the Internet. Upon reception, the external IP packets are stripped away, exposing the original packets for delivery. Encapsulation allows the transport of packets that will not otherwise conform to Internet address standards. Figure 12.4 shows the main components of the Point-to-Point Tunneling Protocol (PPTP). For data transmission using PPTP, tunneling makes use of two basic packet types [22]: (a) data packets and (b) control packets. Control packets are used strictly for status inquiry and signaling information and are transmitted and received over a TCP connection. The data portion is sent using PPP encapsulated in Generic Routing Encapsulation (GRE) protocol. GRE protocol provides a way to encapsulate arbitrary data packets within an arbitrary transport protocol. Although PPTP did not have any provision for authentication or encryption when it was first developed, it has been enhanced recently to support encryption and authentication methods.

12.7.2 Layer-2 Transport Protocol (L2TP)

Similar to PPTP, L2TP is basically a tunneling protocol and does not include any encryption or authentication mechanism. The main difference between PPTP and L2TP is that L2TP combines the data and control channels and runs over the User Datagram Protocol (UDP). The latter is faster for sending packets that are commonly used in real-time Internet communication because it does not retransmit lost packets. On the other hand, PPTP separates the control stream, which runs over TCP, and the data stream, which runs over GRE. Combining these two channels and using high performance UDP makes L2TP more firewall friendly than the PPTP. This is the main advantage as most firewalls do not support GRE.

In PPP, a connection is tunneled using IP. An L2TP access concentrator is the client end of

The PPTP Standard

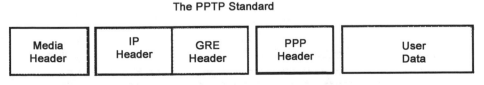

(GRE=Genetic Routing Encapsulation Protocol)

Figure 12.4   The Point-to-Point Tunneling Protocol (PPTP) standard

the connection while an L2TP network server is the server side. The PPP packets are encapsulated in an L2TP header that is encapsulated in IP. These IP packets can traverse the network just like ordinary IP datagrams.

Data transmission in an L2TP can be implemented as a UDP-based IP protocol. The packet is first generated at the client computer. This IP packet is sourced from the client computer and destined for the remote network. The packet is encapsulated in PPP. This packet is then encapsulated in L2TP. UDP header is added to this L2TP packet and is encapsulated in an IP datagram. This IP packet is destined for the Internet Service Provider (ISP) network. The IP packets will again be encapsulated at PPP and terminate at the ISP's network authentication server. This final heavily encapsulated packet will be sent over the circuit switched layer 2 network.

12.7.3 Internet Protocol Security (IPSec)

IPSec is an open standard that is based on network layer 3 security protocol. The latter protects IP datagrams by defining a method of specifying how the traffic is protected and to whom it is sent. In order to protect IP datagrams, the IPSec protocol uses either the Encapsulation Security Payload (ESP) or Authentication Header (AH) protocols [3,21].

The data origin authentication ensures that the received data is the same as that sent and the recipient knows who sent the data (see Figure 12.5). Data integrity ensures data transmission without alteration while relay protection offers partial sequence integrity. Data confidentiality ensures that no one can read the transmitted data which can be possible by using encryption algorithms.

Integrating L2TP with IPSec offers the ability to use L2TP as a tunneling protocol; however, securing the data is achieved using an IPSec scheme. Using L2TP as the tunneling protocol gives the added advantage of increased manageability for end-to-end communications. Moreover, L2TP is a widely available standard; therefore the interoperability between vendors is far better than just IPSec alone [3,21].

The same VPN technology can be used to secure wireless systems. The Access Points (APs) are configured for open access with no WEP encryption, but wireless access is isolated from the enterprise network by a VPN server and a VLAN between the APs and VPN servers. Authentication and full encryption over the wireless network is provided using the VPN servers which also act as gateways to the private network. Clearly, a VPN-based solution has the advantage of being scalable for a very large number of users. Figure 12.6 illustrates the general configuration of a VPN.

12.8 Summary

Almost all wireless networks are at risk of compromise. Unfortunately, fixing the problem is

	Data Origin Authentication	Data Integrity	Relay Protection	Data Confidentiality
AH	✓	✓	✓	
ESP	✓	✓	✓	✓

Figure 12.5 Comparison between Authentication Header (AH) and Encapsulation Security Payload (ESP)

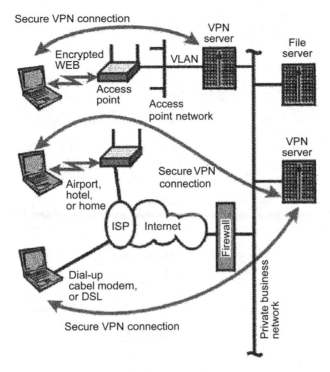

Figure 12.6 A Virtual Private Network (VPN) configuration [17–23]

not a straightforward procedure. It has been found that all IEEE 802.11 wireless networks deployed have security problems [20].

Among the effective interim short-term solutions is the use of a WEP with a robust key management system, VPNs schemes and high-level security schemes such as IPSec. Although these schemes do not completely resolve the problem, they can be used until the IEEE 802.11 standard committee establishes new effective encapsulation algorithms. Basically, there is no wireless technology that is better than the other for all applications. Each has its own advantages and drawbacks. Despite the fact that wireless networks are not completely secure, there ease of use has always been considered a key factor for their amazing widespread success. Biometric-based security schemes have great potential to secure and authenticate access to all types of networks including wireless networks. We are witnessing these days an increasing interest in this technology due to its great potential.

References

[1] Stallings W. *Network Security Essentials: Applications and Standards*, Prentice Hall, Upper Saddle River, NJ, 2000.

[2] Obaidat M. S. and Sadoun B. Keystroke Dynamics Based Authentication, *in Biometrics: Personal Identification in Networked Society*, Jain A., Bolle R. and Pankanti S., eds., Kluwer, Norwell, MA, 1999, pp. 213–230.

[3] Stallings W. *Cryptography and Network Security: Principles and Practice*, Second Edition, Prentice Hall, Upper Saddle River, NJ, 1999.

[4] http://rr.sans.org/wireless/wireless_list.php

[5] http://www.netmotionwireless.com/resource/whitepapers/security.asp

[6] Denning D. *Cryptography and Data Security*, Addison-Wesley, Reading, MA, 1983.

[7] Obaidat M. S. and Sadoun B. Verification of Computer Users Using Keystroke Dynamics, 27(2), April, 1997, 261–269.

[8] Obaidat M. S. An Evaluation Simulation Study of Neural Network Paradigm for Computer Users Identification, *Information Sciences Journal–Applications,* 102(1–4), November, 1997, 239–258.

[9] Obaidat M. S. A Methodology for Improving Computer Access Security, *Computers & Security*, 12, 1993, 657–662.

[10] Obaidat M. S and Macchairolo D. T. An On-line Neural Network System for Computer Access Security, *IEEE Transactions in Industrial Electronics*, 40(2), 1993, 235–241.

[11] Bleha S. and Obaidat M. S. Dimensionality Reduction and Feature Extraction Applications in Identifying Computer Users, *IEEE Transactions Systems, Man and Cybernetics*, 21(2), March/April, 1991.

[12] Bleha S. and Obaidat M. S., Computer User Verification Using the Perceptron, *IEEE Transactions Systems, Man and Cybernetics*, 23(3), May/June, 1993, 900–902.

[13] IEEE 802.11b Wired Equivalent Privacy (WEP) Security at: http://www.wi-fi.com/pdf/Wi-FiWEPSecurity.pdf

[14] Security of WEP Algorithm at: http://www.isaac.cs.berkeley.edu/isaac/wep-faq.html

[15] Walker J. Overview of 802.11 Security. Available at: http://grouper.ieee.org/groups/802/15/pub/2001/Mar01/01154r0P802-15_TG3%

[16] Ukela S. Security in Wireless Local Area Networks, available at: http://www.tml.hut.fi/Opinnot/Til-110-501/1997/wireless_lan.html

[17] Stallings W. *Wireless Communications and Networks*, Prentice Hall, Upper Saddle River, NJ, 2002.

[18] Walker J. Unsafe at any Key Size: An Analysis of the WEB Encapsulation, *Tech. Report 03628E, IEEE 802.11 Committee,* March, 2002. Available at: http//grouper.ieee.org/groups/802/11/Documents/DocumentHolder/0-362.zip

[19] IEEE 802.11 Working Group, available at: http//grouper.ieee.org/groups/802/11/index.html.

[20] Arbaugh W. A., Shankar N and Justin Wan Y. C. Your Wireless Network has No Clothes, available at: http://www.cs.umd.edu/~waa/wireless.pdf

[21] http://www.checkpoint.com/products/vpn1/vpnwp.html

[22] http://www.cisco.com/warp/public/779/largeent/learn/technologies/VPNs.html

[23] http://www.rsasecurity.com/rsalabs/3-6-3.html

13

Simulation of Wireless Network Systems

This chapter deals with simulation of wireless network systems. We introduce the basics of discrete-event simulation as it is the simulation technique that is used for simulating wireless networks. We then review the main characteristics of the commonly used stochastic distributions used for the simulation of wireless networks. The techniques used to generate and test random number sequences are investigated. Then, we introduce the techniques used to generate random variates followed by performance metrics considerations. The chapter concludes with cases studies on the simulation of some wireless network systems.

13.1 Basics of Discrete-Event Simulation

Simulation is a general term that is used in many disciplines including performance evaluation of computer and telecommunications systems. It is the process of designing a model of a real system and conducting experiments with this model for the purpose of understanding its behavior, or of evaluating various strategies of its operation. Others defined simulation as the process of experimenting with a model of the system under study using computer programming. It measures a model of the system rather than the system itself.

A model is a description of a system by symbolic language or theory to be seen as a system with which the world of objects can be expressed. Thus, a model is a system interpretation or realization of a theory that is true. Shannon defined a model as 'the process of designing a computerized model of a system (or a process) and conducting experiments with this model for the purpose either of understanding the behavior of the system or of evaluating various strategies for the operation of the system.'

Based on the above definition of a model, we can redefine simulation as the use of a model, which may be a computer model, to conduct experiments which, by inference, convey an understanding of the behavior of the system under study. Simulation experiments are important aspect of any simulation study since they help to:

- discover something unknown or test an assumption
- find candidate solutions, and provide a mean for evaluating them.

Basically, modeling and simulation of any system involve three types of entities: (a) real system; (b) model; and (c) simulator. These entities are to be understood in their interrelation to one another as they are related and dependent on each other. The real system is a source of raw data while the model is a set of instructions for data generating. The simulator is a device for carrying out model instructions. We need to validate and verify any simulation model in order to make sure that the assumptions, distributions, inputs, outputs, results and conclusions, as well as the simulation program (simulator), are correct [1–10].

Systems in general can be classified into stochastic and deterministic types [1–3]:

- *Stochastic systems.* In this case, the system contains a certain amount of randomness in its transitions from one state to another. A stochastic system can enter more than one possible state in response to an activity or stimulus. Clearly, a stochastic system is nondeterministic in the sense that the next state cannot be unequivocally predicted if the present state and the stimulus are known.
- *Deterministic systems.* Here, the new state of the system is completely determined by the previous state and by the activity or input.

Among the reasons that make simulation attractive in predicting the performance of systems are [1–3]:

- Simulation can foster a creative attitude for trying new ideas. Many organizations or companies have underutilized resources, which if fully employed, can bring about dramatic improvements in quality and productivity. Simulation can be a cost-effective way to express, experiment with, and evaluate such proposed solutions, strategies, schemes, or ideas.
- Simulation can predict outcomes for possible courses of action in a speedy way.
- Simulation can account for the effect of variances occurring in a process or a system. It is important to note that performance computations based solely on mean values neglect the effect of variances. This may lead to erroneous conclusions.
- Simulation promotes total solutions.
- Simulation brings expertise, knowledge and information together.
- Simulation can be cost effective in terms of time.

In order to conduct a systematic and effective simulation study and analysis, the following phases should be followed [1,4,5]. Figure 13.1 summarizes these major steps.

- *Planning.* In the planning phase, the following tasks have to be defined and identified:
  - *Problem formulation.* If a problem statement is being developed by the analyst, it is important that policymakers understand and agree with the formulation.
  - *Resource estimation.* Here, an estimate of the resources required to collect data and analyze the system should be conducted. Resources including time, money, personnel and equipment, must be considered. It is better to modify goals of the simulation study at an early stage rather than to fall short due to lack of critical resources.
  - *System and data analysis.* This includes a thorough search in the literature of previous approaches, methodologies and algorithms for the same problem. Many projects have failed due to misunderstanding of the problem at hand. Also, identifying parameters, variables, initial conditions, and performance metrics is performed at these stages. Furthermore, the level of detail of the model must be established.

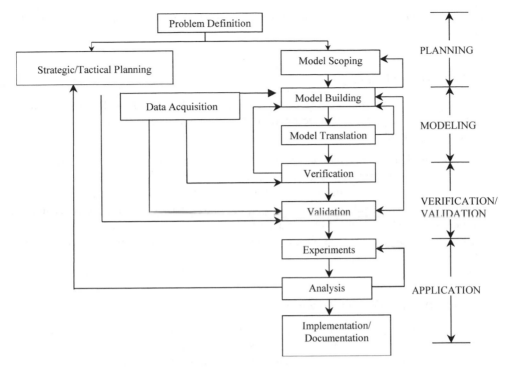

Figure 13.1 Overview of the simulation methodology [1–5]

- *Modeling phase.* In this phase, the analyst constructs a system model, which is a representation of the real system.

    - *Model building.* This includes abstraction of the system into a mathematical relationship with the problem formulation.
    - *Data acquisition.* This involves identification, specification, and collection of data.
    - *Model translation.* Preparation and debugging of the model for computer processing.

Models in general can be of different types. Among these are: (a) descriptive models; (b) physical models such as the ones used in aircraft and buildings; (c) mathematical models such as Newton's law of motion; (d) flowcharts; (e) schematics; and (f) computer pseudo code.

The major steps in model building include: (a) preliminary simulation model diagram; (b) construction and development of flow diagrams; (c) review model diagram with team; (d) initiation of data collection; (e) modify the top-down design, test and validate for the required degree of granularity; (f) complete data collection; (g) iterate through steps (e) and (g) until the final granularity has been reached; and (h) final system diagram, transformation and verification.

In the context of this phase, it is important to point out two concepts:

- *Model scooping.* This is the process of determining what process, operation, equipment, etc., within the system should be included in the simulation model, and at what level of detail.
- *Level of detail.* This is determined based on the component's effect on the stability of the

analysis. The appropriate level of detail will vary depending on the modeling and simulation goals.

13.1.1 Subsystem Modeling

When the system under simulation study is very large, a subsystem modeling is performed. All subsystem models are later linked appropriately. In order to define/identify subsystems, there are three general schemes:

- *Flow scheme*. This scheme has been used to analyze systems that are characterized by the flow of physical or information items through the system, such as pipeline computers.
- *Functional scheme*. This scheme is useful when there are no directly observable flowing entities in the system, such as manufacturing processes that do not use assembly lines.
- *State-change scheme*. This scheme is useful in systems that are characterized by a large number of interdependent relationships and that must be examined at regular intervals in order to detect state changes.

13.1.2 Variable and Parameter Estimation

This is usually done by collecting data over some period of time and then computing a frequency distribution for the desired variables. Such an analysis may help the analyst to find a well-known distribution that can represent the behavior of the system or subsystem.

13.1.3 Selection of a Programming Language/Package

Here, the analyst should decide whether to use a general-purpose programming language, a simulation language or a simulation package. In general, using a simulation package such as NS2 or Opnet may save money and time, however, it may not be flexible and effective to use simulation packages as they may not contain capabilities to do the task such as modules to simulate the protocols or features of the network under study.

13.1.4 Verification and Validation (V&V)

Verification and validation are two important tasks that should be carried out for any simulation study. They are often called V&V and many simulation journals and conferences have special sections and tracks that deal with these tasks, respectively.

Verification is the process of finding out whether the model implements the assumptions correctly. It is basically debugging the computer program (simulator) that implements the model. A verified computer program can in fact represent an invalid model; a valid model can also represent an unverified simulator.

Validation, on the other hand, refers to ensuring that the assumptions used in developing the model are reasonable in that, if correctly implemented, the model would produce results close to these observed in real systems. The process of model validation consists of validating assumptions, input parameters and distributions, and output values and conclusions. Validation can be performed by one of the following techniques: (a) comparing the results of the simulation model with results historically produced by the real system operating under the same conditions; (b) expert intuition; (c) theoretical (analytic) results using queuing theory or other analytic methods; (d) another simulation model; and (e) artificial intelligence and expert systems.

13.1.5 Applications and Experimentation

After the model has been validated and verified, it can be applied to solve the problem under investigation. Various simulation experiments should be conducted to reveal the behavior of the system under study. Keep in mind that it is through experimentation that the analyst can understand the system and make recommendations about the system design and optimum operation. The extent of experiments depends on cost to estimate performance metrics, the sensitivity of performance metrics to specific variables and the interdependencies between control variables [1,4,5].

The implementation of simulation findings into practice is an important task that is carried out after experimentation. Documentation is very important and should include a full record of the entire project activity, not just a user's guide.

The main factors that should be considered in any simulation study are: (a) Random Number Generators (RNGs); (b) Random Variates or observations (RVs); (c) programming errors; (d) specification errors; (e) length of simulation; (f) sensitivity to key parameters; (g) data collection errors in simulation; (h) optimization parameter errors; (i) incorrect design; and (j) influence of initial conditions.

The main advantages of simulation are [4,5]:

- *Flexibility*. Simulation permits controlled experimentation with the system model. Some experiments cannot be performed on the real physical system due to inconvenience, risk and cost.
- *Speed*. Using simulation allows us to find results of experiments in a speedy manner. Simulation permits time compression of a system over an extended period of time.
- Simulation modeling permits sensitivity analysis by manipulating input variables. It allows us to find the parameters that influence the simulation results. It is important to find out which simulation parameters influence performance metrics more than others as proper selection of their operating values is essential for stable operation.
- Simulation modeling involves programming, mathematics, queuing theory, statistics, system engineering and science as well as technical documentation. Clearly, it is an excellent training tool.

The main drawbacks of simulation are [4,5]:

- It may become expensive and time-consuming especially for large simulation models. This will consume long computer simulation time and manpower.
- In simulation modeling, we usually make assumptions about input variables and parameters, and distributions, and if these assumptions are not reasonable, this may affect the credibility of the analysis and the conclusions.
- When simulating large networks or systems, the time to develop the simulator (simulation program) may become long.

- It is usually difficult to initialize simulation model parameters properly and not doing so may affect the credibility of the model as well as require longer simulation time.

13.2 Simulation Models

In general, simulation models can be classified in three different dimensions [3]: (a) a static versus dynamic simulation model, where a static model is representation of a system at a particular time, or one that may be used to represent a system in which time plays no role, such as Monte Carlo models, and a dynamic simulation model represents a system as it evolves over time; (b) deterministic versus stochastic models where a deterministic model does not contain any probabilistic components while a stochastic model has at least some random input components; (c) continuous versus discrete simulation models where a discrete-event simulation is concerned with modeling of a system as it evolves over time by representation in which the state variables change instantaneously at separate points in time, usually called events. On the other hand, continuous simulation is concerned with modeling a system by a representation in which the state variables change continuously with respect to time.

In order to keep track with the current value of simulation time during any simulation study, we need a mechanism to advance simulation time from one value to another. The variable that gives the current value of simulation time is called the simulation clock. The schemes that can be used to advance the simulation clock are [1]:

- *Next-event time advance*. In this scheme, simulation clock is initialized to zero and the times of occurrences of future events are found out. Then simulation clock is advanced to the time of occurrence of the most imminent event in the future event list, then the state of the system is updated accordingly. Other future events are determined in a similar manner. This method is repeated until the stopping condition/criterion is satisfied. Figure 13.2 summarizes the next-event time advance scheme.
- *Fixed-increment time advance*. Here, simulation clock is advanced in fixed increments. After each update of the clock, a clock is made to determine if any events should have occurred during the previous fixed interval. If some events were scheduled to have occurred during this interval, then they are treated as if they occurred the end of the interval and the system state is updated accordingly.

A fixed-increment time advance scheme is not used in discrete-event simulation. This is due to the following drawbacks: (a) errors are introduced due to processing events at the end of the interval in which they occur; and (b) it is difficult to decide which event to process first when events that are not simultaneous in reality are treated as such in this scheme.

The main components that are found in most discrete-event simulation models using the next-event time advance scheme are [1–5]: (a) system state which is the collections of state variables necessary to describe the system at a particular time; (b) simulation clock which is a variable giving the current value of simulated time; (c) statistical counters which are the variables used for storing statistical information about system performance; (d) an initializing routine which is a procedure used to initialize the simulation model at time zero; (e) a timing routine which is a procedure that determines the next event from the event list and then

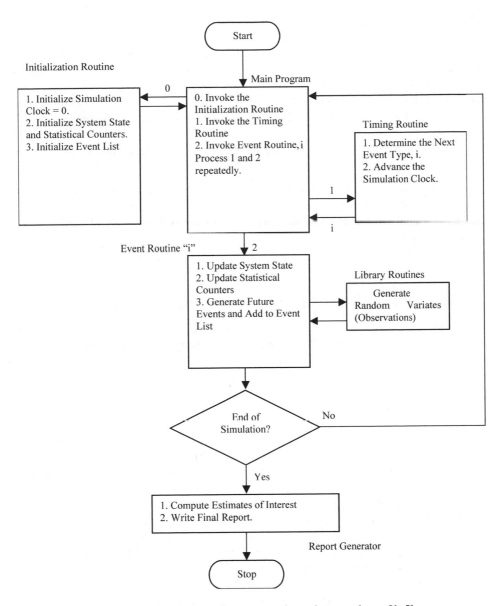

Figure 13.2 Summary of next-event time advance scheme [1–5]

advances the simulation clock to the time when that event is to occur; (f) an event routine which is a procedure that updates the system state when a particular type of event occurs; (g) library routines that are a set of subprograms used to generate random observations from probability distributions; (h) a report generator which is a procedure that computes estimates of the desired measures of performance and produces a report when the simulation ends; and (i) the main program which is a procedure that invokes the timing routine in order to determine the next event and then transfers control to the corresponding event routine to properly update the system state, checks for termination and invokes the report generator when the conditions for terminating the simulation are satisfied.

Simulation begins at time 0 with the main program invoking the initialization routine, where the simulation clock is initialized to zero, the system state and statistical counters are initialized, as well as the event list. After control has been returned to the main program, it invokes the timing routine to find out the most eminent routine. If event i is the most eminent one, then simulation clock is advanced to the time that this event will occur and control is returned to the main program.

The available programming languages/packages for simulating computers and network systems are:

- General purpose languages such as C, C++, Java, Fortran, and Visual Basic.
- Special simulation languages such as Simscript II.5, GPSS, GASP IV, CSIM, Modsim III.
- Special simulation packages such as Comnet III, Network II.5, OPNet, QNAP, Network Simulation 2 (NS-2).

13.3 Common Probability Distributions Used in Simulation

The basic logic used for extracting random values from probability distribution is based on a Cumulative Distribution Function (CDF) and a Random Number Generator (RNG). The CDF has Y values that range from 0 to 1. RNGs produce a set of numbers which are uniformly distributed across this interval. For every Y value there exists a unique random variate value, X, that can be calculated.

All commercial simulation packages do not require the simulationist to write a program to generate random variates or observations. The coding is already contained in the package using special statements. In such a case, a model builder simply: (a) selects a probability distribution from which he desires random variates; (b) specifies the input parameters for the distribution; and (c) designates a random number stream to be used with the distribution.

Standard probability distributions are usually perceived in terms of the forms produced by their Probability Density Functions (pdf). Many probability density functions have parameters that control their shape and scale characteristics. There are several standard continuous and discrete probability distributions that are frequently used with simulation. Examples of these are: the exponential, gamma, normal, uniform continuous and discrete, triangular, Erlang, Poisson, binomial, Weibull, etc. Standard probability distributions are used to represent empirical data distributions. The use of one standard distribution over the other is dependent on the empirical data that it is representing, or the type of stochastic process that is being modeled. It is essential to understand the key characteristics and typical applications of the standard probability distributions as this helps analysts to find a representative distribution for empirical data and for processes where no historical data are available. Next is a brief review of the main characteristics of the most often used probability distributions for simulation [1–4].

- *Bernoulli distribution.* This is considered the simplest discrete distribution. A Bernoulli variate can take only two values, which are denoted as failure and success, or $x = 0$ and

$x = 1$, respectively. If p represents the probability of success, then $q = 1 - p$ is the probability of failure. The experiments to generate a Bernoulli variate are called Bernoulli trials. This distribution is used to model the probability of an outcome having a desired class or characteristic; for example, a packet in a computer network reaches or does not reach the destination, and a bit in a packet is affected by noise and arrives in error. The Bernoulli distribution and its derivative can be used only if the trials are independent and identical.

- *Discrete uniform.* This distribution can be used to represent random occurrence with several possible outcomes. A Bernoulli (1/2) and Discrete Uniform (DU) $(0, 1)$ are the same.

- *Uniform distribution (continuous).* This distribution is also called the rectangular distribution. It is considered one of the simplest distributions to use. It is commonly used if a random variable is bounded and no further information is available. Examples include: distance between source and destination of message on a network, and seek time on a disk. In order to generate a continuous uniform distribution, $U(a, b)$, you need to: generate $u \sim U(0, 1)$ and return $a = (b - a)u$. The key parameters are: $a = $ lower limit and $b = $ upper limit, where $b > a$. The continuous uniform distribution is used as a 'first' model for a quantity that is felt to be randomly varying between two bonds a and b, but about which little else is known.

- *Exponential distribution.* This is considered the only continuous distribution with memoryless property. It is very popular among performance evaluation analysts who work in simulation of computer systems and networks as well as telecommunications. It is often used to model the time interval between events that occur according to the Poisson process.

- *Geometric distribution.* This is the discrete analog of the exponential distribution and is usually used to represent the number of failures before the first success in a sequence of Bernoulli trials such as the number of items inspected before finding the first defective item.

- *Poisson distribution.* This is a very popular distribution in queuing, including telephone systems. It can be used to model the number of arrivals over a given interval such as the number of queries to a database system over a duration, t, or the number of requests to a server in a given duration of time, t. This distribution has a special relation with the exponential distribution.

- *Binomial distribution.* This distribution can be used to represent the number of successes in t independent Bernoulli trials with probability p of success on each trial. Examples include the number of nodes in a multiprocessor computer system that are active (up), the number of bits in a packet or cell that are not affected by noise or distortion, and the number of packets that reach the destination node with no loss.

- *Negative binomial.* It is used to model the number of failures in a system before reaching the kth success such as the number of retransmissions of a message that consists of k packets or cells and the number of error-free bytes received on a noisy channel before the k in-error bytes.

- *Gamma distribution.* Similar to the exponential distribution, this is used in queuing modeling of all kinds, such as modeling service times of devices in a network.

- *Weibull Distribution.* In general, this distribution is used to model lifetimes of components such as memory or microprocessor chips used in computer and telecommunications systems. It can also be used to model fatigue failure and ball bearing failure. It is considered the most widely used distribution to represent failure of all types. It is interesting to point out that the exponential distribution is a special case of the Weibull distribution when the shape parameter α is equal to 1.

- *Normal or Gaussian distribution*. This is also called the bell distribution. It is used to model errors of any type including modeling errors and instrumentation errors. Also, it has been found that during the wearout phase, component lifetime follows a normal distribution. A normal distribution with zero mean and a standard deviation of 1 is called standard normal distribution or a unit normal distribution. It is interesting to note that the sum of large uniform variates has a normal distribution. This latter characteristic is used to generate the normal variate, among other techniques such as the rejection and Polar techniques. This distribution is very important in statistical applications due to the central limit theorem, which states that under general assumptions, the mean of a sample of n mutually independent random variables, that have distribution with finite mean and variance, is normally distributed in the limit $n \rightarrow \infty$.
- *Lognormal distribution*: The log of a normal variate has a distribution called lognormal distribution. This distribution is used to model errors that are a product of effects of a large number of factors. The product of a large number of positive random variates tends to have a distribution that can be approximated by lognormal.
- *Triangle distribution*. As the name indicates, the pdf of this distribution is specified by three parameters (a, b, c) that define the coordinates of the vertices of a triangle. It can be used as a rough model in the absence of data.
- *Erlang distribution*. This distribution is usually used in queuing models. It is used to model service times in a queuing network system as well as to model the time to repair and time between failures.
- *Beta distribution*. This distribution is used when there is no data about the system under study. Examples include the fraction of packets or cells that need to be retransmitted.
- *Chi-square distribution*. This was discovered by Karl Pearson in 1900 who used the symbol χ^2 for the sum. Since then statisticians have referred to it as the chi-square distribution. In general, it is used whenever a sum of squares of normal variables is involved. Examples include modeling the sample variances.
- *Student's distribution*. This was derived by Gosset who was working for a winery whose owner did not appreciate his research. In order not to let his supervisor know about his discovery, he published his findings in a paper under the pseudonym student. He used the symbol t to represent the variable and hence the distribution was called the 'student's t distribution'. It can be used whenever a ratio of normal variate and the square root of chi-square variable is involved and is commonly used in setting confidence intervals and in t-tests in statistics.
- *F-Distribution*. This distribution is used in hypothesis testing. It can be generated from the ratio of two chi-square variates. Among its applications is to model the ratio of sample variances as in the F-test for regression and analysis of variances.
- *Pareto distribution*. This is also called the double-exponential distribution, the hyperbolic distribution, and the power-law distribution. It can be used to model the amount of CPU time consumed by an arbitrary process, the web file size on an Internet server, and the number of data bytes in File Transfer Protocol (FTP) bursts [8].

13.4 Random Number Generation

In order to conduct any stochastic simulation, we need pseudorandom number sequences that are generated using pseudorandom generators. The latter are often called Random Number Generators (RNGs). The majority of programming languages have subroutines, objects, or functions that generate random number sequences. The main requirements of a random number generator are: (a) numbers produced must follow the uniform distribution, since truly random events follow it; (b) the sequence of generated random numbers produced must be reproducible (replicable) as long as the same seed is used, which permits replication of simulation runs and facilitates debugging; (c) routines used to generate random numbers must be fast and computationally efficient; (d) the routines should be portable to different computer platforms and preferably to different programming languages; (e) numbers produced must be statistically independent; (f) ideally, the sequence produced must be nonrepeating for any desired length, however, this is impractical as the period must be very long; (g) the technique used should not require large memory space; and (h) the period of the generated random sequences must be sufficiently long before repeating themselves

The goal of a RNG is to generate a sequence of numbers between 0 and 1 which imitates the ideal characteristics of uniform distribution and independence as closely as possible. There are special tests that can be used to find out whether the generation scheme has departed from the above goal. There are RNGs that have passed all available tests, therefore these are recommended for use.

The algorithms that can be used to generate pseudorandom numbers are [1–4]: (a) linear congruential generators; (b) midsquare technique; (c) Tausworthe technique; (d) extended Fibonacci technique; and (e) combined technique.

13.4.1 Linear-Congruential Generators (LCG)

This is a widely used scheme to generate random number sequences. This technique was initially proposed by Lehmer in 1951. In this technique, successive numbers in the sequence are generated by the recursion relation:

$$X_{n+1} = (aX_n + b) \bmod m, \qquad \text{for } n \geq 0$$

where m is the modulus, a is the multiplier, and b is the increment. The initial value X_0 is often called the seed. If $b \neq 0$, the form is called the mixed congruential technique. However, when $b = 0$, the form is called the multiplicative congruential technique. It should be stated that the values of a, b, and m drastically affect the statistical characteristics and period of the RNG.

Moreover, the choice of m affects the characteristics of the generated sequence.

- *Multiplicative LCG with $m = 2^k$.* This choice of m, provides an easy mode of operation. However, such generators do not have a full period as the maximum period for multiplicative LCG with modulus $m = 2^k$ is only 1/4th of the full period, that is, $2^k - 2$. This period is achieved if the multiplier 'a' is of the form $8i \pm 3$ and the initial seed is an odd integer.
- *Multiplicative LCG with $m \neq 2^k$.* In order to increase the period of the generated sequence, the modulus m is chosen to be a prime number. A proper choice of the multiplier 'a', can

give us a period of $(m - 1)$, which is almost equal to the maximum possible length m. Note that unlike the mixed LCG, X_n obtained from a multiplicative LCG can never be zero if m is prime. The values of X_n lie between 1 and $(m - 1)$, and any multiplicative LCG with a period of $(m - 1)$ is called a full-period generator.

13.4.2 Midsquare Method

This method was developed by John Von Neumann in 1940. The scheme relies on the following steps: (a) start with a seed value and square it; (b) use the middle digits of this square as the second number in the sequence; (c) this second number is then squared and the middle digits of this square are used as the third number of the sequence; and then (d) repeat steps (a) to (d). Although this scheme is very simple, it has important drawbacks: (a) short repeatability periods; (b) numbers produced may not pass randomness tests; and (c) if a 0 is generated then all other numbers generated will be 0. The latter problem may become very serious.

13.4.3 Tausworthe Method

This technique was developed by Tausworthe in 1965. The general form is:

$$b_n = (C_{q-1}b_{n-1}) \text{ XOR } (C_{q-2}b_{n-2}) \text{ XOR } \cdots \text{ XOR } (C_0 b_{n-q})$$

where c_i and b_i are binary variables. The Tausworthe generator uses the last q bits of a sequence. It can easily be implemented by hardware using Linear Feedback Shift Registers (LFSRs).

13.4.4 Extended Fibonacci Method

A Fibonacci sequence is generated by

$$X_n + X_{n-1} + X_{n-2}$$

A Fibonnacci series is modified in order to generate random numbers. The modification is

$$X_n + X_{n-1} + X_{n-2} \bmod m$$

The random number sequences generated using this technique do not have good randomness properties, especially the fact that they have serial correlation.

The seed value in general should not affect the characteristics of the random sequence generated. However, some seed values may affect the randomness characteristics of the RNG. In general, good random number generators should produce good characteristics regardless of the seed value. However, some RNGs may produce shorter sequences and inadequate randomness characteristics if their seed values are not selected carefully. Below are recommended guidelines that should be followed when selecting the seed of a random number generator [4]:

- Avoid using zero. Although a zero seed may be fine for mixed LCGs, it would make a multiplicative LCG or Tausworthe generator stick at zero.
- Do not use even values. If the generator is not a full-period generator such as multiplicative LCG with modulus $m = 2^k$, the seed should be odd. For other cases even values are often as good as odd values. Avoid generators that have too many restrictions.
- Never subdivide one stream. A common mistake is to use a single stream for all variables. For example, if $(r_1, r_2, r_3, \ldots)$ is the sequence generated using a single seed r_0, the analyst may for example, use r_1 to generate interarrival times, r_2 to generate service times, and so forth. This may result in a strong correlation between the two variables.
- Do not use overlapping streams. Each stream of random numbers that is used to generate a specific even should have a separate seed value. If the two seeds are such that the two streams overlap, there will be a correlation between the streams, and the resulting sequence will not be independent. This will lead to misleading conclusions and wrong simulation results.
- Make sure to reuse seeds in successive replication. When a simulation experiment is replicated several times, the random number stream need not be reinitialized, and the seeds left over from the previous replication can continue to be used.
- Never use random seeds. Some simulation analysts think that using random seeds, such as the time of the day, or current date, will give them good randomness characteristics. This is untrue as it may cause the simulation not to be reproduced. Also, multiple streams may overlap. Random seed selection is not recommended. Moreover, using successive random numbers obtained from the generator as seeds is also not recommended.

13.5 Testing Random Number Generators

The desirable properties of a random number sequence are uniformity, independence and long period. In order to make sure that the random sequence generated from a RNG have these properties, a number of tests have been established. It is important to stress that a good RNG should pass all available tests.

The process of testing and validating pseudorandom sequences involves the comparison of the sequence with what would be expected from the uniform distribution. The major techniques for doing so are [1–4,8]: (a) the chi-square (frequency) test; (b) the Kolmogorov–Smirnov (K-S) test; (c) the serial test; (d) the spectral test; and (e) the poker test.

A brief description of these techniques is given below:

- *Chi-square test.* This test is general and can be used for any distribution. It can be used to test random numbers that are independent and identically uniformly distributed between 0 and 1 and for testing random variate generators. The procedure can be summarized as follows:

  1. Prepare a histogram of observed data.
  2. Compare observed frequencies with those obtained from the specified density function.
  3. For k cells, let O_i = observed frequencies, E_i = expected frequencies, then $D =$ Difference $= \sum (O_i - E_i)^2 / E_i$.
  4. For an exact fit D should be 0.
  5. D can be shown to have a chi-square distribution with $(K - 1)$ degrees of freedom,

where k is the number of cells (classes or clusters). We use the significance level α for not rejecting or the confidence level $(1 - \alpha)$ for accepting.

- *Kolmogorov–Smirnov (K-S) test.* This test compares an empirical distribution function with the distribution function, F, of the hypothesized distribution. It does not require grouping/clustering of data into cells as in the chi-square test. Moreover, the K-S test is exact for any sample size, n, while the chi-square test is only valid in an asymptotic sense. The K-S test compares distribution of the set of numbers to a theoretical (uniform) distribution. The unit interval is divided into subintervals and CDF of the sequence of numbers is calculated up to the end of each subinterval. By comparing Ks with those listed in special tables, we can determine if observations are uniformly distributed. In this test, the numbers are normalized and sorted in increasing order. If the sorted numbers are: $X_1, X_2, \ldots X_n$ such that $X_{n-1} \leq X_n$. Then two factors called $K+$ and $K-$ are calculated as follows:

$$K+ = (n)^{0.5} \max\left(\frac{j}{n} - X_j\right)$$

$$K- = (n)^{0.5} \max\left(X_j - \frac{j-1}{n}\right)$$

where n is the number of numbers tested and j is the order of the number under test. If the values of $K-$ and $K+$ are smaller than $K_{[1-\alpha,n]}$ listed in the K-S tables, the observations are said to come from the specified distribution at the α level of significance.
- *Serial test.* This test measures the degree of randomness between successive numbers in a sequence. The procedure relies on generating a sequence of M consecutive sets of N random numbers each. Then the numbers range is partitioned into K intervals. For each group, construct an array of size $(K \times K)$. The array is initialized by zeros. Then, the sequence of numbers is examined from left to right, pairwise. If the left number of a pair is in interval i, while the right number is in interval j, increment the (i,j) element by 1. After this, the final results of M groups are compared with each other and with expected values using the chi-square test.
- *Spectral test.* This test is used to check for a flat spectrum by checking the observed estimated cumulative spectral density function with the K-S test. Basically, it measures the independence of adjacent sets of numbers.
- *Poker test.* The poker test treats the random numbers grouped together as a poker hand. The hands obtained are compared with what is expected using the chi-square technique. For more detailed information on this, see Refs. [1–6,8]

13.6 Random Variate Generation

Random number generators are used to generate sequences of numbers that follow the uniform distribution. However, in simulation we encounter other important distributions such as exponential, Poisson, normal, gamma, Weibull, beta, etc. In general, most simulation analysts use existing programming library routines or special routines built into simulation languages. However, some programming languages do not have built-in procedures for all distributions. Therefore, it is essential that the simulation analysts understand the techniques used to generate random variates (observations). All methods to be discussed start by generating one or more pseudorandom number sequences from the uniform distribution. Then a transform is applied to this uniform variate to generate the nonuniform variates. The main techniques used to generate random variates are as follows [1,4,8].

13.6.1 The Inverse Transformation Technique

This technique can be used to sample from exponential, uniform, Weibull and triangle distributions as well as empirical distributions. Moreover, it is considered the main technique for sampling from a wide variety of discrete distributions.

In general, this technique is useful for transforming a standard uniform deviate into any other distribution. The density function $f(x)$ should be integrated to find the cumulative density function $F(x)$, or $F(x)$ is an empirical distribution. The scheme is based on the observation that given any random variable x with Cumulative Distribution Function (CDF), $F(x)$, the variable $u = F(x)$ is uniformly distributed between 0 and 1. We can obtain x by generating uniform random numbers and computing: $X = F^{-1}(U)$.

Example The Exponential Distribution The Probability Density Function (pdf) is given by

$$f(x) = \begin{cases} \lambda e^{-\lambda x}, & x \geq 0 \\ 0, & x < 0 \end{cases}$$

The Cumulative Distribution Function (CDF) is given by

$$F(x) = \begin{cases} 1 - e^{-\lambda x}, & x > 0 \\ 0, & x < 0 \end{cases}$$

The parameter λ can be interpreted as the average number of arrivals (occurrences) per unit time. It is equal to $\lambda = 1/\beta$ where β is interpreted as the mean interarrival time. We set the CDF $= U = 1 - e^{-\lambda x}$ where U is uniformly distributed between 0 and 1.

$$X = -\frac{1}{\lambda} \ln(1 - U)$$

Since U and $(1 - U)$ are both uniformly distributed between 0 and 1, we will use U instead of $(1 - U)$ in order to reduce the computational complexity. Thus,

$$X = -\frac{1}{\lambda} \ln U$$

which is the required exponential random variate. The last expression is easy to implement using any programming language.

13.6.2 Rejection Method

This method is also called the acceptance-rejection technique. Its efficiency depends upon being able to minimize the number of rejections. Among the distributions that can be generated using this technique are the Poisson, gamma, beta and binomial distributions with large N. The basis for this scheme is that the probability of r being $\leq bf(x)$ is $bf(x)$ itself. That is

$$\text{Prob}[r \leq bf(x)] = bf(x)$$

where r is the standard uniform number. If x is generated randomly in the interval (c, d), and x is rejected if $r > bf(x)$, then the accepted x's will satisfy the density function $f(x)$. In order to use this scheme, $f(x)$ has to be bounded and x valid over some range $(c \leq x \leq d)$. The steps to be taken are: (1) normalize the range of $f(x)$ such that $bf(x) \leq 1$, $c \leq x \leq d$; (2) define x as a uniform continuous random variable $x = c + (d - c)r$; (3) generate a pair of random variables (k_1, k_2); (4) if the pair satisfies the property $k_2 \leq bf(x)$, then set the random deviate to $x = c + (d - c)k_1$; (5) if the test in the previous step fails, return to step 3 and repeat steps 3 and 4.

13.6.3 Composition Technique

This technique is used when the required cumulative distribution function F can be expressed as a combination of other distributions such as $F_1, F_2, F_3, F_4, \ldots$. The goal is to be able to sample from F_i more easily than F.

$$F(x) = \sum p_i F_i(x)$$

Moreover, this technique can be used if the probability density function $f(x)$ can be expressed as a weighted sum of other probability density functions.

$$f(x) = \sum p_i f_i(x)$$

In both cases, the steps used for generation are basically the same: (a) generate a positive random integer i such that $P(I = i) = p_i$ for $i = 1,2,3,\ldots$, which can be implemented using the inverse transformation scheme; (b) return X with the ith CDF $F_i(x)$. The composition technique can be used to generate the Laplace (double-exponential), and the right-trapezoidal distributions.

13.6.4 Convolution Technique

In many cases, the required random variable X can be expressed as a sum of other n random variables (Y_i) that are independent and identically distributed (IID). In other words,

$$X = Y_1 + Y_2 + \cdots + Y_n$$

Therefore, X can be generated by simply generating n random variates Y_i and then summing them. It is important to point out the difference between the convolution and composition techniques. In the convolution scheme, the random variable X itself can be expressed as a sum of other random variables whereas in the composition scheme, the distribution function of X is a weighted sum of other distribution functions. Clearly, there is a fundamental difference between the two schemes.

The algorithm used here is quite intuitive. If X is the sum of two random variables Y_1 and Y_2, then the pdf of X can be obtained by a convolution of the pdfs of Y_1 and Y_2. This is why this method is called the 'Convolution Method.'

The convolution technique can be used to generate the Erlang, binomial, Pascal (sum of m geometric variates), and triangular (sum of two uniform variates) distributions.

13.6.5 Characterization Technique

This method relies on special characteristics of some distributions such as the relationship between Poisson and exponential distributions. Such characteristics allow variates to be generated using algorithms tailored for them.

If the interarrival times of a process are exponentially distributed with mean $1/\lambda$ then the number of arrivals n over a given period T has a Poisson distribution with parameter λT. This means that a Poisson variate can be obtained by continuously generating exponential variates until their sum exceeds T and returning the number of variates generated as a Poisson variate.

13.7 Case Studies

This section presents examples on the simulation of wireless networks. These are the simulation of an IEEE 802.11 wireless LAN, simulation analysis of QoS in IEEE 802.11 WLAN system, simulation comparison of the TRAP and RAP wireless LANs protocols, and Simulation of Topology Broadcast Based on Reverse-Path Forwarding (TBRPF) protocol using an 802.11 WLAN-based Mobile Ad hoc NETwork (MANET) model.

13.7.1 Example 1: Performance Evaluation of IEEE 802.11 WLAN Configurations using Simulation

In this example, we used simulation modeling to evaluate the performance of wireless LANs under different configurations. In wireless LANs, as opposed to wired LANs, different transmission results can be observed for different transmission rates due to radio propagation characteristics where the signal decay is far greater than on cables. This leads to new and interesting operational and modeling phenomena and issues such as hidden node, capture effect and spatial reuse.

We used the network simulation (ns) package for this task. *ns* is not a visualization tool and is not a Graphical User Interface (GUI) either. It is basically an extension of *oTcl* (*Object Tcl*); therefore it looks more like a scripting language which can output some trace files [11,12]. However, a companion component called *nam* (for Network AniMator) allows the user to have a graphical output. *ns* can simulate: (1) topology: wired, wireless, and satellite; (2) scheduling/dropping algorithms: FIFO, Drop Tail, RED, SFQ, CBQ, etc.; (3) transport protocols: TCP (all flavors) and UDP; (4) routing algorithms: static and dynamic routing, and MPLS; (5) applications: FTP, HTTP (web-caching), Telnet, and traffic generators based on probabilistic distributions (CBR, Pareto, exponential); (6) multicast traffic and routing algorithms; (7) various error models for link failures. *ns* uses $C++$ for per-packet action (TCP implementations, for instance) and *oTcl* for control (topology, scenario design).

In this case study, we present the performance evaluations of IEEE 802.11 standard/Direct Sequence (DS) using simulation modeling with a transmission rate of 2, 5 and 11 Mbps. The model used is an optimized model for the IEEE 802.11 MAC scheme. This optimization tries to maximize the speed of the simulation and will sometimes lead to a slight simplification or approximation in the modeling [13].

We considered the cases of having 2, 5, 10, 15 or 20 nodes in the WLAN system with data rates of 2, 5 and 11 Mbps. The traffic is generated with large packets of size 150 bytes (12,000

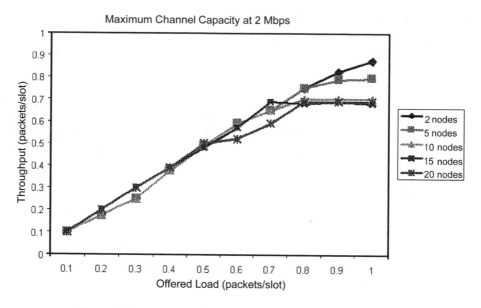

Figure 13.3   Throughput versus offered load for a 2 Mbps WLAN

bits) and the network was simulated for different load conditions with a load ranging from 10% to 100% of the channel capacity. The resulting simulation allows us to find out the maximum channel capacity of the IEEE 802.11 standard. The results are given in Figures 13.3–13.5.

As shown in the figures, the channel throughput decreases as the number of nodes increases. This is a general result of the CSMA scheme. We can also see that the normalized

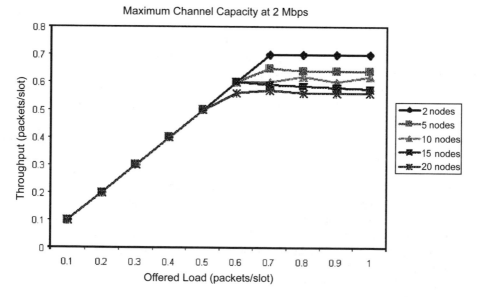

Figure 13.4   Throughput versus offered load for a 5 Mbps WLAN

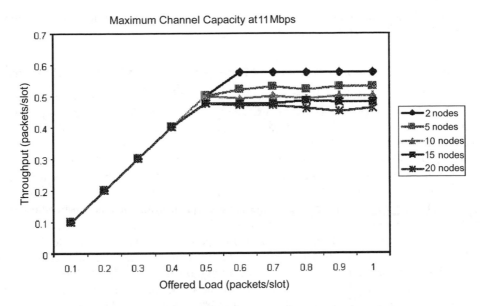

Figure 13.5   Throughput versus offered load for a 1 Mbps WLAN

channel throughput decreases as the data transmission rate increases. This phenomenon can be explained by the fixed overhead in the frames.

The broadcast mode of operation was also evaluated using simulation modeling. Basically, we studied the scenario where all the nodes send a broadcast traffic and investigated the success rate of these transmissions. The results are shown in Figure 13.6.

The simulation results show that the collision rate is more than 10% for a load greater than

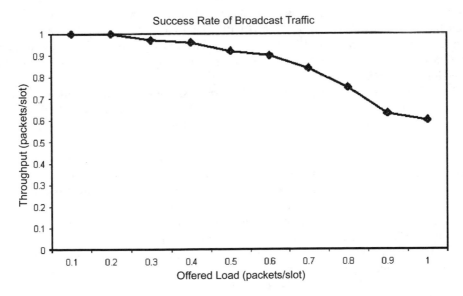

Figure 13.6   Throughput versus offered load in the broadcast mode of operation with ten stations

50% of the channel capacity. This poor performance for broadcast traffic is a well-known issue for IEEE 802.11 WLAN standard.

13.7.2 Example 2: Simulation Analysis of the QoS in IEEE 802.11 WLAN System

This study deals with the analysis of quality of service (QoS) in the IEEE 802.11 Wireless Local Area Network (WLAN). We analyze multimedia traffic in a WLAN with special emphasis on the Quality of Service (QoS). The analysis is based on a simulation of video, voice and data traffic in a WLAN environment.

The usability of the wireless medium to send voice and video packets has been studied in this example. Although IEEE 802.11 has been deployed predominantly for data, it has not been used to transmit voice traffic, primarily because of the infancy of voice packet and the requirements that voice traffic places on the network [14]. Data packets can handle larger delays than voice and video packets, which allow them to be buffered and transmitted in a best effort manner. On the other hand, multimedia packets are more sensitive to delay. To achieve good quality of voice/video service, at least 99% of the packets should arrive within 200 ms. This 200 ms refers to the complete delay in the path including the Internet routing delays. We simulated a data combined multimedia network operating in a WLAN environment. The network has 20 wireless stations capable of transferring data, voice and video. Simulation analysis was conducted to understand the effect of increasing the network load on the number of voice media channels that can be supported.

The IEEE 802.11 standard covers the Medium Access Control (MAC) sublayer and the physical layer of the OSI reference model. Here, we focus on the MAC layer. The 802.11 MAC algorithms are also called the Distributed Foundation Wireless MAC (DFWMAC), which provide a distributed access control mechanism with an optional control on top of that. Figure 13.7 illustrates the MAC architecture – a general description of this architecture is available in Refs. [14–17]. This protocol supports two types of services: Distributed Coordination Function (DCF) and the Point Coordination Function (PCF). The lower sublayer is called the Distributed Control Function (DCF). The DCF uses a contention algorithm to provide access to all traffic and it is often used. The top layer is the Point Coordination Function (PCF); it is an optional layer and uses a centralized MAC algorithm to provide contention free service.

The DCF uses a simple Carrier Sense Multiple Access with Collision Avoidance (CSMA-CA) algorithm. When a station has MAC frames to transmit, it listens to the medium, and if the medium is idle for a duration greater than the Distributed coordination Function Interframe Space (DIFS) then the packet is sent. During the transmission of the packet, the medium is busy for the transmission time of the packet, which depends on the packet length and the medium bandwidth. Once the current packet transmission is complete, only then can the next packet be sent. If a station wants to send data and if the medium is busy, then the station enters a back off period where it polls the medium. If the medium is idle for a period of 'DIFS time', it decrements its back off counter. When the back off counter reaches zero, the station again tries to send the packet. If it is not successful, it doubles the back off counter value and restarts the process. If the back off value reaches a maximum value then the packet is dropped and a message is sent to the upper layer indicating dropping of the packet. Figure 13.8 describes the DCF access mechanism [17].

In the back off mechanism, the counters are set to a random number between 0 and the

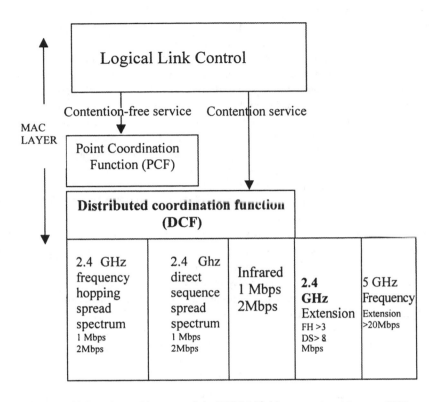

Figure 13.7   The architecture of the IEEE 802.11 protocol architecture [19]

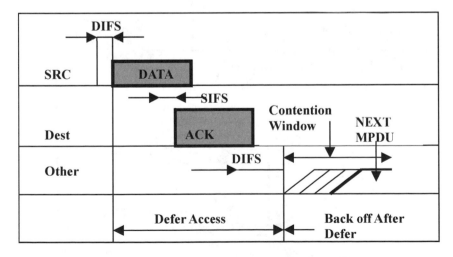

Figure 13.8   Basic DCF access mechanism [18]

minimum back off counter (C_{wmin}, value = 31) [15,16]. If the medium is idle for DIFS duration then the counter is decremented. Once the counter goes to 0 then the packet is resent. If another collision occurs then the counter value is doubled ($2 \times C_{wmin}$) and a random number is chosen between 0 and ($2 \times C_{wmin}$). The same process is repeated until the value is incremented to the maximum back off counter (C_{wmax}, value = 1023). The value stays at C_{wmax} and if the packet is not sent within the 'packet lifetime,' (a fixed value chosen for the packets' lifetime), then the packet is discarded and an error message is sent to the upper layer.

Next, we describe the simulation model. Data packets can handle larger delays than voice/video packets. Therefore they can be buffered and transmitted in a best effort manner. On the other hand, voice/video packets are more sensitive to delay. To achieve good quality of voice service, at least 99% of the voice packets should arrive within 200 ms [3]. This 200 ms refers to the complete delay in the path including the Internet routing delays. Since other switching delays can also contribute to the delay, it was assumed that the WLAN MAC delay should be less than 100 ms. Figure 13.9 shows the DCF state machine.

We have simulated a data combined with voice network and data combined with video network operating in a WLAN environment. The network has 20 wireless stations transferring data and numerous voice/video stations are also connected. Simulations were done to understand the effect of increasing the network data load on the quality of the voice/video channels and the number of voice/video channels that can be supported. Increasing the data load on the network causes more contention in the network, which causes greater delays to the packets transmitted thereby reducing the quality of the voice traffic.

We also simulated the effect of increasing the load and its effect on the quality of the voice/

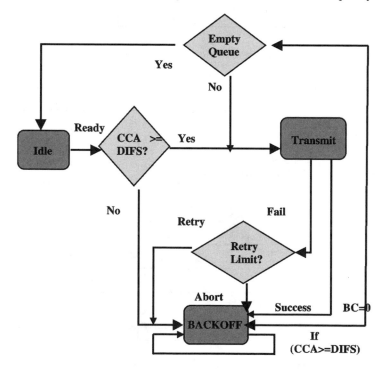

Figure 13.9   DCF state machine [18]

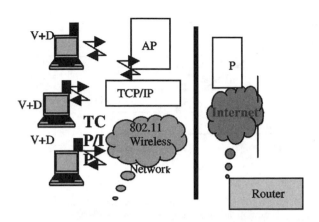

Figure 13.10 A BSS including data, voice and video stations [20]

video traffic. The quality of the voice/video traffic was measured by the percentage of voice/ video packets, which have a delay greater than N ms ($N = 50$ and 100). The network manager can, therefore, easily trade off QoS with the number of wireless stations connected.

This simulation was based on stations transmitting and receiving data at 11 Mbps. The voice/video stations are also assumed to be operating at 11 Mbps. Figure 13.10 shows a BSS including data, voice and video stations [20]. We considered a network that consists of 20 data stations in a Basic Service Set (BSS). The data packet length is 3200 bits, the voice packet length is 2016 bits while the video packet length is 11,680 bits. The back off period was used as described in the IEEE 802.11b standard. The minimum back off counter was chosen to be 31 and the maximum back off counter was chosen to be 1023. The data packets were generated according to the uniform distribution. All data packets were assumed to be of the same length. The same was assumed about the voice/video packets. Voice traffic was carried using the User Datagram Protocol (UDP). Each UDP packet was formed out of 252 bytes of speech samples, an 8-byte UDP header, 20-byte IP header and a 14- byte Data Link Control (DLC) header appended to it [20]. Voice packets were modeled as arriving at the destination at a rate of one every 30 ms. Real player uses G2 protocol, which generates packets of payload of 1460 bytes. The packets were generated every 21.5 ms. As real player server uses HTTP protocol for transmission, the TCP, IP and DLC overheads are also added to the payload. The final packet of size 1542 bytes was sent every 21.5 ms. The data packet arrival rate depends on the load of the network. The load of the network is assumed to be 100% for 11 Mbps throughput. Therefore, a load of 10% means that 1.1 Mbps of data throughput is being consumed by the data stations. As the load of the network increases, the data packet arrival rate also increases. The arrival rate is given by

$$\text{Arrival rate} = \frac{\text{WLAN rate} \times \text{load}}{\text{Data Packet length} \times 8 \times \text{number of data stations}}$$

where the mean arrival time $= 1/\text{arrival rate}$. The actual arrival time was chosen in the interval between (0, mean arrival time $\times 2$).

A uniform random number generated the arrival time of the data packets. When collision occurs on the network, each station enters a back off period during which it polls the physical medium. The value for back off was chosen between (0, back off counter). After the back off

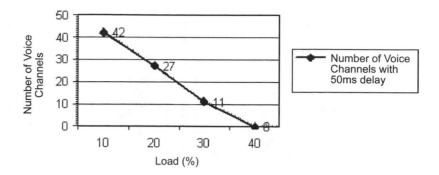

Figure 13.11   Number of voice channels versus load for a delay of 50 ms

duration has expired the station tries to retransmit, if the retransmission fails then the station doubles the back off duration and reenters back off. This process continues until either the packet is transmitted or the lifetime of the packet has exceeded. The lifetime of the packet refers to the time duration within which the packet should reach the destination. When the packet is not transmitted, an error message is sent to the upper layer. This simulation assumes that a short preamble mechanism is used for MAC transmission. In this mechanism, the short preamble is sent at 1 Mbps followed by the Physical Layer Convergence Procedure (PLCP) header. The transmission time for a MAC packet in the simulation also includes the DIFS time out and the ACK request, which is sent after a time interval of SIFS (short interframe space). We have DIFS duration = 50 μs, transmission time for PCLP preamble = 72/1,000,000 s, transmission time for PLCP header = 48/2,000,000 s, transmission time for data = packet length/transmission rate, transmission time for medium = 200 m/200 m/μs, SIFS duration = 10 μs and ACK transmission time = (72 μs + 24 μs + 14)/transmission rate. Thus, total transmission time = {50 + 2(72 + 24 + 1) + 10 + (packet length + 14)}/transmission rate. Load of network = bandwidth used by data stations)/(11 Mbps) × 100%.The simulation program was run for 20 s and various simulation experiments were conducted to reveal some of the characteristics and behavior of the network.

The wireless medium is a shared medium. It is used by both TCP/IP traffic and voice and video traffic. TCP/IP is more immune to delays and jitter, and is also called the 'background load' on the network. This load is expressed as a percentage of the available data rate (11 Mbps). As the background load increases, the ability of the 802.11 MAC to allow high priority traffic like voice reduces. Figure 13.11 shows the effect of increasing the load on the network, on the number of voice channels, which can be supported. The delay allowed in this experiment was 50 ms. With all other parameters remaining constant, we can see that the number of simultaneous voice lines which can be supported decreases from 42 for a lightly loaded network (with 10% load) to 0 for a heavily loaded network. As the medium utilization of the network depends on the number of data and voice stations connected to an access point, i.e. the BSS size, reducing the BSS size would increase the number of simultaneous voice lines. Decreasing the BSS size has other issues, namely more cost and co-channel interference.

The requirement for voice is to have a total end-to-end delay of less than 200 ms. As each part of the network contributes to the delay, it is necessary to limit the delay in each leg of the network. To achieve a better quality of voice medium a multitier QoS has been proposed

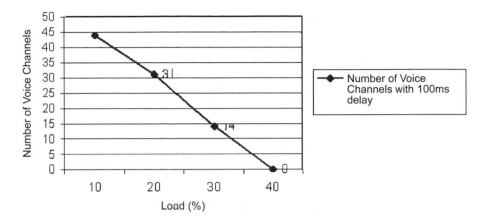

Figure 13.12   Number of voice channels versus load for a delay of 100 ms

which will allow the network manager to trade quality of voice calls with the number of voice calls. In Figure 13.12 we see how relaxing the requirement of delay and increasing it to 100 ms increases the number of voice channels that can be supported.

Thus, to allow improved QoS for voice channels, suitable QoS modifications are necessary according to the 802.11e specification. Proposals have been submitted to the IEEE 802.11 working group E for improvements to the DCF block.

Figure 13.13 shows the number of voice channels versus the percentage of voice packets below 100 ms. Figure 13.14 shows a comparison of the number of video channels supported to the background load for the desired QoS of 100 ms. As the background load of the wireless network increases, the number of simultaneous video streams that can be supported reduces. At 40% of background load, no video streams can be supported.

If the maximum packet delay is reduced to 50 ms, then the number of simultaneous video channels supported does not change substantially. The plot in Figure 13.15 shows the drop in quality as the number of video channels increases. The background load is held constant at 35%. The quality of channel for six simultaneous video streams is so bad that only 16% of the

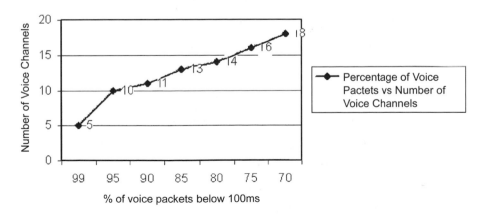

Figure 13.13   Number of voice channels versus the percentage of voice packets below 100 ms

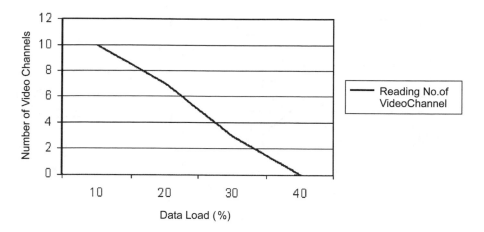

Figure 13.14   Number of video channels versus load for a delay of 100 ms

packets arrive within 50 ms. Figure 13.16 shows the number of video channels versus the percentage of video packets within 100 ms.

Figure 13.17 shows the availability of the number of voice channels in a network where 35% of the effective bandwidth is used by TCP/IP devices. The vertical axis shows the percentage of packets arriving before 50 ms. It compares the different MAC algorithms and shows how EDCF is beneficial in this environment as a means for differentiating services.

13.7.3 Example 3: Simulation Comparison of the TRAP and RAP Wireless LANs Protocols

In this example, we compare the performance of the TDMA-based Randomly Addressed Polling (TRAP) protocol, a wireless networking protocol proposed in [21] and the Randomly Addressed Polling (RAP) protocol [22] using simulation modeling. TRAP employs a variable-length TDMA-based contention stage with the length based on the number of active

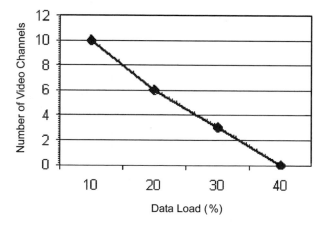

Figure 13.15   Number of video channels versus load for a delay of 50 ms

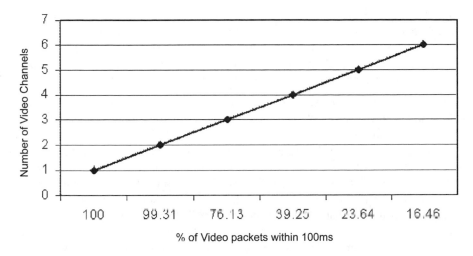

Figure 13.16 Number of video channels versus percentage of video packets within 100 ms

stations. At the beginning of each polling cycle, the base station invites all active mobile stations to register their intention to transmit via transmission of a short pulse. The base station uses the aggregate received pulse in order to obtain an estimate of the number of contending stations and schedules the contention stage to contain an adequate number of time slots for these stations to successfully register their intention to transmit. Then, it transmits a READY message carrying the number of time slots P. Each mobile node (station) calculates a random address in the interval $[0...P - 1]$, transmits its registration request in the respective time slot and then the base station polls according to the random addresses received.

RAP is a protocol designed for infrastructure WLANs. In RAP, the cell's base station initiates a contention period in order for active nodes to inform of their intention to transmit packets. For each polling cycle, contention is resolved by assigning addresses only to the active stations within the cell at the beginning of the cycle. All active mobile nodes generate a random number and transmit it simultaneously to the Base Station (BS) using Code Division Multiple Access (CDMA) or Frequency Division Multiple Access (FDMA). The number

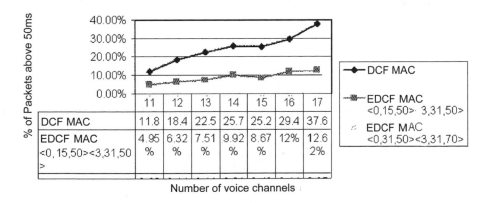

Figure 13.17 Quality of service versus MAC protocol (voice)

transmitted by each station identifies this station during the current cycle and is known as its random address. RAP uses a fixed number of random addresses P with values of P around 5 suggested [23]. This limitation stems from the requirement for orthogonal transmission of the random addresses. For a RAP WLAN that consists of N active mobile stations under the coverage of a base station, the protocol consists of the following stages:

- *Contention invitation stage.* Whenever the base station is ready to collect packets from the mobile nodes, it transmits a *READY* message, which may be piggybacked in a previous downlink transmission.
- *Contention stage.* Here, each active mobile node generates a random number R, ranging from 0 to $P - 1$. All active nodes transmit their random numbers simultaneously to the base station using CDMA or FDMA. The number transmitted by each station identifies this station during the current cycle and is known as its random address.
- *Polling stage.* Suppose that at the lth stage $(1 \leq l \leq L)$ the base station receives the largest number of distinct addresses and these are, in ascending order, $R_1, R_2, \ldots, R_n$. These numbers are used to poll the mobile nodes. When the base station polls mobile nodes with R_k, nodes that transmitted R_k as their random address at the lth stage transmit packets to the base station. This means that if two or more nodes have transmitted the same random address at the lth stage, a collision would occur. If $n = N$, however, no collision occurs. If the base station successfully receives a packet from a mobile node, it sends a positive acknowledgment (ACK). Acknowledgment packets are transmitted right before polling the next mobile node. If a mobile node receives an ACK, it assumes correct delivery of its packet, otherwise it waits for the current polling cycle to complete and retries during the next cycle.

TRAP employs a variable-length TDMA-based contention stage, which lifts the requirement for a fixed number of random addresses. The TDMA-based contention stage comprises a variable number of slots, with each slot corresponding to a random address. However, a mechanism is needed in order for the base station to select the appropriate number of slots (equivalently, random addresses) in the TDMA contention stage. To this end, at the beginning of each polling cycle, all active mobile nodes (stations) register their intention to transmit via transmission of a short pulse. All active stations' pulses are added at the base station, which uses the aggregate received pulse to estimate the number of active stations. The time slots will obviously be of fixed length, thus a mobile station that generates a random address p, $0 \leq p < P$, will transmit its random address at slot p. Based on this approach, the proposed protocol works as follows [21]:

- *Active stations estimation.* At the beginning of each polling cycle, the base station sends an *ESTIMATE* message in order to receive pulses from active stations. After the base station estimates the number of active stations N based on the aggregate pulse received, it schedules the TDMA-based contention stage to comprise an adequate number of random addresses $P = kN$, where k is an integer, for the active stations to compete for medium access.
- *Contention invitation stage.* The base station announces it is ready to collect packets from the mobile nodes, and transmits a *READY* message, containing the number of random addresses P to be used in this polling cycle.
- *Contention stage.* Each active mobile node generates a random number R, ranging from 0

to $P - 1$. Active nodes transmit their random numbers at the appropriate slot of the TDMA-based contention scheme. As in RAP, stations can generate addresses up to q times in a single contention stage and the contention stage may be repeated L times, with each active station generating a random address for each stage. Clearly, if two or more mobiles select the same random address, their random address transmissions collide and are not received at the base station. Thus, the random addresses received correctly at the base station are always distinct, with each number identifying a single active station.

- *Polling stage*. Suppose that at the lth stage ($1 \leq l \leq L$), the base station received the largest number of random addresses and these are, in ascending order, $R_1, R_2, \ldots, R_n$. The base station polls the mobile nodes using those numbers. When the base station polls mobile nodes with R_k, nodes that transmitted R_k as their random address at the lth stage transmit packets to the base station.

- If a base station successfully receives a packet from a mobile node, it sends a positive acknowledgment (ACK). Acknowledgment packets are transmitted right before polling the next mobile node. If a mobile node receives an ACK, it assumes correct delivery of its packet, otherwise, it waits for the current polling cycle to complete and retries during the next cycle.

In order to compare the performance of TRAP against RAP, we used a discrete-event simulator coded in C. The simulator models N mobile clients, the base station and the wireless links as separate entities. Each mobile station uses a buffer to store the arriving packets. The buffer length is assumed to be equal to Q packets. Any packets that arrive while the buffer is full are dropped. The packet interarrival times are assumed to be exponentially distributed. The arrival rate is assumed to be the same for all mobile stations. The condition of the wireless link between any two stations was modeled using a finite state machine with two states. Such structures can efficiently approximate the bursty-error behavior of a wireless channel [24] and are widely used in WLAN modeling [25,26]. The model comprises two states: (a) state G, denotes that the wireless link is in a relatively 'clean' condition and is characterized by a small Bit Error Rate (BER), which is given by the parameter *GOOD BER*; (b) state B, denotes that the wireless link is in a condition characterized by increased BER, which is given by the parameter *BAD BER*.

It was assumed that the background noise is the same for all stations and thus the principle of reciprocity stands for the condition of any wireless link. Therefore, for any two stations A and B, the BER of the link from A to B and the BER of the link from B to A are the same. The time spent by a link in states G and B is exponentially distributed, but with different mean values, given by the parameters *TIME GOOD* and *TIME BAD*, respectively. The status of a link probabilistically changes between the two states. When a link is in state G and its status is about to change, the link transits to stage B. When a link is in state B and its status is about to change, the link transits to stage G. By changing the model's parameter values, the protocols can be simulated for a variety of environments and conditions. The main assumptions considered in this study are:

- No data traffic is exchanged between the base station and the mobiles.
- The effect of adding a physical layer preamble was not included in our simulations.
- It is assumed that no error correction is used. Whenever two packets collide, they are assumed to be lost.

The performance metrics considered are the mean throughput, and delay. The number of mobile stations, N, under the coverage of the base station, the buffer size, Q, and the parameter, *BAD BER*, were taken as follows: (a) network N_1: $N = 10$, $Q = 5$, *BAD BER* $= 10^{-6}$; (b) network N_2: $N = 10$, $Q = 5$, *BAD BER* $= 10^{-3}$; (c) network N_3: $N = 50$, $Q = 5$, *BAD BER* $= 10^{-6}$; (d) network N_4: $N = 50$, $Q = 5$, *BAD BER* $= 10^{-3}$.

All other parameters remain constant for all simulation results: *GOOD BER* $= 10^{-10}$, *TIME GOOD* $= 30$ s, *TIME BAD* $= 10$ s, $L = 2$, *PRAP* $= 5$, $k = 2$, *RETRY LIMIT* $= 3$. The variable *RETRY LIMIT* sets the maximum number of retransmission attempts per packet. If the number of retransmissions of a packet exceeds this value (either due to collisions or channel errors) the packet is dropped. At the MAC layer, the size of all control packets for the protocols is set to 160 bits, the *DATA* packet size is set to 6400 bits and the overhead for the orthogonal transmission of the random addresses in RAP is set to five times the size of the poll packet. The wireless medium bit rate was set to 1 Mbps. The propagation delay between any two stations was set to 0.05 ms.The delay versus throughput characteristics for both the TRAP and RAP wireless protocols are shown in Figures 13.18–13.21. From these graphs, it is obvious that TRAP is superior to RAP in cases of medium- and high-load conditions. This superiority is due to the ability of TRAP to dynamically adjust the number of available random addresses according to the number of active mobile stations.

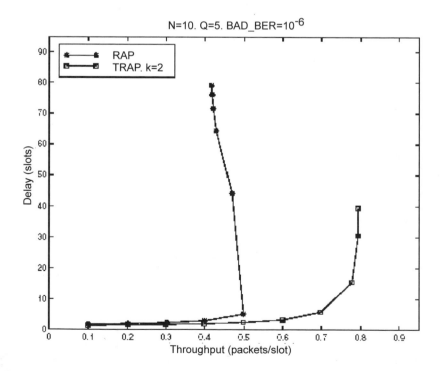

Figure 13.18   The delay versus throughput characteristics of RAP and TRAP when applied to network N_1

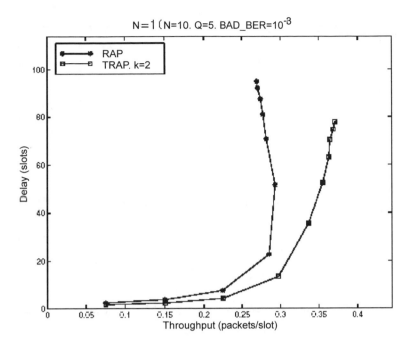

Figure 13.19   The delay versus throughput characteristics of RAP and TRAP when applied to network N_2

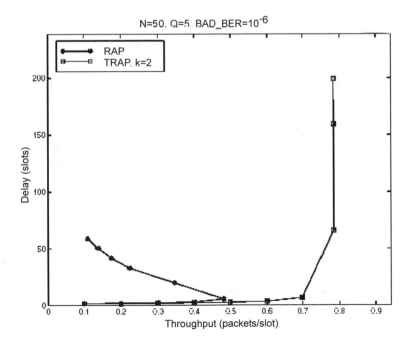

Figure 13.20   The delay and throughput characteristics of RAP and TRAP when applied to network N_3

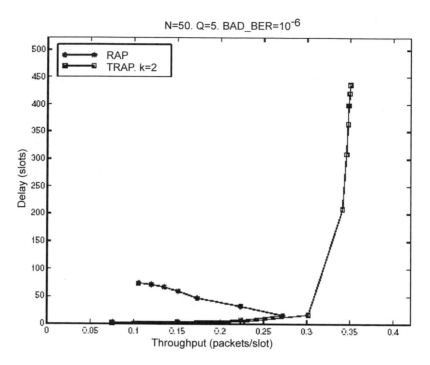

Figure 13.21   The delay and throughput characteristics of RAP and TRAP when applied to network N_4

13.7.4 Example 4: Simulation Modeling of Topology Broadcast Based on Reverse-Path Forwarding (TBRPF) Protocol Using an 802.11 WLAN-based MANET Model [27]

In this example, we describe a comparative simulation study of the topology broadcast based on reverse-path forwarding mobile routing protocol [28] and the Open Shortest Path First-2 (OSPF2) [29]protocol using the OPNET [30,31] simulation package. The study used a new model of 802.11wireless LAN (WLAN) designed for the Mobile Ad hoc Networking (MANET) protocol. The model consists of an 802.11b Wireless Local Area Network (WLAN) with enhancements to the physical layer, Media Access Control (MAC) layer, and propagation model to facilitate the design and study of the proposed MANET protocols. TBRPF is a link-state protocol used to turn wireless point-to-point networks into routed mobile networks that can react efficiently to node mobility. TBRPF is a proactive link state protocol that performs neighbor discovery though sending out periodic 'HELLO' packets using a protocol known as TND. TND can send out shorter HELLO messages than OSPF2 because its messages only have the addresses of newly discovered neighbors that have not yet been added to the neighbor table. Nodes periodically broadcast a HELLO with their own address. When a node receives a HELLO from a node it does not have in its routing table, it sends out HELLO messages with that new node's address in the 'newly discovered neighbors' section of the HELLO. When a node receives three new HELLO messages from a neighbor with its own address in the message, it discovers that it has bidirectional communications with the new neighbor and adds the new neighbor to its neighbor table. Once a neighbor has been added to the neighbor table, TBRPF no longer broadcasts that neighbor's address in its HELLOs. This allows TBRPF to generate shorter messages than OSPF2's HELLO which always includes the addresses of all known neighbor nodes. In dense networks, OSPF's HELLO protocol can become expensive in terms of network overhead.

The main difference between TBRPF and OSPF is that TBRPF uses reverse-path forwarding of link state messages through the minimum hop broadcast tree instead of using flooding broadcasts from each node as used in OSPF2 to send link state updates throughout the network. By using an improved version of the Extended Reverse Path-Forwarding (ERPF) algorithm as its topology broadcast method, it can be shown that TBRPF can scale to larger networks or handle more dynamic networks than traditional link state protocols that use flooding for topology broadcast. In a WLAN network, all links to a node's immediate neighbors are inherently broadcast links. When a message is sent out from a central node, all nodes within the transmission range of that node can hear the message. The TBRPF model takes advantage of the ability to broadcast updates to all of a link's immediate neighbors. The original ERPF algorithm was not designed to be reliable for calculating the reverse path in dynamic networks. TBRPF has two important modifications that distinguish it from the original ERPF algorithm. The first modification is the use of sequence numbers so updates can be ordered. The second change is that TBRPF is the first protocol where the computation of minimum hop trees is based on the network's topology information received along the broadcast tree rooted at the source of the routing information. By using minimum-hop trees instead of shortest-path trees (based on link costs) TBRPF generates less frequent changes to the broadcast trees and therefore generates less routing overhead to maintain the trees. Another important feature of TBRPF FT that distinguishes it from other link state MANET protocols, such as optimized link state routing (OLSR), is that each TBRPF node has a full topology map of the network so it can calculate alternate or disjointed paths quickly when topology changes occur. Full topology information can also be useful for calculating the most efficient multicasting routes or meeting other QoS objectives in the future.

In general, a network can be represented by a graph consisting of vertices (router nodes) and bidirectional edges (unicast links between nodes). This means that we can write $G = \{V,E\}$. Protocols such as OSPF2 that use flooding send topology updates down all unicast edges and have a best and worst case complexity of (Big O) $\Theta(E)$ for all messages. TBRPF sends updates down a reverse path consisting of a minimum hop spanning tree, which spans between the vertices so most messages have a complexity of V, however, there is a special case in TBRPF that can have a worst case complexity $\Theta(V^2)$. In general, we can consider a WLAN as a partial broadcast network where many nodes are within radio range of each other. In WLANs, all nodes within the radio broadcasting range of a particular node can hear any messages sent by that node. If a networking protocol is optimized to take advantage of the partial broadcast topologies of WLAN networks, it can greatly reduce the cost required to update the network during a topology update since the broadcast can effectively send a message down many or all of a nodes edges at once. In a partial broadcast network medium, such as WLAN, the complexity of TBRPF varies conversely with the network's density. In TBRPF's worst case, complexity will be $\Theta(V^2)$. This occurs in topologies where every node must generate a topology update to every other node. OSPF can also take advantage of a broadcast medium to improve its performance by broadcasting from each node to all its neighbors, rather than broadcasting down each edge individually between nodes. OSPF2 will require $\Theta(V^2)$ messages to converge the network as all nodes must transmit their routing

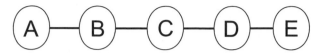

Figure 13.22   Sparse string network

tables and repeat (flood) the transmissions of all other nodes tables. In OSPF2 all network topology changes after convergence will generate a constant complexity $\Theta(V)$ because the flooding mechanism causes all nodes to repeat any node's topology update broadcast.

The worst-case scenario for TBRPF is a minimally sparse network, such as the string network shown in Figure 13.22. This topology can generate the worst case performance of $\Theta(V^2)$. Let us consider the case of convergence where only the two end nodes are the leaves in the broadcast tree. Essentially every node must generate a NEW_PARENT message and send it to every node down the string except for the end leaf. When the network is converged, any addition or deletion anywhere in a string affects the entire minimum hop broadcast tree so each node must propagate NEW_PARENT and CANCEL_PARENT messages one hop at a time throughout the entire network for a complexity of $\Theta(V^2)$. This is TBRPF's worst case, but luckily this is a rare case in most MANET networks. In a link state update which does not generate a change to the broadcast tree, such as the loss of a leaf node at the end of the string, the worst complexity in a string will always be $\Theta(V)$ but the best and average cases are also essentially V since all nodes except the leaf nodes are part of the minimum hop reverse path tree and must repeat all routing messages. If a link-cost change is propagated from a leaf node, the message can be propagated in $(|V - 1|)$ updates as the update is sent down every edge in the network and the final node does not transmit the message. If an update that does not change, the minimum hop tree is generated by a nonleaf node, that node can broadcast to two neighbors at once, but then each node on the reverse path can transmit to only one neighbor at a time thus only lowering the complexity to $(|V - 2|)$. In a string topology OSPF2's complexity to converge is $\Theta(V^2)$ and for all other topology broadcast events it will be $\Theta(V)$ as in any other topology. In string networks OSPF may outperform TBRPF for some events due to the complexity of updating the minimum hop tree.

The best possible topology for TBRPF is a maximum density network, as shown in Figure 13.23 where all nodes are fully connected to all other nodes. TBRPF's minimum hop broadcast spanning tree in a fully connected network has 1 root node and $(V - 1)$ leaf nodes. This

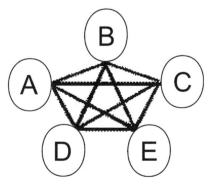

Figure 13.23   Fully connected network

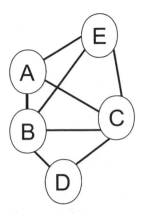

Figure 13.24   Typical MANET network configuration

allows TBRPF to update the network with a single broadcast for a constant complexity of $\Theta(V)$.

While a string topology shows a limitation of TBRPF and a fully connected broadcast topology shows TBRPF at its most efficient, most MANET networks are neither strings nor fully connected, but are rather ad hoc collections of nodes such as those depicted in Figure 13.24.

In this typical network, the complexity for TBRPF is less than V since the minimum hop broadcast tree's longest path is two hops while the shortest broadcast path (from B, or C to all nodes) requires only one broadcast. OSPF2 broadcast flooding will require the constant work of six broadcasts as each node sends out an update. OSPF2 is not able to use the partial broadcast topology to its full advantage like TBRPF does.

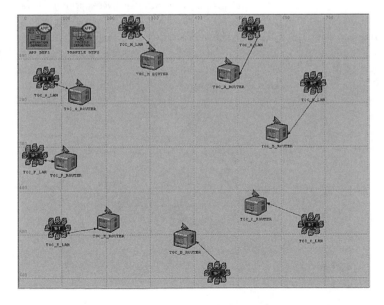

Figure 13.25   OPNET modeling environment

Figure 13.26   String test case

In order to test the difference in mobile networking protocols, we created a customized MANET WLAN model using MIL3's OPNET 7.0 (for more information on OPNET see www.mil3.com/opnet) modeling environment as pictured in Figure 13.25.

We added several extensions to bring the OPNET WLAN code up to the latest 802.11b standard. We also added code to allow us to model new routing, power adaptation, path loss, and security extensions to 802.11. The WLAN node considered here is a mobile router that may be attached to end nodes, rather than a mobile end node itself. We are assuming line of sight path loss for all topologies and are not taking into account multipath losses or the effects of intervening obstacles between nodes. We have not used the standard OPNET path loss equation because we have found that the low antenna heights of WLAN radios installed in wearable and portable computers cause more severe path loss than the free space path loss equation typically used in modeling. To compensate for the unusually low antenna heights of most WLAN MANETs, our model uses the equation

$$P_{\text{loss}} = 7.6 + 40\log_{10}d - 20\log_{10}h_t h_r$$

that was introduced in Ref. [8] to determine the link ranges viable for wearable WLAN radios. Tree basic topologies were created to test our node models. The topologies are the string, a fully connected, and a typical MANET network. In each topology, we stimulate topology broadcasts by changing the transmit power of nodes A and E to cause the network to drop and reinstate these leaf and nonleaf nodes.

Figures 13.26–13.28 show the string test, fully connected, and typical configurations, respectively. Although TBRPF is a draft IETF protocol, the working implementation code for the latest version is the intellectual property of Stanford Research Institute (SRI) International. It has several enhancements to the original IETF draft code, such as a more efficient 'HELLO' protocol, that add to its bandwidth efficiency in MANETs. The implementation of OSPF2 for our model came from the standard library of models that come with OPNET. The OSPF2 implementation follows RFC 2328 standards.

Tables 13.1 and 13.2 show the routing control traffic measured throughout the network and

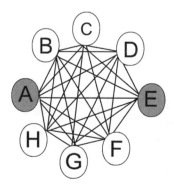

Figure 13.27   Fully connected case

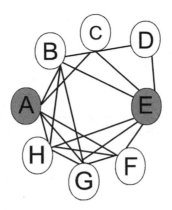

Figure 13.28   Typical MANET network

Table 13.1   Peak bits/second

Topology	OSPF2	TBRPF	Reduction (%)
String	25,920	20640	20
Full	26,000	6440	75
Typical	26,160	6200	76

Table 13.2   Average bits/second

Topology	OSPF2	TBRPF	Reduction (%)
String	1125	625	45
Full	6374	648	90
Typical	3032	468	85

the reduction of traffic from OSPF2 to TBRPF using the three test cases: string, full, and typical.

An analysis of these data shows that TBRPF's broadcast down the minimum hop reverse-path spanning tree greatly reduces the average network management cost over flooding in all situations, even in a worse case scenario of a string topology. However, at peak times, TBRPF did exhibit $\Theta(V)^2$ behavior which made it almost as costly as OSPF2. In partial broadcast and full broadcast topologies, the cost savings is significant with an average MANET network using approximately 85% less bandwidth.

The simulation showed that when compared to OSPF2 in a partial broadcast 802.11 WLAN network, the TBRPF protocol used 85% less bandwidth to maintain packet routing. Military and commercial users are already experimenting with TBRPF but tuning and debugging implementation of the code has always been cumbersome. This model that runs in the popular OPNET environment, allows accelerating the development and integration of TBRPF and other protocols in MANET radios which can be integrated into laptops, pocket PCs, cell phones, and wearable computers.

13.7 Summary

This chapter deals with the basics of simulation modeling and its application to wireless networking. We started by introducing the fundamentals of discrete-event simulation, the basic building block of any simulation program (simulator), and simulation methodology. Then we surveyed the commonly used distributions, their major characteristics, and applications. We presented the techniques used to generate and test random numbers. Then the techniques used to generate random variates (observations) are explained and the variates that can be generated by each of these techniques are investigated. Finally, we concluded by presenting four examples on the simulation of wireless network systems. These examples cover the performance evaluation of a simple IEEE 802.11 WLAN, simulation of QoS in IEEE 802.11 WLAN system, simulation comparison of the TRAP and RAP wireless LANs protocols, and Simulation of Topology broadcast Based on Reverse-Path Forwarding (TBRPF) protocol using an 802.11 WLAN-based Mobile Ad hoc NETwork (MANET) model.

References

[1] Law A. M. and Kenton W. D. *Simulation Modeling and Analysis*, Third Edition, McGraw-Hill, 2000.

[2] Banks J., Carson I. S., Nelson B. L. and Nicol D. M. *Discrete-Event System Simulation*, Third Edition, Prentice Hall, Upper Saddle River, NJ, 2001.

[3] Pooch U. and Wall I. *Discrete-Event Simulation - A Practical Approach*, CRC Press, FL, 1993.

[4] Jain R. *The Art of Computer Systems Performance Evaluation*, Addison-Wesley, New York, 1991.

[5] Obaidat M. S. Simulation of Queueing Models in Computer Systems, in *Queueing Theory and Applications*, Ozekici S, ed., Hemisphere, New York, 1990, pp. 111–151.

[6] Sadiku M. and Dyas M. *Simulation of Local Area Networks*, CRC Press, FL, 1995.

[7] Chun Chan W. *Performance Analysis of Telecommunications and Local Area Networks*, Kluwer, Boston, MA, 2002.

[8] Trivedi K. *Probability and Statistics with Reliability. Queueing and Computer Science Applications*, Second Edition, Wiley, 2002.

[9] Obaidat M. S. (Guest Editor) Special Issue on High Speed Networking: Simulation Modeling and Applications. *Simulation Journal. SCS*, 64(1), 1995.

[10] Obaidat M. S. (Guest Editor). Special Issue on Modeling and Simulation of Computer Systems and Networks: Part I: Networks, *Simulation Journal, SCS*, 68(1), 1997.

[11] ns manual - http://www.isi.edu/nsnam/ns/doc/index.html

[12] Using the ns simulator - http://dpnm.postech.ac.kr/research/01/ipqos/ns/

[13] Muhlethaler P. and Najid A. An Efficient Simulation Model for Wireless LANs Applied to the IEEE 802.11 Standard, INRIA Research Report No. 4182, April, 2001.

[14] Zahedi A. et al. Capacity of Wireless LAN with Voice and Data Services; *IEEE Transactions on Communications*, 48(7), July, 2000.

[15] IEEE 802.11 Standard, Wireless LAN Medium Access Control (MAC) and Physical Layer (PHY) Specification, June, 1997.

[16] IEEE Std 802.11e/D1, Draft Supplement for IEEE 802.11, 1999 Edition, March, 2001.

[17] Stallings W. *Wireless Communications and Networks*, Prentice Hall, Upper Saddle River, NJ, 2002.

[18] O'Hara B. and Patrick A. *IEEE 802.11 Handbook - A Designer's Companion*, Standards Information Network IEEE Press, January, 1999.

[19] Specification of the Bluetooth System.http://www.bluetooth.com/

[20] Cali F., Conti M. and Gregori E. IEEE 802.11 Wireless LAN: Capacity Analysis and Protocol Enhancement, in *Proceedings of IEEE INFOCOM'98*, 1998, San Francisco, CA, pp. 142–149.

[21] Nicopolitidis P., Papadimitriou G. I., Obaidat M. S. and Pomportsis A. S. TRAP: a High Performance Protocol for Wireless Local Area Networks, *Computer Communication Journal*, May, 2002.

[22] Chen K.-C. and Lee C.-H. RAP-A Novel Medium Access Control Protocol for Wireless Data Networks, in *Proceedings of IEEE GLOBECOM*, TX, USA, 1993, pp. 1713–1717.

[23] Chen K.-C. Medium Access Control of Wireless LANs for Mobile Computing, *IEEE Network*, September/October, 1994, 50–63.

[24] Gilbert E. Capacity of a Burst Noise Channel, *Bell System Technology Journal*, 39, 1960, 1253–1265.

[25] Bhaqwat P., Bhattacharya P., Krishna A. and Tripathi S. Enhancing Throughput Over Wireless LANs Using Channel State Dependent Packet Scheduling, in *Proceedings of IEEE INFOCOM' 96*, 1996, pp. 1133–1140.

[26] Bhaqwat P., Bhattacharya P., Krishna A. and Tripathi S. Using Channel State Dependent Packet Scheduling to Improve Tcp Throughput Over Wireless LANs, *ACM/Baltzer Wireless Networks*, 1997, 91–102.

[27] Green D. and Obaidat M. S. Modeling and Simulation of a Topology Broadcast Based on Reverse-Path Forwarding (TBRPF) Mobile Ad–hoc Network, in *Proceedings of the, 2002 International Symposium on Performance Evaluation of Computer and Telecommunications Systems, SPECTS*, July, 2002, San Diego, CA.

[28] Bellur B., Ogier R and Templin, F., Topology Broadcast based on Reverse–Path Forwarding (TBRPF), *IETF MANET Working Group Draft*, September, 2001.

[29] Moy J. OSPF Version 2, IETF Network Working Group Draft RFC 2328, April, 1998.

[30] OPNET Technologies. *OPNET Modeler Brochure" and "OPNET Radio Module;*, OPNET Technologies, Bethesda, MD, 2002.

[31] www.mil3.com/opnet

14

Economics of Wireless Networks

14.1 Introduction

The field of mobile wireless communications is currently one of the fastest growing segments of the telecommunications industry. Wireless devices have nowadays found extensive use and have become an indispensable tool on the everyday life of many people, both the professionally and personally. To gain insight into the momentum of the growth of the wireless industry, it is sufficient to state the tremendous growth in the number of worldwide subscribers of wireless systems. This figure has risen from only one mobile subscriber per 100 inhabitants worldwide in 1990 to 26 subscribers per 100 inhabitants in 1999 and the growth continues. This increasing number of subscribers is obviously reflected in monetary terms as well. For example, in the United States alone the wireless industry has grown from a 7.3 billion dollar industry in 1992 to a 40 billion dollar industry in 2000 [1]. Over the same period, the revenues from mobile services grew at an annual rate of 28%, a lot higher than the 7% rate achieved by the telecommunications industry, excluding the wireless sector. It can be easily seen that with such growth rates, it is just a matter of time before the use of wireless systems surpasses that of wireline systems. In fact, this transition has already taken place in some countries, such as Korea, where wireless telephony has replaced fixed telephony as the primary means of telecommunication [1].

As far as the near future is concerned, it is estimated that the growth of the wireless industry will continue, although at a slower rate [1]. Although there are predictions that bring the number of worldwide wireless subscribers to 2 billion by 2010, this may be difficult to achieve due to economic and social issues. The increase in the number of subscribers between 1984 and 2000 is mainly due to market penetration in societies of developed countries. In order for the number of subscribers to reach 2 billion by 2010, either almost everyone in these societies will have to possess a mobile phone, or the market has to open up in undeveloped countries as well. The latter, however, has obvious difficulties due to the lower income of people in these countries.

From the above discussion, it is logical to expect a decline in the growth rate of worldwide wireless subscribers [1]. However, this is not necessarily bad news for the wireless industry. The fact remains that cellular phones will continue to be used by very many people, who will form the base for the next major step in the wireless industry. This step is the integration of the wireless world with another area of high market penetration: the Internet. Although voice telephony will continue to be a significant application, the wireless–Internet combination will shift the nature of wireless systems from today's voice-oriented wireless systems towards data-centric systems. As far as the market opportunities are concerned, it is logical to expect a bright future, since the combination of the two fastest growing segments in the telecommunications industry will exponentially multiply market opportunities and revenues. A first step towards data orientation of wireless systems is the 2.5G standards, such as GPRS, which are deployed today in various parts of the world. However, a more radical approach will be taken by the next generations of wireless networks [6,7]:

- *Third Generation (3G)* wireless networks will be commercially deployed in the very near future, offering data rates up to 2 Mbps. Such speeds are enough for supporting wireless data applications.
- *Fourth Generation (4G)* and beyond wireless networks. These will evolve towards an integrated system, which will produce a common packet-switched (possibly IP-based) platform for wireless systems, offering support for high-speed data applications and transparent integration with the wired networks.

14.1.1 Scope of the Chapter

This chapter discusses a number of economic issues relating to wireless networks. It is organized as follows. Section 14.2 discusses the economic benefits of wireless networks. Section 14.3 discusses the changing economics of the wireless industry due to the above-mentioned movement towards the 'wireless Internet'. Section 14.4 provides a discussion on the expected growth for the demand of wireless data. Section 14.5 discusses charging issues [2] for wireless networks. Finally, Section 14.6 presents a summary.

14.2 Economic Benefits of Wireless Networks

Due to their ability to reduce overall networking costs, wireless networks can produce significant economic benefits for operators, compared to wired networks. It is significant to state that a wireless network requires less cabling than a wired network, or no cabling at all. Despite the fact that this obviously results in significant costs savings, since no installation of wires or fiber optics is needed, this fact is also extremely useful in several other situations:

- *Network deployment in difficult to wire areas.* Such is the case for cable placement in rivers, oceans, etc. Another example of this situation is the asbestos found in old buildings. Inhalation of asbestos particles is very dangerous and thus either special precautions must be taken when deploying cables or the asbestos must be removed. Unfortunately, both solutions increase the total cost of cable deployment.
- *Prohibition of cable deployment.* This is the situation in network deployment in several cases, such as historical buildings.
- *Deployment of a temporary network.* In this case, cable deployment does not make sense, since the network will be used for a short time period.

Another economic advantage enjoyed by wireless is that, contrary to wired networks, network capacity can be quickly and fully reused. The unused capacity of each wireless access point is ready to serve any newly arriving subscriber in the area, without the need for wires to reach the subscriber and technicians to install services.

14.3 The Changing Economics of the Wireless Industry

The movement towards integration of wireless networks and the Internet has reached a point which marks a change for the business of the wireless industry. The evolution from a voice-oriented to a data-oriented market will be the reason for introduction of new services and revenues as well as major changes in the industry's value chain. Furthermore, the wireless industry is likely to move from a vertical integration model to a horizontal integration model. Vertical integration refers to the situation of one or more companies covering the entire range of layers that are needed to offer services to the consumer. On the other hand, horizontal integration follows a layered approach, where the products of multiple companies are needed in order to offer services to the consumers. Although in most cases horizontal integrators lost out to vertical integrators, there are exceptions where horizontal integration dominates the market. This exception is expected to characterize the wireless industry as well. Overall, the trend towards data-oriented wireless systems is expected to change the economics of the wireless industry. In the following, we summarize the main factors affected by this change [1].

14.3.1 Terminal Manufacturers

14.3.1.1 Movement Towards Internet Appliances

It is expected that current wireless terminals will be substituted by Internet-enabled ones, such as Internet-enabled pagers, phones, digital assistants, etc. Thus, terminal manufacturers will face a new challenge in the design and implementation of their products. Whereas today the main target of terminal manufacturers is reduction in size and battery power consumption, in the future the target will also be terminals that support high-speed data services. It is likely that terminals will be classified into a number of categories, with each category addressing a different part of the consumer base. Thus, terminal categories will possibly be characterized by different device costs and capabilities.

14.3.1.2 Increasing Sales Figures

Mobile terminals are expected to continue to enjoy a sales increase despite the previously mentioned expectation for a reduction in the growth rate of the customer base. This is to be expected, since people are likely to change their terminals every couple of years in order to be able to keep up with the new services offered by mobile carriers. This fact already characterizes the mobile industry, with a simple example being the upgrade from a GSM to a GPRS phone in order to be able to use the higher data rates offered by GPRS. This evolution towards terminals of higher capabilities will be a challenging task due to the added complexity induced by the extra functionality. As a measure of comparison, we mention that the volume of software in a GPRS phone exceeds the volume of that in a standard GSM phone by ten times.

14.3.1.3 Lower Prices

Mobile terminals will continue to be based on silicon technology. This will continue to lower terminal sizes and prices. The evolution of silicon-based technology will also result in lower levels of power consumption. Thus, average battery lifetime is expected to increase.

14.3.1.4 Increased Competition from Asian Manufacturers

Due to the fact that Japan used a different 2G standard from the rest of the world, Japanese firms were left out of the international competition for 2G terminals. As a result, this has left a space open for American and European companies. However, this fact is not expected to continue in the future; rather, Japanese companies are expected to be the strongest competitors in the era of 3G wireless systems, especially Wideband CDMA (W-CDMA), which will soon be commercially deployed. This momentum of Japanese companies can be realized by the fact that many of the first trial 3G system deployments were made in Japan and by the announcement of the world's biggest operator, Vodafone, that 80% of its 3G terminals will be Japanese.

14.3.2 Role of Governments

14.3.2.1 Revenue due to Spectrum Licensing

Governments are actually very interested in the wireless telecommunication market from the point of view of economical benefits for themselves. This can be seen in the case of 3G spectrum auctions, which turned out to be very profitable for some governments. Such was the case with 3G spectrum auctions in Great Britain, which eventually created a revenue of about 40 billion dollars for the British government, ten times more than was expected. The fact that governments are likely to get a lot of money through spectrum licensing can be made clearer by stating that, compared to the 40 billion dollar revenue for the British government due to 3G spectrum, the total revenue to all European countries for 2G spectrum was about ten times less. The huge prices of 3G spectrum clearly show a difficult competitive environment for the mobile carriers.

14.3.2.2 License Use

Licensing spectrum parts to specific companies does not mean selling the spectrum; rather, the spectrum parts are leased for a certain period of time. Different governments lease spectrum for different time periods and some of them also restrict its use to only certain services. For example, the Federal Communications Commission (FCC), the national regulator inside the United States, licenses spectrum to operators without limiting them on the type of service to deploy over this spectrum. On the other hand, the spectrum regulator of the European Union does impose such a limitation. This helps growth of a specific type of standard, an example being the success of GSM in Europe.

14.3.2.3 Governments Can Affect the Market

Since governments control the way spectrum is used, they can control the number of licenses and thus the number of competing carriers. By increasing or decreasing this number, governments can affect the growth rate of the market and the competitiveness of the carriers. Finally, another way of affecting the market comes through privatization of telecommunication companies, which is a general trend around the world.

14.3.3 Infrastructure Manufacturers

14.3.3.1 Increased Market Opportunities

Due to the deployment of the next generations of wireless networks in the near future, the infrastructure of the mobile market is likely to rapidly increase in size. It is estimated that until 2006, this market will grow to a 200 billion dollars, four times the size it had achieved in 1999. Such conditions obviously promise a bright future for the infrastructure manufacturers.

14.3.3.2 Increased Entry Barriers

The increased complexity of infrastructure equipment for the next generations of wireless networks and the increased demand for such equipment is likely to favor companies which already enjoy a large market share. Furthermore, manufacturers of equipment for data networks are likely to enter this market.

14.3.4 Mobile Carriers

14.3.4.1 Market Challenges

The mobile carriers will face the greatest challenges in the new era of the wireless industry. They will have to adapt to the reducing growth rates of the subscriber base and the declining prices. The latter is a result of the maturing market and is due to the competition between carriers and the low prices of fixed line services. Furthermore, mobile carriers will have to adapt to the movement towards the wireless Internet and find ways to make profit from it. Of course this also means a risk for carriers, as they will have to spend a lot of money on investments (such as 3G licenses, new infrastructure and equipment, etc.) hoping that the wireless Internet finds the necessary popularity among the subscribers so that the carrier eventually gets its money back. This adoption of the wireless Internet as a primary means of revenue means that mobile carriers need to play a number of additional roles in order to stay competitive. These additional roles are that of the Internet Service Provider (ISP), the portal, the application service provider and the content provider. These roles are summarized below:

- *The ISP role.* The mobile carriers will have to carefully examine the case of the fixed Internet world. In that case, local telephone companies in North America lost the opportunity of becoming major ISPs and America On Line (AOL) emerged as the dominant player in the field. Thus, mobile carriers will want to ensure that the same does not happen with the wireless Internet. This means reduction of wireless Internet prices; however, it will be difficult to reach the prices of the wired Internet due to the fact that the wireless bandwidth is a scarce and expensive resource. Finally, it remains to be seen whether ISPs of the fixed Internet world will enter the wireless Internet arena. In this case, they are likely to take a substantial part of the market due to their experience and preservation of their subscriber base.

- *The portal role.* Mobile carriers will also have to run their own portals to the wireless Internet world. In this case, it is logical to expect that portals already flourishing on the wired Internet will have a big advantage over those of mobile carriers. The same of course holds for the case of mobile carriers that are associated with successful portals of the wired Internet. In that case, mobile carriers will have the advantage of gaining from the knowledge and customer base of the successful fixed-Internet portal.
- *The application service provider role.* In the 3G generations and beyond of wireless networks, many new services will appear. Thus, mobile carriers are potential providers of these new services, which may constitute a significant portion of revenue. Examples of such services are location-based services.
- *The content provider role.* Mimicking the world of fixed Internet, mobile carriers will also have to prepare content for their portals.

14.3.4.2 Few Carriers

The cost of the equipment for the rollout of the new services is estimated to be 2–4 times higher than the cost of 2G equipment. This means that a reduced number of carriers is likely to characterize each market. This number is estimated to be between two and four carriers for each country's market. (Actually, it has been proved through game theory that the maximum number of carriers that does not slow down profitability is 4 [1]). In cases where a larger number of competitive carriers appear, the chances are that those with the largest subscriber base will probably acquire the biggest part of the market. This means that the market is divided between those carriers with obvious advantages to their revenues. Smaller carrier companies obviously will not be able to survive the competition and they will be forced to merge in order to stay competitive. Overall, the market for mobile Internet will resemble an oligopoly, with a streak of strategic behavior from competing carrier companies. This means that the prices of products of a company affect those of its competitors. In such an environment, companies implicitly come to a common agreement regarding their prices. This kind of agreement is known as self-enforcing, since the competitors abide by it due to the fact that this is in their interest. Such a market, where a company chooses its strategy given the strategies of its competitors in order to maximize its profit is said to be in a Nash equilibrium.

14.3.4.3 Bundled Products

In most cases, consumers appear to prefer bundled products. Carriers associated with telecom operators, especially for data services, will have a relative advantage.

14.3.4.4 Changing Traffic Patterns

Increased intra-country mobility, especially within the European Union where a common standard (GSM) is used, increases traffic related to roaming between countries. In some small countries, traffic due to roaming will actually constitute more than half of the traffic exchanged.

14.3.4.5 Different Situation in each Country

Due to the different factors that dominate the telecommunications scene and the society of each country, it is difficult to make predictions on successful carriers. In the United States, the wireless market is affected by the large distances, lack of spectrum, increased competition, large subscriber base, Internet popularity and a divergence of standards. In the European Union, however, the scenario is somewhat different: Internet use is not that widespread, a single standard exists (GSM) and, as mentioned above, roaming traffic is an important part of the total traffic.

14.4 Wireless Data Forecast

As stated, wireless data will become a significant part of the traffic over future mobile wireless data. It is interesting to note the similarity of today' situation regarding the wireless Internet with that of the wired Internet in the early 1990s. In those years, Internet was characterized by lower data rates (due to low-speed (up to 9.6 kbps) dial-up modems) and applications far from today's user-friendly ones, such as the inconvenient Mosaic web browser. Furthermore, information was available mostly in text format and graphics were of low resolution. However, speeds increased (reaching 56 kbps for dial-up and 128 kbps for ISDN) as did usability (an example being the introduction of Netscape's and Internet Explorer's graphical interfaces) thus raising the popularity and penetration of the Internet. Specifically, it enjoyed a tremendous evolution with traffic per user rising from one MB per month in 1991 to 200 MB per month in 1999.

A somewhat similar situation with that of the early days of Internet characterizes today's wireless data scene: low data rates, abbreviated user interfaces (e.g. those of the Short Message Service (SMS) and Wireless Application Protocol (WAP)), text-like output and low-resolution graphics. As the capabilities and usability of wireless networks increases, a growth similar to that of fixed Internet will be observed for the wireless Internet as well.

14.4.1 Enabling Applications

A number of capacity-demanding data applications are expected to be used over wireless networks. These will offer compelling value to the consumer and due to their popularity are expected to increase wireless data traffic. Some of these applications are briefly highlighted below [5–7]:

- *Video telephony and videoconferencing.* These will be typical mobile multimedia applications. They will offer users the ability to participate in virtual meetings and conferences through their wireless terminals. Moreover, they will offer the ability to access multimedia content, such as CD-quality music and TV-quality video feeds, from service platforms and the Internet.
- *Internet browsing.* This will be a significant application. It will be greatly enabled by the emergence of XML, which will enable internet content to be more accessible by wireless devices without the need to offer web content separately for wireless devices, as is the case with the Wireless Access Protocol (WAP).
- *Mobile commerce.* These will offer the ability to make on-line purchases and reservations upon demand without having to be in front of an Internet-connected PC. Market analysts

predict that e-commerce will be a multitrillion dollar industry by 2003. Introducing e-commerce to the mobile platform will be an important source of operator revenues.

- *Multimedia messaging.* These applications will offer support for multimedia-enhanced messages such as voice mails and notifications, video feeds software applications and multimedia data files.
- *Geolocation.* Geolocation determines the geographical location of a mobile user. There are two types of geolocation techniques, one based on the handset and the other on the network. The first one uses the GPS system to determine user location while in the second one the replicas of the signals from the same handset at different base stations are combined in order to determine user location. Some obvious applications employing geolocation technology include mobile map service and identification of user location for emergency calls. In fact, geolocation technology has already been deployed in Japan and Korea generating over one million position references per day [5].

14.4.2 Technological Alternatives and their Economics

There are a number of candidate technologies for offering data transfer in wireless networks. In this section we summarize some of these technologies.

- *cdma2000.* This is a fully backwards-compatible descendant of IS-95 (cdmaOne) utilizing the same 1.25 MHz carrier structure of cdmaOne. Cdma2000 offers both voice and data at rates up to 2 Mbps. It uses two spreading modes, 1X and 3X. The 1X mode uses a single cdmaOne carrier providing average data rates up to 144 kbps, while 3X is a multicarrier system. 1X and 3X are the two modes currently standardized, although modes such as 6X, 9X and 12X may be standardized in the future.
- *High Data Rate (HDR).* This is an enhancement of 1X for data services. HDR uses more modulation, thus offering higher speeds than 1X.
- *Wideband CDMA (WCDMA).* WCDMA introduces a new 5 MHz-wide channel structure, capable of supporting voice and average data at speeds up to 2 Mbps.
- *General Packet Radio Service (GPRS).* GPRS is a packet-switched overlay over 2G networks. Its operation is based on allocation of more slots to a user within a GSM frame. GPRS terminals support a variety of rates, ranging from 14.4 to 115.2 kbps, both in symmetric and asymmetric configurations.

It is estimated [5] that, based on a cost per megabyte scenario, CDMA-based technologies have an economic advantage over GPRS due to the limited capacity of the latter. Of the cdma-based technologies, HDR is the most advantageous for supporting data traffic, as it has a two to three times cost advantage over cdma2000 1X and WCDMA. This advantage of HDR is due to its optimization for data traffic.

14.5 Charging Issues

A fundamental issue in the wireless market is the way carriers charge their customers. Although customers are certainly attracted to new and exciting technologies, most of them will make their choice of carrier based on the charges. Thus, it can be seen that charging policies have the potential to greatly impact the success of mobile carriers.

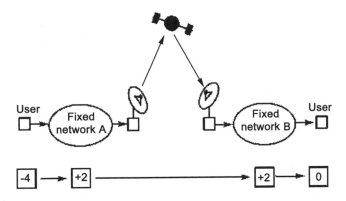

Figure 14.1   Charging on an international call

In both fixed and mobile telephony worlds, carriers can send bills only to their own customers. This of course means that there must exist a way for users to be charged for calls terminating at the network of a different carrier. In order to illustrate this scenario, Figure 14.1 shows the charges (in monetary units) when a user of carrier A makes a call to a telephone belonging to a different carrier B. It can be seen that the user pays for the usage both of carrier A and B. Since most countries originally had only one phone company (typically owned by the government), such a situation arose in international calls trough fixed telephony networks. The way the user of a phone company A was charged for making a call to phone B was defined through a set of regulations, known as interconnect agreements, between the national phone companies. Obviously, both companies profited from international calls.

Since the scheme of the interconnect agreements required each carrier to form a separate agreement with every other carrier, the International Telecommunications Organization (ITU) devised the international accounting rate system. This actually allowed carriers to charge as much as they wanted for calls terminating on their own network. Since charging for this type of service did not affect their own customers, most carriers decided to charge a lot. This situation, which resulted in high prices for international calls, began to change in the 1990s, when multiple fixed telephony carriers began to appear within the market of the same country. These carriers were interconnected with others of the same country in order to allow users of competing carriers to call each other. The calls between telephones of different carriers were charged in a way similar to that presented in Figure 14.1. Some of these new carriers also set up connections with carriers of neighboring countries by bypassing the accounting rate system. In order to be competitive, they offered lower charges for international calls and thus prices for such calls began to fall.

14.5.1 Mobility Charges

In most cases the price for placing a call through a mobile carrier is a lot higher than that through a fixed telephone carrier. This is due to the fact that (a) mobile carriers have paid a significant amount of money to obtain spectrum licenses and (b) they frequently spend money in installing new infrastructure. The actual price for a mobile telephone call is not constant but rather depends on the policy of the carrier, the time at which the call is placed, the user's contract, etc. However, despite the fact that mobile calls cost more than fixed ones, these

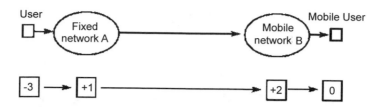

Figure 14.2   Calling party pays

prices follow a declining rate for a number of reasons, such as competition between carriers and the target of making mobile telephony a direct competitor of the fixed system.

Another interesting issue is the charge for the case of a user that places a call that ends at the network of a mobile carrier. Here, there are two approaches:

- *Calling Part Pays (CPP)*. This approach, shown in Figure 14.2, is mostly used in European countries. It can be seen from the figure that the caller pays for usage of both the fixed and the mobile networks resulting in a free call for the receiving party. Thus, calling a mobile phone from a fixed one is more expensive than a call placed between two fixed telephones. In order to provide fairness to the callers, mobile numbers are preceded by special codes, which let the caller know that the charge for such a call will be higher than that for a call to a fixed telephone.
- *Receiving (called) Party Pays (RPP)*. This approach, shown in Figure 14.3, is mostly used in the United States and Canada. It can be seen from the figure that the called party pays for usage of the mobile network. Thus, calling a mobile phone from a fixed one costs the calling party the same amount of money as a call placed between two fixed telephones. This approach is driven by the fact that in the United States consumers are accustomed to the fact that local calls are free, thus paying for a call to a mobile phone being in the same area would seem strange to them.

An advantage of the CPP approach is that it brings no burden on the owners of the mobile phone. Since the calling party is the one that is charged by the call to the mobile carrier, owners of mobile phones can freely give their number to whomever they want. Such a situation of course does not apply to the RPP approach. In that case, people are reluctant to give their mobile phone numbers since they will pay for all incoming calls. In some cases they even close their phones in order to avoid receiving unwanted calls. Finally, the CPP approach is much more likely to be used in marketing. This is because most of the time mobile carriers advertise themselves based on the cost of placing a call from a mobile phone, which is continuously declining due to competition. Thus, people tend to prefer carriers who

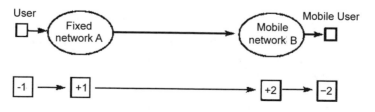

Figure 14.3   Receiving party pays

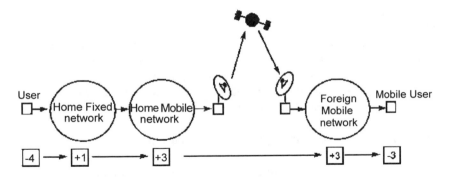

Figure 14.4 Charges for a call placed to a roaming user.

impose the lower cost for making calls and do not pay attention to the charges imposed on others for calling them.

14.5.2 Roaming Charges

Figure 14.4 shows the case of a call placed from a fixed telephone to a user of a mobile carrier, who has moved to the operating area of a mobile carrier located in a different country. This situation is known as roaming and imposes relatively high charges on the receiving party. As shown in the figure, the RPP approach is in effect in roaming situations. This is because it would be unfair to charge the caller for usage of the foreign mobile network since he/she has no way of knowing that the called party is roaming to a foreign network. Thus, the cost of the call for the calling party is just the sum of the cost of using the fixed network and the cost of using the home mobile network, meaning that the charge for the calling party is what it would be if the called party was not roaming. The extra cost of using the foreign mobile network is charged to the called party. This charge is usually a lot bigger than the amount of money that is charged to customers of the foreign network, a fact that may make roaming an expensive service.

14.5.3 Billing: Contracts versus Prepaid Time

Once the charges for utilizing network resources are summed up, the mobile carriers have to send bills to the customers in order to get their money. There are two main approaches here: contracts and prepaid billing. A contract is essentially leasing of a connection to the network of the carrier. In most situations, users that sign such contracts also get the mobile handset free of charge. The mobile operators of course eventually get back the cost of the handset, since the contract forces the user to pay a monthly rental charge for his/her connection irrespective of the fact that he/she may not use the connection at all. Of course the user is also charged for both the calls made and generally for all the services used. Obviously, when the paid rental charges total to the cost of the handset, all the money paid by the user from this point onwards is pure profit to the mobile carrier.

Contracts have the disadvantage of limiting the user to a specific carrier for a certain amount of time. This means that in order to get a new phone for free, customers get stuck with the same contract for quite some time (about a year most of the time). Thus, another approach appeared; that of 'prepaid' time. This approach was first applied by Telecom Portugal (TMN) in 1995. According to this, users pay in advance for both their handsets and the calls they make. Handsets can be bought from electronics stores and they are usually 'loaded' with a certain amount of credits, which translate into speaking time (and obviously credits for using other network services, such as the Short Message Service (SMS)). Once the user of the phone has exhausted all the credits, he/she can recharge the phone by entering special code numbers that can be found on special cards sold by stores, automatic teller machines, etc.

The prepaid approach has found significant acceptance in Europe [8]. One year after its introduction in Portugal, revenues for mobile services for TMN grew by 65% and in 1997 TMN experienced a 130% increase in its customer subscriber base due to the popularity of prepaid mobile products. In 1999, over 85% of TMN customers used prepaid services. Similar penetration rates of prepaid services also hold for other countries, such as Spain. Overall, prepaid mobile services constituted more than 67% of European mobile subscribers in 1999. In the same year, the doubling of the subscriber base in Spain and Greece was made possible due to the prepaid approach. However, it has not yet gained significant momentum in the United States market where it is primarily restricted to older analog phones and it constitutes approximately 6% of the overall subscriber base.

The small acceptance of the prepaid approach in the United States can be attributed to the fact that in this market, RPP is used, thus users of contracts are sure that they will be able to receive a call at all times. This would not be true with a CPP that has run out of credits. In Europe, however, CPP is used and thus anyone can call a mobile phone irrespective of the fact that it may have run out of credits. In fact, this has created a significant revenue problem for mobile carriers. This is because many people choose to never recharge their phone and therefore use it only for receiving calls. In order to deal with this problem, most carriers block calls to a phone that has not been recharged for a specific time period (most of the time this is some months). Moreover, carriers generally offer better charges for contract users in order to promote such subscriptions over prepaid ones. This is due to the fact that contract users are always able to place calls, thus the chances of revenue to the carrier from a contract subscriber are better than that from a prepaid subscriber.

The advantages of the prepaid approach are that:

- since no monthly charged is employed, customers have greater control of their costs,
- from the operators point of view prepaying is beneficial since they get their money in advance and are not burdened with the overhead and cost of producing bills for prepaid customers,
- prepaid is beneficial for users who would otherwise not have a credit rating sufficient to qualify for a contract mobile subscription. Such an example is the case of Australia, where the introduction of prepaid mobile services gave access to a vary large number of people. These people would otherwise not have access to mobile services due to the fact that they could not meet the credit checks. This accounts for about 40% of all the people that want a mobile phone in Australia.

14.5.4 Charging

There are three main motivations for charging in mobile wireless networks. These are briefly highlighted below:

- Recovery of the investment in infrastructure equipment.
- Generation of profit for the mobile operators and service providers.
- Controlling network congestion by providing service levels of different prices.
- For the case of noncommercial organizations, such as schools and universities, congestion control through a charging scheme is used for social reasons. In this case, charging may be based on 'tokens' and thus not reflected in monetary terms.

Charging methods largely depend on the structure of the network. The majority of wireless networks until the 3G era were primarily designed for voice traffic and are thus of a circuit-switched network. Nevertheless, the movement towards the next generation of wireless networks is towards a packet-switched network. In a circuit-switched network, a dedicated path is assigned between the communicating sides for the entire duration of the connection. Of course, the entire capacity of a link is not necessarily dedicated to a single connection but can rather be time or frequency-multiplexed in order to serve more connections. Circuit switching introduces some overhead for link establishment, however, after this takes place, the delay incurred by switching nodes is insignificant. Thus, circuit switching can support isochronous services such as voice, which is the primary reason that circuit switching has been widely utilized in earlier cellular systems.

However, circuit switching is efficient for data traffic, since in such cases the circuit will be idle most of the time. Packet switching solves this problem by routing packets between the communicating parties with each packet following a possibly different path. Each packet carries a control header, which contains information that the network needs to deliver the packet to its destination. In each switching node, incoming packets are stored and the node has to pick up one of its neighbors to hand it the packet. This decision entails a number of factors, such as cost, congestion, QoS, etc., and depends on the routing algorithm used. A benefit of using packet switching for data services is that bandwidth is used more efficiently, since links are not occupied during idle periods. Furthermore, in a packet-switched network, priorities can be used. Packet switching has emerged as an efficient way of handling asynchronous data in cellular systems. Examples of this approach are the CDPD and GPRS standards in 2G networks. The rising significance of data traffic over wireless systems makes the importance of using packet switching in such systems even greater. This can be realized by the fact that the next generations of wireless systems (4G and beyond) are envisaged to be an integrated a common packet-switched (possibly IP-based) platform.

14.5.4.1 Charging Methods

Below we describe some methods for charging in mobile networks [3]. Most of these methods have already been proposed for the Internet but are equally applicable in the case of mobile networks.

- *Metered charging.* This model is already used by many ISPs, mobile and fixed telephony carriers. The model charges the subscriber with a monthly fee irrespective of the time he spends using the network services. However, most of the time this fee also includes some 'free' time of network use. When the user has spent this time, he/she is charged for the extra time using the network. This method is used in 2G networks for charging voice traffic. The method of charging for voice calls is quite straightforward: The duration of the call is proportional to the call's cost. Nevertheless, sometimes charges decrease for increased network usage. Metered charging is well suited to voice calls which are typically circuit-switched, since the user pays for the period of time the circuit is used, that is essentially the duration of the voice call. Furthermore, it adds little network overhead and is transparent to customers since it does not require configuration in their devices. However, this model is not suitable for charging for data services that are expected to be offered by the 'wireless Internet'.

- *Packet charging.* This method is used for charging in packet-switching networks. It is more suitable for data than metered charging. This is because the user is not charged based on time but rather on the number of packets that he/she exchanges with the network. Thus, this method obviously calls for a system that is able to efficiently count the number of packets that belong to a specific user and produce bills based on these measurements. The disadvantage of packet charging is the fact that its implementation might be difficult and thus costly, since the cost of counting packets for each user might increase the complexity to the network, either due to increased traffic or additional infrastructure requirements. This results in increased network overhead; however, the overhead to subscribers remains minimal as the method is transparent to them.

- *Expected capacity charging.* This method involves (a) an agreement between the user and the carrier regarding the amount of network capacity that will be received by the user in the case of network congestion and (b) a charge for that level of service. However, users are not necessarily restricted to the agreed capacity. In cases of low network congestion, a user might receive a higher capacity than the agreed one without additional charges. Nevertheless, the network monitors each user's excess traffic and when congestion is experienced, this traffic is either rejected or charged for. The advantage of this method is that it enables mobile carriers to achieve a more stable long-term capacity planning for their network. Expected capacity charging is less complex than packet charging both in terms of network and subscriber overhead.

- *Paris-metro charging.* In this method, the network provides different traffic classes, with each class characterized by different capabilities (such as capacity) and hence a different charge. Thus, users can assign traffic classes to their different applications based on the desired performance/cost ratio. For example, a user may decide to assign a higher traffic class to business e-mails and a lower one to personal messaging services. Furthermore, in cases of congestion of a certain traffic class, a user may decide to change the traffic class of his/her congested connections in order to improve performance. Such a switching between traffic classes might also be initiated by the network itself in order to provide self-adaptivity. From the above discussion, it is obvious that Paris-metro charging is useful for providing network traffic prioritization in wireless data networks. Another advantage of the method is that it provides customers with the ability to control the cost of their network connections. However, the disadvantages of the method are (a) an increase in the mathematical complexity of the network's behavior and thus cost of implementation and (b) the fact that users need to be familiar with the process of assigning traffic classes to their connections introduces some overhead for them. The latter problem could be solved by an automatic assignment method. However, this would require extensions to the network protocols and thus increase network complexity and protocol overheads.

- *Market-based reservation charging.* This method entails an auctioning procedure for acquiring network resources. Users place monetary bids and based on these bids the network assigns appropriate connections to users. An advantage of this method is the fact that users are in control of the quality of service they receive from the network. For example, business users will be more likely to accept a higher charge for their connections than customers that use the network for recreational activities. However, the disadvantages of this method are that (a) due to the bidding procedure, customers are never sure regarding the quality of service they receive from the network, (b) the auctioning approach adds to network overhead, (c) users must make bids, thus the method is not transparent to them and familiarization with it is required. Furthermore, market-based reservation charging raises the issue of unfairness since some customers may not be able to receive the desired performance. It is generally agreed that this method is not suitable for the wireless Internet.

Figure 14.5 summarizes some characteristics of the above charging methods. These methods may be used in combination to produce flexible charging schemes for covering a diverse range of requirements. For example, a mobile operator could offer free calls up to a certain limit and only charge a monthly rental fee for the subscription. To gain revenue from users that make many calls and exceed the above limit, the operator can choose to use metered charging for the extra calls. Customers that read their e-mail and access the Web through their mobile phones could be charged using a fixed charge for reception of text and packet charging for e-mail attachments. By utilizing such a combination of charging schemes, the required charging policy for a diverse range of customer types, ranging from teenagers and students to business users, can be achieved.

Once a charging method has been decided for a user and put into operation, the network is responsible for capturing information relating to this user's traffic. This information may comprise a number of charge sources, such as charges for using the network, accessing the

Charging method	Implementation cost	Network overhead	Transparence to subscribers
Metered charging	Medium-High	Low	Low
Packet charging	High	High	High
Expected capacity charging	Medium-High	Medium	Medium
Paris-Metro Charging	Medium	High	Low
Market-based reservation charging	Medium-High	High	Low

Figure 14.5 Comparison of charging models

pages of a web server, using certain applications, accessing other networks, etc. Then this information must be processed and finally used to produce the bills sent to the customer.

14.5.4.2 Content-based Charging

A different approach to the problem of how to charge a customer for utilizing the network is content-based charging. The novelty of this approach is that users are not charged based on usage, but rather on the type of content they access. Some examples of the significance of content-based charging follow:

- Content-based charging has been applied in Japan by NTT DoCoMo and experience showed that customers are willing to pay extra for certain simple services such as stock quote information.
- Another example is the case of the Short Message Service (SMS): since this service consumes extremely few network resources, it is a significant point of revenue for operators due to the facts that (a) the price of an SMS message is around 0.1 dollar and (b) SMS is a very popular service.
- Another example is that of on-line games [4] through the wireless Internet. Although such applications are both popular and impressive, they require little amount of information exchange between terminals, since graphic display is local to the devices. Thus, the traffic exchanged between devices conveys only game-state information (such as player positions and ball trajectory in sports games) and perhaps instant-messages exchanged between the players. It is obvious that for such an application, users would easily accept a charge significantly higher than that corresponding to the amount of exchanged traffic. The usefulness of this fact to both operators and application providers is obvious.

14.6 Summary

Wireless networks constitute an important part of the telecommunications market. In the United States alone the wireless industry has grown from a 7.3 billion dollar industry in 1992 to a 40 billion dollar industry in 2000. Despite the fact that the growth of wireless network subscriptions is expected to decline, the industry is up against a new challenge: that of integration with the Internet. The result of this integration, the wireless Internet, is expected to significantly increase the demand for wireless data services and provide a new revenue source for wireless telecommunication companies. This chapter overviews several economic aspects of wireless networks, including economic benefits of wireless networks, facts that affect the economics of the wireless industry, a forecast for the growth of wireless mobile data services and several charging issues for wireless data services.

References

[1] Hugh S. M. A., Down K., Clements J. and McCarron M. *Global Wireless Industry Report: Part 1: the Changing Economics of the Wireless Industry*,
available at http://www.totaltele.com/ whitepaper/docs/wireless111600.pdf
[2] Franzen H. Charging and Pricing in Multi-Service Wireless Networks, Master Thesis, Department of Micro-electronics and Information Technology Royal Institute of Technology of Sweden, 2001.

[3] Cushnie J., Hutchison D. and Oliver H. Evolution of Charging and Billing Models for GSM and Future Mobile Internet Services, in *Proceedings Of QofIS Symposium*, 2000, pp. 313–323.

[4] Value-Based Billing for Wireless Internet Services, *Portal Overview*, available at http://www.asiatele.com/internet/wireless.pdf

[5] The Economics of Wireless Mobile Data, *Qualcomm White Paper*, available at http://www.qualcomm.com/main/whitepapers/WirelessMobileData.pdf

[6] Nicopolitidis P., Papadimitriou G. I., Obaidat M. S. and Pomportsis A. S. Third Generation and Beyond Wireless Systems, *Communications of the ACM*, 2002, in press.

[7] Nicopolitidis P., Papadimitriou G. I., Obaidat M. S. and Pomportsis A. S. 3G Wireless Systems and beyond: A Review, in *Proceedings of IEEE ICECS*, 2002, in press.

[8] Beaubrun R. and Pierre S. Technological Developments and Socio–Economic Issues of Wireless Mobile Communications, *Telematics and Informatics*, 18, 2001, 143–158.

Further Reading

[1] Dornan A. *The Essential Guide to Wireless Communications Applications*, Prentice Hall, Upper Saddle River, NJ, 2001.

Index

a street microcell that has the form of a grid comprising square buildings, there exist two possible situations.

- If a LOS exists between the transmitter and the receiver (e.g. receiver A in Figure 2.9), then the path loss model comprises two parts. Up to a certain breakpoint, the exponent n is around 2, as in free-space loss. However, beyond this breakpoint the signal strength decreases more steeply with a value of n around 4. Andersen et al. [3] mention that the breakpoint is given by $2\pi h_b h_m/\lambda$, where h_b is the antenna height of the base station and h_m is the antenna height of the mobile station.
- If a LOS does not exist between the transmitter and the receiver (e.g. receiver B in Figure 2.9), then the path loss is greater for the receiver. Up to the intersection of the two streets, the exponent n is around 2, however beyond the intersection n takes values between 4 and 8.

Various propagation models for street microcells have been proposed based on ray-optic theory. The preliminary two ray model calculates received signals for LOS channels by taking into account a direct ray and a ground-reflected ray. Enhancements of this model use more rays for greater accuracy. Hence, the four-ray model also assumes two rays that stem from reflection by nearby buildings, the six-ray model assumes double reflected rays by buildings, etc. Generally, model using a large number of rays is more accurate than a model assuming a smaller number of rays. Other methods also exist that try to take into account corner diffraction of signals and partially overlapping microcells.

2.3.2.3 Indoor propagation and its differences to outdoor propagation

Indoor propagation has attracted significant attention due to the rising popularity of indoor voice and data communication systems, such as wireless local area networks (WLANs), cordless telephones, etc. Although the phenomena that govern indoor propagation are the same as those that govern outdoors (reflection, diffraction, scattering), there are several differences [3] between indoor and outdoor environments:

- *Dependence on building type.* Radio propagation is more difficult to predict in indoor environments and on a number of factors relating to the building (architecture, materials used for building construction, the way which people move throughout the building, whether windows and doors are open or closed). Thus, several characteristics of a building directly impact propagation of signals within the building. A great number of measurements have been performed and researchers have classified buildings into various types, with buildings in each type inducing different propagation behavior to signals. The types of buildings mentioned in the literature [3] are homes in suburban areas, homes in urban areas, office buildings with fixed walls, open office buildings with movable soft panels of height less than the ceiling dividing the office area, factories, grocery stores, retail stores and sports arenas. Inside buildings, two types of transmitter/receiver path exist, based on whether the transmitter is visible to the receiver: LOS paths and obstructed (OBS) paths. Buildings types are summarized in Figure 2.10, which also gives values for n and σ for transmission at the specified frequency in these environments [3]. The above discussion implies that the path loss model of Equation (2.6) is also good for indoor channels too; a typical r_0 value is 1 m.

Building type	Frequency (MHz)	n	σ dB
Retail stores	914	2.2	8.7
Grocery stores	914	1.8	5.2
Office with fixed walls	1500	3.0	7.0
Office with soft panels	900	2.4	9.6
Office with soft panels	1900	2.6	14.1
LOS Textile/chemical factories	4000	2.1	7.0
OBS Textile/chemical factories	4000	2.1	9.7

Figure 2.10   Values for exponent n and σ for various building types

- *Delay spread.* Inside a building, objects that cause scattering are usually located much closer to the direct propagation path between the transmitter and the receiver. Thus, delay spread due to multipath propagation is typically smaller in indoor systems. Buildings that have few metal and hard partitions have rms delay spreads between 30 and 60 ns, whereas for larger buildings with more metal this number can be as large as 300 ns.
- *Propagation between floors.* Typically, there will be a reuse of frequencies between different floors of a building in an effort to increase spectrum efficiency. Thus, inter-floor interference will significantly depend on the inter-floor propagation characteristics. This makes prediction of propagation between floors an important factor. Although this problem is quite difficult some general rules exist: (a) the type of material that separates floors impacts signal attenuation between the floors; solid steel planks induce more signal attenuation than planks that are produced by pouring concrete over metal layers; (b) buildings with a square footprint induce greater attenuation than buildings with a rectangular footprint due to signals traveling between floors; (c) the greatest path loss of a signal crossing floors occurs when the signal passes from the originating floor to an adjacent one. After this point, propagation to the next floors is characterized by smaller path losses for each floor crossed by the signal. This phenomenon is probably due to diffraction of radio energy across the sides of the building and arrival at distant floors of signal energy scattered from nearby buildings. For separation of one floor, Andersen et al. [3] mention a typical loss of 15 dB with an additional loss of 6–10 dB occurring for the next four floors. For floors further away, the overall path loss increases by a few dB for each floor.
- *Outdoor to indoor signal penetration.* Indoor environments are often affected by signals originating from other buildings or outdoor systems. This phenomenon should be taken into account since it could generate problems in cases where such systems use the same frequencies. Although exact models for this phenomenon do not exist, Andersen et al. [3] make some general remarks. It appears that outdoor to indoor signal attenuation decreases for the higher floors of a building. This is due to the fact that at such floors a LOS path with the antenna of the outdoor system may exist. In some reports, however, this is accompanied by an attenuation increase for floors higher than a certain level, possibly due to shadowing by nearby buildings. Moreover, signal penetration into buildings is reported to be a function of signal frequency with attenuation decreasing for an increasing signal frequency.